Lecture Notes in Computer Science 6235

Commenced Publication in 1973
Founding and Former Series Editors:
Gerhard Goos, Juris Hartmanis, and Jan van Leeuwen

Alexey Vinel   Boris Bellalta
Claudio Sacchi   Andrey Lyakhov
Miklós Telek   Miquel Oliver (Eds.)

Multiple Access Communications

Third International Workshop, MACOM 2010
Barcelona, Spain, September 13-14, 2010
Proceedings

 Springer

Volume Editors

Alexey Vinel
Saint-Petersburg Institute for Informatics and Automation of the
Russian Academy of Sciences (SPIIRAS)
St. Petersburg, 199178, 14 line, 39, Russia
Email: vinel@ieee.org

Boris Bellalta
Miquel Oliver
Universitat Pompeu Fabra, Dept. of Information and Communication Technologies
Roc Boronat 138, 08018 Barcelona, Spain
E-mail: {boris.bellalta;miquel.oliver}@upf.edu

Claudio Sacchi
University of Trento, Dept. of Information Engineering and Computer Science (DISI)
Via Sommarive 14, I-38050 Povo (Trento), Italy
E-mail: sacchi@disi.unitn.it

Andrey Lyakhov
Institute for Information Transmission Problems, RAS
Bolshoy Karetny per., 127994 Moscow, Russia
E-mail: lyakhov@iitp.ru

Miklós Telek
Budapest University of Technology and Economics, Dept. of Telecommunications
P.O. Box 91, 1521 Budapest, Hungary
E-mail: telek@hit.bme.hu

Library of Congress Control Number: 2010933252

CR Subject Classification (1998): C.2, H.4, K.6.5, D.4.6, D.2, E.3

LNCS Sublibrary: SL 5 – Computer Communication Networks
and Telecommunications

ISSN   0302-9743
ISBN-10   3-642-15427-1 Springer Berlin Heidelberg New York
ISBN-13   978-3-642-15427-0 Springer Berlin Heidelberg New York

springer.com

© Springer-Verlag Berlin Heidelberg 2010
Printed in Germany

Typesetting: Camera-ready by author, data conversion by Scientific Publishing Services, Chennai, India
Printed on acid-free paper   06/3180

Preface

It is our great pleasure to present the proceedings of the Third International Workshop on Multiple Access Communications (MACOM) that was held in Barcelona during September 13–14, 2010.

In 1961, Claude Shannon established the foundation for the discipline now known as "multi-user information theory" in his pioneering paper "Two-way Communication Channels," and later Norman Abramson published his paper "The Aloha System—Another Alternative for Computer Communications" in 1970 which introduced the concept of multiple access using a shared common channel. Thereafter, for more than 40 years of study, numerous elegant theories and algorithms have been developed for multiple-access communications.

During the 1980s and 1990s the evolution of multiple-access techniques proceeded in conjunction with the evolution of wireless networks. Novel multiple access techniques like code division multiple access (CDMA) and orthogonal frequency division multiple access (OFDMA) provided increased spectral efficiency, dynamicity and flexibility in radio resource allocation with intrinsic anti-multipath and anti-interference features. In this first decade of the 21st century, multiple-access techniques, derived from advanced wireless transmission methodologies based on the diversity concept (e.g., MC-CDMA, MIMO-OFDMA and SC-FDMA), opened the road to a renewed idea of multiple access. Today multiple-access communications involve many challenging aspects not only limited (like in the past) to physical layer design. Medium access control (MAC) techniques play a crucial role in managing the radio resources that users will exploit to transmit their data streams. Recent developments in software radios and cognitive radios have led to a significant impact also on spectrum management and access paradigms. Old paradigms of multiple-access management were based on locked and exclusive reservation policies of spectrum resources. Cognitive radios allow us to sense the spectrum and to occupy free bandwidth portions following opportunistic approaches.

Technical contributions to all these topics were presented and discussed in MACOM 2010 and are included in this book. We received more than 50 submissions for the conference, including 10 invited papers. After the review process, 21 high-quality full papers were accepted for presentation at the workshop, together with 6 posters. Almost every submission received at least three reviews from the members of the Technical Program Committee and/or external reviewers. Our gratitude goes to the Technical Program Committee and external reviewers for their efforts.

MACOM 2010 also included four distinguished keynote speakers: Dimitry Osipov, Alexander N. Dudin, David Malone and, especially for his relevance in the history of multiple-access communications, Norman Abramson. Additionally, a special session devoted to the IC0906 COST action WINEMO (Wireless

Networking for Moving Objects) was planned to promote interactions and further discussions with researchers from all European countries working in these fields.

Finally, we would like to take this opportunity to express our gratitude to the sponsors and supporters, together with the local organizers, who help to make MACOM 2010 a very successful event.

September 2010   A. Vinel
                                                                 B. Bellalta
                                                                 C. Sacchi
                                                                 A. Lyakhov
                                                                 M. Telek
                                                                 M. Oliver

Organization

MACOM 2010 was organized by the TTCN group (Telecommunication Technologies and Computer Networks Group) from SPIIRAS (St. Petersburg Institute for Informatics and Automation of RAS), Russia, and the NeTS group (Networking Technologies and Strategies Research Group) from UPF (Universitat Pompeu Fabra), Spain.

Executive Committee

Conference Chair	Alexey Vinel, SPIIRAS, Russia
General Co-chairs	Boris Bellalta, Universitat Pompeu Fabra, Spain
	Adolf Finger, Dresden University of Technology, Germany
TPC Chairs	Andrey Lyakhov, IITP RAS, Russia
	Claudio Sacchi, University of Trento, Italy
	Miklos Telek, Technical University of Budapest, Hungary
Local Chair	Miquel Oliver, Universitat Pompeu Fabra, Spain
Local Organization	Cristina Cano, Universitat Pompeu Fabra, Spain
	Anna Sfairopoulou, Universitat Pompeu Fabra, Spain
Publicity Chair	Min Chen, Seoul National University, Korea
Industrial Chairs	Sergey Balandin, Nokia Research Center, Finland
	Javier Del Ser, Tecnalia-Telecom, Spain

Steering Committee

Khalid Al-Begain	University of Glamorgan, Pontypridd, UK
Ernst Gabidulin	MIPT, Russia
Vitaly Gutin	ETU "LETI", Russia
Angel Lozano	Universitat Pompeu Fabra, Spain
Felix Taubin	SUAI, Russia
Victor Zyablov	IITP RAS, Russia
Vladmir Vishnevsky	IRE RAS, Russia
Bernhard Walke	RWTH Aachen University, Germany

Technical Program Committee

Sergey Andreev	SPIIRAS, Russia
Konstantin Avrachenkov	INRIA Sophia Antipolis, France
Florin Avram	Universite de Pau, France
Abdelmalik Bachir	Imperial College London, UK
Jaume Barcelo	Universidad Carlos III de Madrid, Spain
Boris Bellalta	Universitat Pompeu Fabra, Spain
Giuseppe Bianchi	University of Rome Tor Vergata, Italy
Thomas Michael Bohnert	SAP Research, Switzerland
Torsten Braun	University of Bern, Switzerland
Raffaele Bruno	IIT-CNR, Italy
Peter Buchholz	TU Dortmund, Germany
Andrea Cattoni	Aalborg University, Denmark
Eduardo Cerqueira	Federal University of Para, Brazil
Matteo Cesana	Politecnico di Milano, Italy
Periklis Chatzimisios	TEI of Thessaloniki, Greece
Young-June Choi	Ajou University, South Korea
Claudio Cicconetti	University of Pisa, Italy
Ermanna Conte	University of Padova, Italy
Andrea Conti	ENDIF University of Ferrara, WiLAB University of Bologna, Italy
Roberto Corvaja	University of Padova, Italy
Leandro D'Orazio	Siemens S.p.A., Italy
Tugrul Dayar	Bilkent University, Turkey
Alexandre de Baynast	European Microsoft Innovation Center, Germany
Javier Del Ser	TECNALIA-Telecom, Spain
Alexander Dudin	Belarusian State University, Belarus
Alexey Dudkov	University of Turku, Finland
Tamas Elteto	Budapest University of Technology and Economics, Hungary
Marc Emmelmann	Technical Univesity of Berlin, Germany
Stanislav Filin	NICT, Japan
Lorenzo Favalli	University of Pavia, Italy
Istvan Frigyes	Budapest University of Technologies, Hungary
Olga Galinina	Speech Technology Center, Saint Petersburg, Russia
Fabrizio Granelli	University of Trento, Italy
Gaoning He	Telecom ParisTech, France
Geert Heijenk	University of Twente, The Netherlands
Andras Horvath	University of Turin, Italy
David Hunter	University of Essex, UK
Gang Uk Hwang	KAIST, Korea
Eduard Jorswieck	Dresden University of Technology, Germany
Markku Juntti	University of Oulu, Finland

Valentina Klimenok	Belarusian State University, Belarus
Jarkko Kneckt	Nokia Reseach Center, Finland
Vinay Kolar	RWTH Aachen University, Germany
Yevgeni Koucheryavy	Tampere University of Technology, Finland
Andrey Lyakhov	IITP RAS, Russia
David Malone	NUI Maynooth, Ireland
Sebastian Max	RWTH Aachen University, Germany
Michela Meo	Politecnico di Torino, Italy
Enzo Mingozzi	University of Pisa, Italy
Dmitri Moltchanov	Tampere University of Technology, Finland
Qiang Ni	Brunel University, UK
Dmitry Osipov	IITP RAS, Russia
Alexander Pechinkin	Institute of Informatics Problems, RAS, Russia
Aleksi Penttinen	TKK Helsinki University of Technology, Finland
Vicent Pla	Universitad Politecnica de Valencia, Spain
Javier Rodriguez Fonollosa	Universitat Politecnica de Catalunya, Spain
Claudio Sacchi	University of Trento, Italy
Zsolt Saffer	Budapest University of Technology and Economics, Hungary
Alexander Safonov	IITP RAS, Russia
Matilde Sanchez Fernandez	Universidad Carlos III de Madrid, Spain
Christian Schlegel	University of Alberta, Canada
Bruno Sericola	INRIA Rennes - Bretagne Atlantique, France
Pablo Serrano	Universidad Carlos III de Madrid, Spain
Vsevolod Shneer	Technical University of Eindhoven, The Netherlands
Susanna Spinsante	Universitá Politecnica delle Marche, Italy
Dirk Staehle	University of Wuerzburg, Germany
Miklos Telek	Technical University of Budapest, Hungary
Andrea Tonello	University of Udine, Italy
Andrey Turlikov	SUAI, Russia
Rob van der Mei	Centrum voor Wiskunde en Informatica, The Netherlands
Benny Van Houdt	University of Antwerp, Belgium
Maria-Angeles Vazquez-Castro	Universidad Autonoma de Barcelona, Spain
Alexey Vinel	SPIIRAS, Russia
Hongyi Wu	University of Louisiana at Lafayette, USA
Mikhail Yakimov	IITP RAS, Russia
Gennady Yanovsky	SUT, Russia
Mei Yu	Tianjin University, China
Yunpeng Zang	RWTH Aachen University, Germany
Yan Zhang	Simula Research Laboratory and University of Oslo, Norway

Referees

Sergey Andreev	Tugrul Dayar	Ivan Pustogarov
Konstantin Avrachenkov	Alexandre De Baynast	Harri Saarnisaari
Florin Avram	Leandro D'Orazio	Claudio Sacchi
Jaume Barcelo	Alexey Dudkov	Zsolt Saffer
Boris Bellalta	Tams Eltet	Alexander Safonov
Giuseppe Bianchi	Stanislav Filin	Christian Schlegel
Marko Boon	Istvan Frigyes	Rainer Schoenen
Pavel Boyko	Sudarshan Guruacharya	Bruno Sericola
Raffaele Bruno	Andras Horvath	Anna Sfairopoulou
Peter Buchholz	Eduard Jorswieck	Jelena Skulic
Cristina Cano	Valentina Klimenok	Susanna Spinsante
Trang Cao Minh	Yevgeni Koucheryavy	Dirk Staehle
Andrea Cattoni	Andrey Lyakhov	Dimitrios Stratogiannis
Periklis Chatzimisios	David Malone	Gang Uk Hwang
Eduardo Cerqueira	Sebastian Max	Rob Van Der Mei
Andrea Conti	Dmitri Moltchanov	Benny Van Houdt
Roberto Corvaja	Qiang Ni	Alexey Vinel
Eugenio Costamagna	Dmitry Osipov	Mikhail Yakimov
Salvatore D'Alessandro	Massimiliano Panizza	

Technical Sponsors

- Wireless Networking for Moving Objects (WINEMO). IC0906 COST action.
- Foundations and Methodologies for Future Communication and Sensor Networks (COMONSENS). CONSOLIDER-INGENIO 2010.
- The Atomic Redesign of the Internet Future Architecture (TARIFA). The i2CAT Foundation.
- Tecnalia-Telecom : technological corporation.
- A.S. Popov's Society.

Sponsoring Institutions

- Universitat Pompeu Fabra.
- St. Petersburg Institute for Informatics and Automation, RAS.
- Spanish Ministry of Science and Innovation (TEC2008-06055/TEC).
- Agència de Gestió d'Ajuts Univ. i de Recerca (AGAUR), Generalitat de Catalunya.

Table of Contents

Wireless Mesh Networks and WIMAX

Advanced Topics in Wireless Networks

Mobile Ad-Hoc Networks

Physical Model Based Interference Classification and Analysis

Artem Krasilov

Institute for Information Transmission Problems
of the Russian Academy of Science,
B.Karetny lane 19, 127994 Moscow, Russia
krasilov@iitp.ru

Abstract. Interference of links in wireless mesh networks has a great impact on the network performance. Most of previous studies of this problem are based on unrealistic interference models which in some cases lead to erroneous results. In this paper, based on realistic physical interference model we give a full classification of interference cases for two links, provided that stations composing these links are arranged on one line and data flows have the same direction. We show that in some cases channel capacity distribution between links is severe unfair and analyze how the probabilities of appearance of these cases depend on the parameters of physical layer of IEEE 802.11 and relative disposition of stations. Also, we discuss possible mechanisms to minimize these probabilities.

Keywords: CSMA/CA; IEEE 802.11; mesh; interference; physical model.

1 Introduction

It is widely known that the commercial success of Wi-Fi technology is connected with its simplicity (in particular, simplicity of the CSMA/CA method) and support of high data transmission rate. This success, in turn, has led to attempts of using IEEE 802.11 standard in applications for which it was not developed initially. So, IEEE 802.11s amendment [1], which specifies the operation of a multi-hop wireless networks (mesh networks) is almost accepted, and devices implementing this new technology increasingly appear on the market.

According to IEEE 802.11s amendment, the CSMA/CA method is used to access the wireless medium in a mesh network. The CSMA mechanism works as follows. A station that has some data to transmit senses the channel for a given amount of time and, only if the channel is detected idle, the station is allowed to transmit. To reduce the probability of collision on the channel, the Collision Avoidance (CA) mechanism is used. This mechanism is based on backoff procedure: stations defer their transmissions for a random interval, which is measured by integer number of slots. In single-hop networks where all stations can sense transmissions of each other, a collision occurs only if stations choose the same slot to start their transmissions. So, CSMA/CA method works efficiently

A. Vinel et al. (Eds.): MACOM 2010, LNCS 6235, pp. 1–12, 2010.

in single-hop networks. In multi-hop networks, most of stations are hidden from each other and the probability of collision is significantly higher. For example, if stations A and B are in the transmission range of station C and, with that, A and B are hidden form each other, collision of A and B occurs when one of them starts to transmit during a transmission of the other. Stations A and B cannot sense the transmissions of each other and so they cannot prevent collision. All that results in the increase of packet loss probability and channel capacity reduction. Many papers based on simulations [2][3], analytical models [4] and also experimental results [5] show that there are two problems in multi-hop networks due to the interference: the unfairness problem when channel capacity distribution between links becomes unfair, and the oscillation problem when throughput of a link strongly fluctuate in time. However, in these works analysis is made only for a restricted number of topologies and they do not consider the general case of link disposition. To describe the interference phenomenon in multi-hop networks two models can be used: protocol and physical (see Section 2). In [6],which uses the protocol model a classification of possible cases of two links interference is given and the analysis of channel capacity distribution between them in each case is made. It is shown that in certain cases of links disposition the channel capacity distribution between links becomes severe unfair, which may lead to the collapse of one link (starvation effect). But as we will show further, the protocol model is oversimplified and do not take into consideration features of operation of physical layer of IEEE 802.11, specifically capture effect, which in turn lead to erroneous results. More accurate model describing interference phenomenon is the physical model. On the basis of the physical model in [7], a set of conditions which cause unfair channel capacity distribution between links is derived and possible mechanisms to solve the unfairness problem are discussed. However, they do not give a classification of interference cases and do not analyze the degree of unfairness in each case.

In this paper, we study the unfairness problem. On the basis of the physical model, we give a full classification of interference cases for two links, provided that stations composing these links are arranged on one line and data flows have the same direction (unfairness problem appears only when flows have the same direction). For each case, via the simulation, we determine the degree of unfairness and give the qualitative explanation to obtained results. Thereafter, we reveal cases in which the channel capacity distribution becomes severe unfair and determine the probabilities of their appearance among all cases when links interact. Also, we analyze how these probabilities depend on parameters of physical layer of 802.11 and a relative disposition of stations.

2   Interference Models

Nowadays there are two commonly recognized interference models: protocol model and physical model.

2.1 Protocol Model

In the protocol model, for every station two areas are defined: Carrier Sense (CS) range and Transmission (TX) range. CS range of station A is the area where other stations are able to detect PHY activity when A transmits. TX range of station A is the area within which other stations are able to correctly decode frames transmitted by A. In the simplest case with omni-directional antenna, in the absence of obstacles like walls, TX range is a circle, and CS range is a circle with the radius bigger than the radius of the TX range circle.

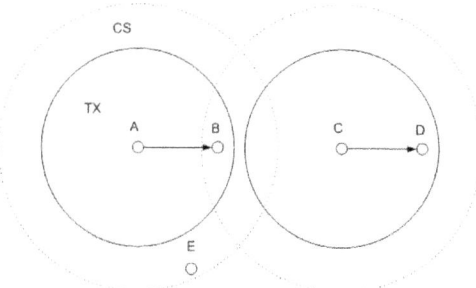

Fig. 1. Illustration of the protocol model

A transmission between stations A and B is successful when B is in the TX range of A and B is not in CS range of any station transmitting at the same time. If station A transmits, all stations which are in CS-range of A sense the channel busy and thus defer their own transmissions. In Fig.1, if A transmits E senses the channel busy and defers its own transmission, while C senses the channel idle and can start its transmission that causes collision on B.

The main advantage of the protocol model is its conceptual simplicity. Previous papers based on the protocol model reveal that in some cases of links disposition network capacity distribution between links is unfair. As shown in [6] the worst case is the case when the receiver of one link is in CS range (not in TX range) of the transmitter of another link and both transmitters are hidden for each other (see Fig.1). In this case, when link $C \to D$, works in saturation it gets almost all the total channel throughput while throughput of link $A \to B$ is extremely low, which means the total collapse of link $A \to B$ (starvation effect). However the results obtained in full-scale experiment [5] strongly differ from the results obtained in simulations and analytical computations which use the protocol model, because the protocol model do not take into consideration features of the operation of physical layer of IEEE 802.11, specifically the capture effect. In this paper, we refer as the capture effect the capability of a station to receive a frame despite the transmission of other station(s). More accurately the capture effect will be defined below.

2.2   Physical Model

The physical model takes into consideration features of operation of the IEEE 802.11 physical layer. In other words, it more accurately describes the process of frame transmission in terms of such physical quantities as signal strength, sensitivity, Signal-Noise-Interference Ratio (SNIR) and so on.

In the absence of obstacles, if the distance between transmitter and receiver is d_0, the average signal strength detected by the receiver is

$$P_{RX}(d_0) = \overline{P}_{RX} \left(\frac{\overline{d}}{d_0} \right)^\theta , \tag{1}$$

where $\overline{P}_{RX}$ is the reference signal strength measured at the distance $\overline{d}$ (usually 1 meter), θ is the path loss exponent ($\theta = 2$ for a free-space LOS (Line-Of-Sight) model and $\theta = 4$ for a ground reflection model). Note that reference signal strength $\overline{P}_{RX}$ depends directly on transmission power P_0.

In the physical model, the aggregate energy detected by a receiver consists of signal (from intended transmitter), interference (from an unwanted transmitter(s)) and background noise. A station can receive a packet with acceptable error rate only if two conditions are satisfied: 1) the received desired signal is greater than the threshold (denoted by P_R, i.e. receiver sensitivity), and 2) the Signal-Noise-Interference Ratio (SNIR) is above the threshold (denoted by S_0).

$$\begin{cases} P_{RX}(d_0) \geq P_R, \\ \dfrac{P_{RX}(d_0)}{P_N + \sum\limits_{i \neq 0} P_{RX}(d_i)} \geq S_0, \end{cases} \tag{2}$$

where P_N is the strength of the background noise, and $P_{RX}(d_i)$ denotes the signal strength from interference source i at distance d_i. 802.11 networks support multiple data rates, and a higher data rate typically requires a higher threshold S_0.

Fig.2 shows a segment of a typical mesh network with a reference transmission from a station TX to a station RX and three other neighboring stations (A, B, and C), where the same transmission power is used by every station. Let's define:

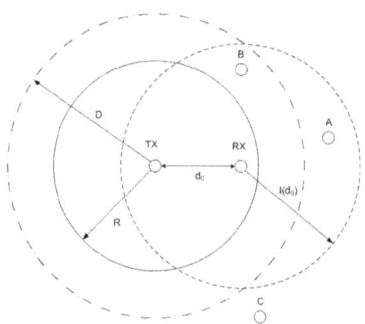

Fig. 2. Illustration of the physical model

R: Radius of the Transmission range, given by

$$R = \overline{d} \left(\frac{\overline{P}_{RX}}{max(P_R, S_0 P_N)} \right)^{1/\theta}. \tag{3}$$

Transmission range of station TX is the area where all other stations are able to decode correctly frames transmitted by TX in the absence of interference, i.e. $\sum_{i \neq 0} P_{RX}(d_i) = 0$.

D: Radius of the Physical Carrier Sensing range, given by

$$D = \overline{d} \left(\frac{\overline{P}_{RX}}{P_C} \right)^{1/\theta}, \tag{4}$$

where P_C denotes the Physical Carrier Sensing (PCS) threshold. If a station detects a signal with energy greater than P_C, it senses the channel busy. Within the Physical carrier sensing range, any station is able to detect an existing transmitter via physical carrier sensing.

I: Radius of Interference range, given by

$$I(d_0) = d_0 \left[\frac{1}{\frac{1}{S_0} - \frac{P_N}{P_{RX}} (\frac{d_0}{d})^\theta} \right]^{1/\theta}. \tag{5}$$

A single transmitter within Interference range of the receiver RX disrupts the reception of the desired transmission (TX-RX). For example, if station A starts to transmit during TX-RX transmission RX is not able to decode correctly the frame from TX. On the contrary, a transmission of station C does not disrupt TX-RX transmission as long as $SINR \geq S_0$ (capture effect).

3 Classification of Interference Cases

Consider a wireless network that consists of four stations (STA1-STA4) composing two active links: link $1 \rightarrow 2$ and link $3 \rightarrow 4$ working in saturation. All stations are arranged on one line, as illustrated in Fig.3. The distance between STA1 and STA2 as well as between STA3 and STA4 equals L. Also we suppose that all stations transmit with the same power and data rate and have the same receiver sensitivity P_R and carrier sensing (PCS) threshold P_C. It means that all stations have the same radius R of Transmission range and the same radius D of Physical Carrier Sensing range.

Let X be the distance between STA1 and STA3. Further we consider only positive values of X because when X is negative the situation is the same if swap links. When X increases from 0 to infinity qualitative changes in interaction between each pair of stations occur. Let us find values of X_i (frontier points) in which these changes occur. Consider the interaction of STA1 and STA4 in detail:

- $0 \le X \le X_1 = R - L$: STA1 and STA4 are in the Transmission range (TX range) of each other. It means that in absence of interference, STA4 is able to decode correctly frames transmitted by STA1 and vice versa.
- $X_1 < X < X_2 = D - L$: STA1 and STA4 are in the Carrier Sensing range (CS range) of each other but not in the TX range. This implies that STA4 is able to detect PHY activity when STA1 transmits, but unable to decode correctly frame transmitted by STA1, and vice versa.
- $X > X_2$: STA1 and STA4 do not hear each other.

Similarly, for pair of stations (STA1,STA3) and (STA2,STA4) we obtain points $X_3 = R$ and $X_4 = D$, and for (STA2,STA3) points $X_5 = R + L$ and $X_6 = D + L$.

If transmissions of STA1 and STA3 overlap in time STA4 can decode correctly a frame transmitted by STA3 when STA1 is not in the Interference range (IF range) of STA4, that is when $X > X_7 = I(L) - L$ (capture effect). Similarly, STA2 can decode correctly frame transmitted by STA1 when STA3 is not in the IF range of STA2, that is when $X > X_8 = I(L) + L$.

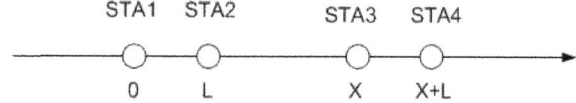

Fig. 3. Considered network

So, we have obtained eight points X_i. Denote by $X_{max} = \max_i \{X_i\}$. When $X \in (0, X_{max})$, links interact between each other. Points X_i break this interval into the subintervals (X_i, X_j) where X_i and X_j are two consecutive points. In each such subinterval (X_i, X_j), a certain interference case occurs. According to (3) - (5), the sequence in which points arrange in interval $(0, X_{max})$ essentially depends on the chosen data rate (threshold S_0) and the distance L. In the next section to confirm our assumptions, via simulation, we analyze in detail the following network configuration: data rate 6 Mbps, distance L=90 m.

4   Analysis of Interference Cases

4.1   Simulation Set-Up

In this section we introduce the simulation environment. To analyze interference cases we use ns-3 network simulator [8]. The physical and link layer of the simulator accurately follows the complete IEEE 802.11 standard. The values of adjustable parameters of physical layer used in simulation are given in Tab.1. All powers in Tab.1 are measured in dBm. The background noise in ns-3 is thermal noise in radio-frequency signal chain given by $P_N = kT\Delta f \dot{N} F$, where $kT\Delta f$ is the Nyquist noise, NF is the noise figure [9] which is specified by the performance

Table 1. Simulation model parameters

Parameter	Value
PHY layer standard, bandwidth, Δf	802.11a, 20 MHz
Transmitter power, P_0	16 dBm
Receiver sensitivity, P_R	-98 dBm
Physical Carrier Sensing threshold , P_C	-98 dBm
Path loss exponent, θ	4
Noise figure, NF	7 dB
Background noise, P_N	-94 dBm

of a radio receiver. In our case we obtain $P_N[dBm] = kT\Delta f[dBm] + NF[dB] = -101dBm + 7dB$.

MAC layer parameters are set up in accordance with IEEE 802.11a specification [10]. The RTS/CTS mechanism is turned off. Stations transmit on basic data rate 6 Mbps with a fixed packet size of 80 bytes on MAC layer.

Threshold S_0 is not an adjustable parameter and its value depends on data rate, packet size and internal features of a radio receiver. To determine its value we study performance of a point-to-point link. In the first simulation, we configured a network of two stations: one sender S and one receiver R. We vary the S-R separation distance and measure the effective throughput provided by the MAC layer at the receiver. The results in Fig.4 show the throughput W against separation distance d and packet delivery rate $(1 - PER)$ against $SINR$. As the value of threshold S_0 we use the value of SNIR when packet delivery rate equals 0.9. Given the value of S_0 which equals 0.3 dB and the values of parameters from Tab.1, we calculate values of R and D which consequently equal 124 m and 160 m.

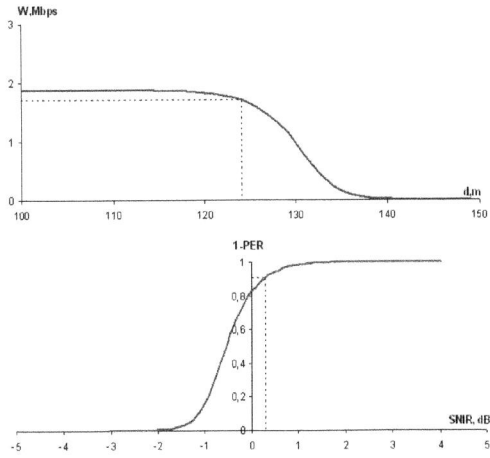

Fig. 4. Performance of point-to-point link

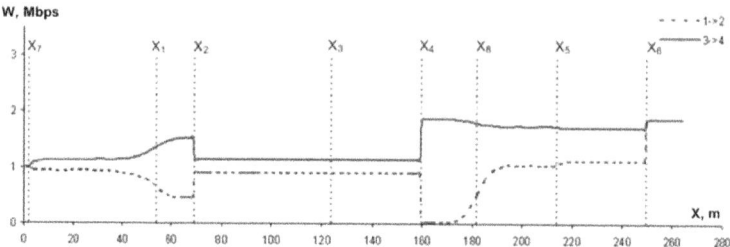

Fig. 5. Interference experiment results

Table 2. The values of frontier points X_i

X_1	X_2	X_3	X_4	X_5	X_6	X_7	X_8
54	70	124	160	214	250	2	182

4.2 Interference Experiment

In the interference experiment we configure a network of four stations as described in section 3. The value of distance L is fixed during the whole experiment and equals to 90 m. The values of frontier points X_i calculated for these network configuration are shown in Tab.2. In the experiment we vary the X distance and measure throughput W of both links $1 \rightarrow 2$ and $3 \rightarrow 4$. The results in Fig.5 show that changes in throughput of both links occur only in frontier points X_i as we supposed in section 3. In points 1,5,7 and 8 we observe smooth changes in the throughput because packet delivery rate (see fig.4)is a smooth function of SNIR in the neighborhood of point S_0 rather than step function as supposed in the physical model.

Let us consider in detail operation of stations in each interval (X_i, X_j) (interference case) and give a qualitative explanation to the obtained results. The network configuration for each case is illustrated schematically in Fig. 6. Solid line means that stations are in the TX range of each other, dashed line in the CS range but not in the TX range. Leader line indicates that a station is in the IF range of the receiver of a neighboring link. For example, in Fig.6a STA1 is in the IF range of the receiver of link $3 \rightarrow 4$. It means that if station STA1 starts to transmit during a transmission of a frame between STA3 and STA4, STA4 is not able to decode correctly the frame from STA3.

a) $X \in (0, X_7)$ – all stations are in the TX range of each other, that is in absence of interference, all frames transmitted by any station can be received by any other station. In this case both links are in equal conditions and hence get equal throughput. Throughput of each link approximately equals half of the throughput of point-to point link.

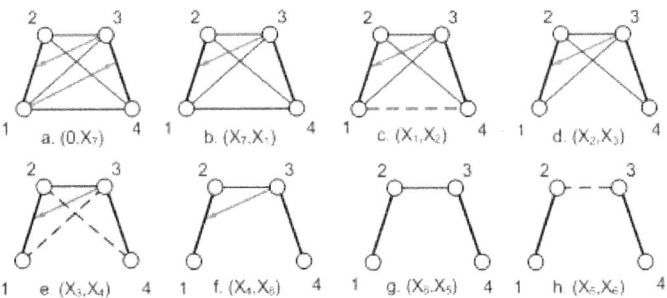

Fig. 6. Interference cases

b) $X \in (X_7, X_1)$ – STA1 is out of the IF range of STA4. It means that when STA1 and STA3 transmit DATA frames simultaneously, STA4 receives a DATA frame from STA3 successfully because of capture effect but STA2 cannot receive the DATA frame from STA1. The difference in reaction on collisions leads to different throughputs for the considered links, because: (i) STA3 can overcome collisions, and its contention window is always minimal; (ii) STA1 increases its contention window after collisions and has fewer chances to get the channel. So, link $1 \rightarrow 2$ throughput is less than link $3 \rightarrow 4$ throughput, i.e. the channel resource distribution becomes unfair.

c) $X \in (X_1, X_2)$ – STA1 and STA4 are in the CS range but not in the TX range of each other. It means that STA4 cannot receive ACK frames from STA1 correctly. After the ACK frame form STA4, STA1 waits for EIFS before resuming decrementing its backoff counter, while STA3 waits only for $DIFS < EIFS$. Thus, STA3 has more chances to win in the contention for the channel, making the channel capacity distribution more unfair.

d) $X \in (X_2, X_3)$ – STA1 and STA4 do not hear each other at all. It means that the phenomenon considered in previous case do not take place here. Practically STA1 and STA3 work as in case b. Note, that throughputs of both links get their previous values.

e) $X \in (X_3, X_4)$ – STAs 1 and 3, as well as 2 and 4, are in the CS range but not in the TX range of each other. In this case, when STA1 starts to transmit its DATA frame to STA2 first, STA3 cannot decode it correctly. This means that STA3 waits for EIFS after the end of the STA3s DATA frame transmission before it can start to count the backoff. During this interval STA2 transmits ACK frame to STA1, and STA1 waits for DIFS ($EIFS = SIFS + T_{ACK} + DIFS = 94\mu s$). So, STA1 and STA3 always start counting their backoffs simultaneously. The situation is the same if STA3 starts its DATA transmission first. We obtain that case e) do not differ from case d), and the channel resource distribution remains the same.

f) $X \in (X_4, X_8)$ – STA1 and STA3 do not hear each other at all, but STA3 is in the IF range of STA2. This situation causes severe unfairness between links $1 \rightarrow 2$ and $3 \rightarrow 4$. It can be explained in the following way. When STA1 transmits a DATA frame to STA2, STA3 does not hear it and can start its own transmission. This causes collision on STA2 but not on STA4. The next transmission of STA1 with high probability is also unsuccessful. It makes STA1s contention window very large, and STA1 has small chances to access the channel, while the contention window of STA3 is of minimal size because its transmission is always successful. As a result, link $3 \rightarrow 4$ gets almost all the total throughput while the throughput of link $1 \rightarrow 2$ is dramatically low, which means the total collapse of link $1 \rightarrow 2$.

g) $X \in (X_8, X_5)$ – STA3 is out of the IF range of STA2. In this case we consider dramatic growth of link $1 \rightarrow 2$ throughput. It caused by capture effect: if STA1 starts to transmit before STA3, STA2 is able to decode correctly frame from STA1 despite the transmission of STA3. Simulation results show that the throughput of link $1 \rightarrow 2$ approximately equals half of the throughput of point-to point link, while the throughput of link $3 \rightarrow 4$ slightly decreases. It caused by growth of the number of ACK frames transmitted by STA2. If STA2 starts to transmit while STA3 sensing the channel activity, STA3 must defer its transmission for period of time $T_{ACK} + DIFS$.

h) $X \in (X_5, X_6)$ – STA2 and STA3 are in the CS range but not in the TX range of each other. This case is similar to the previous one. The difference is that STA3 cannot correctly decode ACK frames transmitted by STA2 and consequently STA3 defer its transmission for $EIFS - DIFS$ more than in previous case before it can start to count the backoff. Results show that throughput of link $1 \rightarrow 2$ slightly increases while throughput of link $3 \rightarrow 4$ decreases.

i) $X \in (X_6, +\infty)$ – links $1 \rightarrow 2$ and $3 \rightarrow 4$ do not interfere at all therefor throughput of each link equals the throughput of point-to point link.

4.3   Analysis of Case f

As we have seen above, in case f) channel capacity distribution between links becomes severe unfair: one link gets almost all the total channel throughput while other link do not work at all. Consider this case in detail. Case f) occurs when the transmitter of one link is in the IF range of another link $(X < I(L) + L)$and transmitters of both links are hidden from each other $(X > D)$. If the value X is uniformly chosen from interval $(0, D + L)$, the conditional probability p_f of the occurrence of case f) among all cases when links interact is given by

$$p_f = \frac{1(L + I(L) - D) \cdot (L + I(L) - D)}{D + L},$$ (6)

where $1(A)$ is a function, returning one if condition A is satisfied and zero otherwise. Conditional probability p_f characterizes the portion of cases when one link repress another provided that links interact. It depends on three parameters: (i) data rate (threshold S_0), (ii) Physical Carrier Sensing threshold P_C, (iii) distance L. Note, that L do not exceed R and R do not exceed D. Using formulas from section 2, we analyze how the probability p_f depends on L and P_C. Fig.7a show that p_f increases with the growth of L. The maximum value of p_f equals 0.5 when $L = R = D$ ($P_C = -93.7dBm$). For each distance L, there exists a limiting value $\widetilde{P_C}$ when p_f becomes zero, see Fig.7b . When P_C equals to $\widetilde{P_C}$, the CS range of the transmitter covers the IF range of the receiver ($D = L + I(L)$). So, if this condition is satisfied it is enough to protect a link from interference with other links. Other stations that is out of the IF range of the receiver can transmit in parallel.

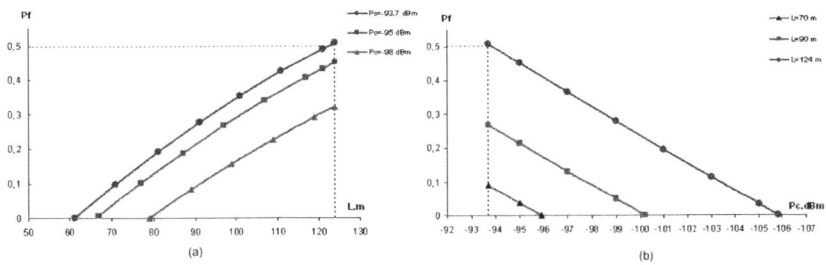

(a)   (b)

Fig. 7. Probability of case f

5   Conclusion

In this paper, we explore the problem of mutual interference between links in IEEE 802.11 networks. On the basis of the physical model, we have given a full classification of interference cases for two links, provided that stations composing these links are arranged on one line and data flows have the same direction. We have shown that in case when the transmitter of one link is in the IF range of another link and transmitters of both links are hidden from each other (case f)), channel capacity distribution between links becomes severe unfair: one link repress another. In case when the transmitter one link is out of the IF range of another link (cases g) and h))due to the capture effect, both links can work in parallel. Obtained results show that in order to protect a link from interference with other links, the CS range of the transmitter should cover the IF range of the receiver. In mesh networks, for each pair transmitter-receiver this condition could be satisfied by means of using: (i) Power Control mechanisms (ii) Adapting Physical Carrier Sensing mechanisms. Results obtained in this paper can be used in designing these mechanisms.

In future work, we will study how the parameters of physical layer, data rate, relative disposition of stations and direction of data flows influence on the appearance of one or another interference case, and investigate the interference phenomenon in general case when stations are arranged on a plane.

References

1. IEEE P802.11s/D3.0, Draft Amendment to Standard. Mesh Networking (March 2009)
2. Xu, S., Saadawi, T.: Does the IEEE 802.11 MAC Protocol Work Well in Multihop Wireless Ad Hoc Networks? IEEE Comm. Magazine 39(6), 130–137 (2001)

3. Huang, X.L., Bensaou, B.: On Max-Min Fairness and Scheduling in Wireless Ad Hoc Networks: Analytical Framework and Implementation. In: Proc. ACM Mobi-Hoc, pp. 221–231 (October 2001)

4. Garetto, M., Shi, J., Knightly, E.W.: Modeling Media Access in Embedded TwoFlow Topologies of Multihop Wireless Networks. In: MobiCom 2005, Cologne, Germany (2005)

5. Lyakhov, A., Pustogarov, I., Safonov, A., Yakimov, M.: Starvation Effect Study in IEEE 802.11 Mesh Networks. In: Proceeding of Third IEEE International Workshop on Enabling Technologies and Standards for Wireless Mesh Networking,

6. Lyakhov, A., Safonov, A., Yakimov, M., Gudilov, A.: Analysis of Direct Transmissions in IEEE 802.11 Networks. In: DCCN 2007, VINITI, Moscow, Russia, vol. 1, pp. 108–115 (2007)

7. Jiang, L.B., Liew, S.C.: Improving Throughput and Fairness by Reducing Exposed and Hidden Nodes in 802.11 Networks. IEEE Trans. Mobile Comput. 7(1), 34–49 (2008)

8. The ns-3 network simulator, http://www.nsnam.org/

9. Gilbert, B.: Noise Figure and Logarithmic Amplifiers. Analog Dialogue 42-06 (June 2008)

10. IEEE Std 802.11-2007, Revision of IEEE Std 802.11-1999. IEEE Standard for Information Technology—Telecommunications and information exchange between systems—Local and metropolitan area network—Specific requirements—Part 11: Wireless LAN Medium Access Control (MAC) and Physical Layer (PHY) specifications. IEEE Computer Society (June 2007)

Dynamic Parameter Adjustment in CSMA/ECA

Jaume Barcelo[1], Boris Bellalta[2], Cristina Cano[2],
Anna Sfairopoulou[2], and Miquel Oliver[2]

[1] Universidad Carlos III de Madrid, Av. de la Universidad 30, 28911 Leganes, Spain
jaume.barcelo@uc3m.es
[2] NeTS Research Group, DTiC, Universitat Pompeu Fabra
{boris.bellalta,cristina.cano,anna.sfairopoulou,miquel.oliver}@upf.edu

Abstract. CSMA/ECA is a modification of CSMA/CA that delivers higher performance by reducing the number of collisions. This paper presents an algorithm to adapt a configuration parameter of the protocol to ensure high performance for any given number of competing stations. Performance is measured in terms of efficiency and fairness. The proposed algorithm has been tested by means of simulation to show that it takes half a second for the system to re-adjust the configuration and suppress the collisions when the number of contending stations suddenly increases from zero to twenty.

Keywords: CSMA, WLAN, IEEE 802.11, throughput, fairness.

1 Introduction

Wireless Local Area Networks (WLAN) have flourished in homes, campuses and enterprises. The vast majority adhere to the IEEE 802.11 [1] protocol that relies on Carrier Sense Multiple Access with Collision Avoidance (CSMA/CA) to share the channel time among the participating stations. It is desired that only a single station transmits at a given time, since overlapping transmissions can easily result in packet loss.

In CSMA/CA, the stations sense the channel for ongoing transmissions before attempting to transmit. If a station senses the channel busy, it will refrain from transmitting and it will delay its transmission a random amount of time. A caveat of CSMA/CA is that only a fraction of the channel time is devoted to successful transmissions while the remainder is wasted in the form of empty channel or collisions. In this article we will study Carrier Sense Multiple Access with Enhanced Collision Avoidance (CSMA/ECA) which is a variant of CSMA/CA that reduces the number of collisions.

Since CSMA/CA curtails the overall network performance, it has deserved many research efforts. Some relevant previous work is revised in Sec. 2. Then, in Sec. 3 , the CSMA/CA protocol is detailed, with particular emphasis on the backoff mechanism. It is explained that using a deterministic backoff value after successful transmissions presents several advantages and the modification of the protocol is called CSMA/ECA. Just as it happens in CSMA/CA, it is also

A. Vinel et al. (Eds.): MACOM 2010, LNCS 6235, pp. 13–24, 2010.

possible to increase the performance of CSMA/ECA by adjusting its contention parameters. Sec. 4 investigates the impact of parameter selection on the performance of CSMA/ECA in different situations. These results are then used in Sec. 5 to suggest an algorithm that dynamically adjusts the configuration of the contending stations to adapt to a changing scenario. Finally, Sec. 6 concludes the article.

2   Related Work

The CSMA/CA medium access control is easy to implement and can be executed in a distributed fashion. These properties make CSMA/CA an ideal candidate for the medium access control of WLAN. It is not an overstatement that the IEEE 802.11 suite of protocols that rely on CSMA/CA has enjoyed a tremendous success in a spectrum of different scenarios. Nevertheless, the research community was concerned from the very beginning by the fact that CSMA/CA does not make an efficient use of the channel time. In [2] the maximum performance that can be delivered using CSMA/CA is studied. The exact values of the efficiency depend on the parameter setting, but in any case it can be concluded that the efficiency of CSMA/CA in IEEE 802.11 is far from 100%. Furthermore, this efficiency significantly degrades as the number of contending stations increases.

A valuable insight into the underlying behaviour of CSMA/CA is provided in [3], where the protocol is modelled as a Markov Chain and key parameters such as transmission probability and conditional collision probability are computed.

Many authors have proposed solutions to improve the performance of CSMA/CA. As an example, the authors of [4] propose that the stations advertise their backoff values to prevent collisions. This approach requires a modification of the headers of the packet and therefore backward compatibility is compromised. The idea of gathering information about other stations intentions to transmit is very promising, and we will use that very same principle in our approach. The difference lies in the fact that in CSMA/ECA this information is distributed implicitly by using a deterministic backoff after successful transmissions, without requiring an additional field in the packet headers.

CSMA/ECA is also related to Reservation Aloha [5] since both approaches separate successful transmissions by a deterministic number of slots. Nevertheless, there is no reservation mechanism in CSMA/ECA.

The adjustment of the contention parameters of CSMA/CA is also a good alternative to improve performance. In [6], an advanced filtering method is proposed to accurately estimate the number of contending nodes in a network and then this information is used to adjust the contention parameters. The approach is fully distributed and it does not require a central entity.

When a central entity is available, it can be used to compute the contention parameters and then distribute them to the remainder of stations using beacon control frames. The solution proposed in [7] is of particular interest because it does not need to estimate the number of contenders. In our approach, we will also make use of a central entity and propose an algorithm that does not need an estimation of the number of contending stations.

Note that both [6] and [7] use CSMA/CA. As a consequence, they are limited in terms of performance by the upper bound presented in [2]. In order to obtain better results, it is necessary to change the protocol. CSMA/ECA proposes a subtle change to the protocol that substantially increases its efficiency.

The performance of CSMA/ECA has already been extensively studied. In [8], simulation is used to assess its performance as perceived by the stations. Specifically, throughput and conditional collision probability plots are provided. It is also shown that CSMA/ECA is advantageous for both elastic (TCP-like) and rigid (UDP-like) traffic flows. Some of these results are captured by an analytical model that is presented in [9].

The ability of CSMA/ECA to handle traffic differentiation is addressed in [10]. That paper also hints the need to adjust the contention parameters of CSMA/ECA in order to accommodate a large number of contenders. This idea is further developed in the present paper.

Some practical issues regarding CSMA/ECA, such as the coexistence with legacy networks, are presented in [11]. It is concluded that CSMA/CA and CSMA/ECA can fairly coexist in the same network since the new protocol is backward compatible with the legacy one. This is an interesting property because it allows for an incremental deployment of CSMA/ECA. The present paper does not address mixed scenarios with new and legacy stations and focuses on a greenfield scenario in which all the stations follow the CSMA/ECA rules.

Previous studies on CSMA/ECA have not considered the possibility of parameter adjustment. When the number of contenders is large and the contention parameters are not adjusted, CSMA/ECA cannot completely suppress the occurrence of collisions. And those collisions prevent an effective use of the channel time. The present paper introduces parameter adjustment for CSMA/ECA and therefore allows for collision-free operation for any number of contending stations.

The main contributions with respect to previous work are as follows. First, the study of fairness among individual CSMA/ECA stations. Second, the assessment of the effect that parameter selection has on performance in a variety of situations. And third, an algorithm that allows the dynamic adjustment of the network parameters to adapt to different network conditions.

3 Carrier Sense Multiple Access with Enhanced Collision Avoidance

When a IEEE 802.11 station joins the CSMA contention for the channel time, it senses the channel for a distributed inter-frame space (DIFS). If the channel is sensed idle, the station transmits. However, if the channel is sensed busy, the transmission will be deferred. A backoff counter is initialized to:

$$B \sim \mathcal{U}[0, CW - 1], \tag{1}$$

where $\mathcal{U}$ represents the uniform random distribution and CW is the value of the contention window, that takes its minimum value CW_{min} for the first transmission attempt.

This backoff counter is decremented every T_e seconds while the channel is sensed idle. Thus T_e is the duration of an empty slot. The backoff counter countdown is frozen while the channel is sensed busy.

When the backoff counter reaches zero, the station transmits. The transmission is successful if the receiver correctly decodes and acknowledges the packet and the sender decodes the acknowledgement control packet. Otherwise, the transmission is a failure and a new random backoff value is selected after doubling the contention window. The contention window value is doubled after each unsuccessful attempt, until reaching its maximum value CW_{max}. There is also a maximum number of retransmissions R.

After a packet has been serviced (either successfully transmitted or discarded due to excessive retransmissions), the contention window is set to its minimum value CW_{min} and the station has to backoff before transmitting another packet. The reason is to prevent that a single station captures the channel by continuously submitting packets.

Although the actual protocol behaviour involves the transmission of signaling packets (ACKs) and different kinds of inter-frame spaces (SIFS and DIFS), it is a common approach in the study of contention protocols to obviate these details (See Fig. 1) . In the following, we consider a simplified model in which we discriminate three kinds of slots, namely: empty, success and collision. The actual content of the successful and collision slots is irrelevant for our discussion.

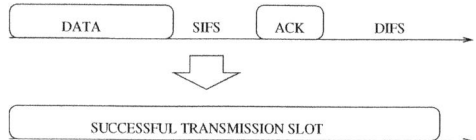

Fig. 1. The behaviour of the protocol is simplified to the slot level

We will assume that the simultaneous transmission of two or more packets results always in a collision, and all the involved packets are lost.

Using the simplification that the channel time can be represented as a succession of slots, we can differentiate three cases in which the backoff is used.

– (i) When a station joins the contention and senses the channel busy.
– (ii) After unsuccessful transmissions.
– (iii) After successful transmissions.

The selection of the backoff should be random for cases (i) and (ii). The reasons are as follows.

Case (i): We have to consider the possibility that two stations join the contention while a third one is transmitting. Then, if the two stations that joined the contention used the same backoff, they would collide in their first transmission attempt. Fig. 2 shows an example in which two stations join the contention

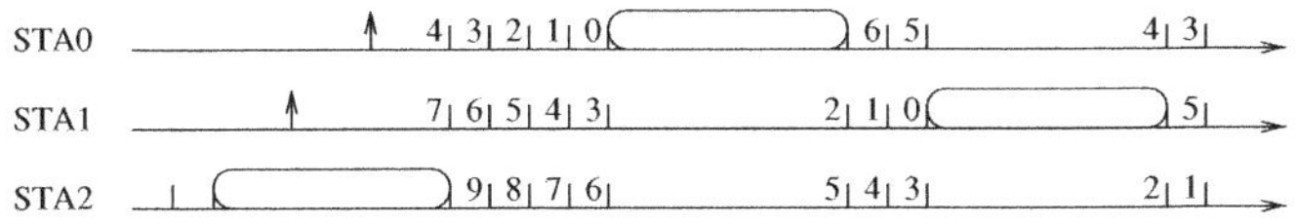

Fig. 2. Example of CSMA/CA. A random backoff is used when the channel is sensed busy after joining the contention.

(represented as an small arrow) while a third one is transmitting. The two new stations choose the initial backoff randomly.

Case (ii): Collisions easily result in unsuccessful transmissions. If the stations that collided chose the same backoff value, they would collide in their next transmission attempt. Fig. 3 shows two stations that collide and then choose random backoffs values and successfully transmit.

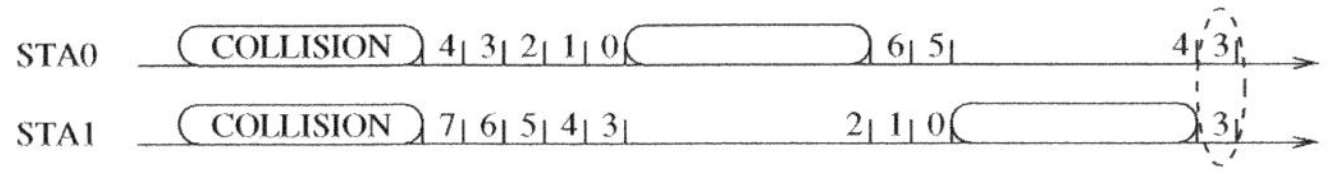

Fig. 3. Example of CSMA/CA. A random backoff is used after successful and unsuccessful transmissions.

In CSMA/CA, all the backoff values are selected randomly. In contrast, CSMA/ECA uses a deterministic backoff after successes (case (iii)). The behaviour of the protocols in each of the three identified cases is summarized in Table 1.

Table 1. Randomness in contention protocols

	CSMA/CA	CSMA/ECA
(i) initial	random	random
(ii) after collision	random	random
(iii) after success	random	deterministic

CSMA/ECA has an advantage over CSMA/CA: two stations that successfully transmitted in their last transmission attempt will not collide among them in their next transmission attempt. Consequently, after all stations have consecutively successfully transmitted, the system operates in a collision-free mode, substantially improving the network performance.

The difference between CSMA/CA and CSMA/ECA can be easily explained by means of an example. Consider Fig. 3 which represents CSMA/CA contention and notice that, after the two stations have successfully consecutively transmitted, it is possible that they share the same backoff value. This will result in a new collision in the stations' next transmission attempt.

Compare this behaviour with the one presented by CSMA/ECA which is depicted in Fig. 4. Consider the second successful transmission that occurs in the figure. After a successful transmission, CSMA/ECA uses a deterministic backoff. Since one and only one station transmitted in the successful slot, the backoff value of the station that just transmitted will be strictly greater than the backoff values from the other stations that successfully transmitted in previous slots. Therefore, collisions are not possible among stations that have successfully transmitted in their last transmission attempt.

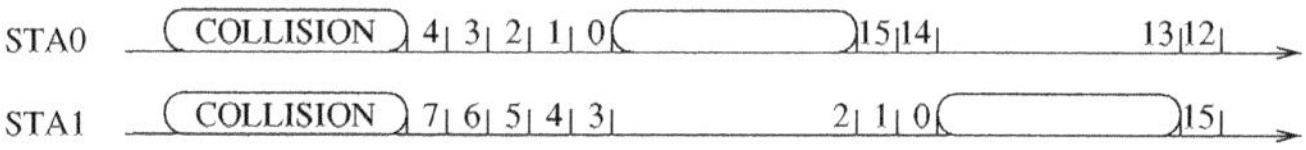

Fig. 4. Example of CSMA/ECA. A random backoff is used after unsuccessful transmissions. Then a deterministic backoff is used after successes.

The proposed protocol, CSMA/ECA, exhibits the following properties:

- Reduced number of collisions when compared with CSMA/CA.
- Possibility of collision-free mode of operation when the number of simultaneous contenders is lower than the deterministic backoff value that it is used after successes.
- The collision-free mode of operation is reached after a transient state, in which the stations perform a random search for empty slots until all the stations consecutively successfully transmit.

In order to make the backoff value deterministic, we use for CSMA/ECA a backoff value V which is equal to the expectancy of the backoff that is used in CSMA/CA. Therefore, V is a function of CW_{min}, which can be dynamically adjusted by means provided in the standard.

$$V = \lceil E\left[\mathcal{U}[0, CW_{min} - 1]\right] \rceil, \tag{2}$$

where $\lceil \rceil$ is the ceiling operator.

4   Parameter Selection and Performance

In the remainder of the article, we will focus on finding the right values for V and CW_{min}. We will use two metrics to assess the performance of the system: efficiency and fairness. As in our previous work, we define efficiency as the fraction of time devoted to successful transmissions (See (2) and (3) in [11]). This metric is also a function of the length of the successful, collision and empty slots. We will assume the default parameter settings in IEEE 802.11b and a frame length of 1500 bytes.

For illustration purposes, simulation results are presented for a scenario with $\varsigma = 8$ contending stations. We have used a custom simulator[1] in Octave [12]

---

[1] Source code is available upon request to the first author.

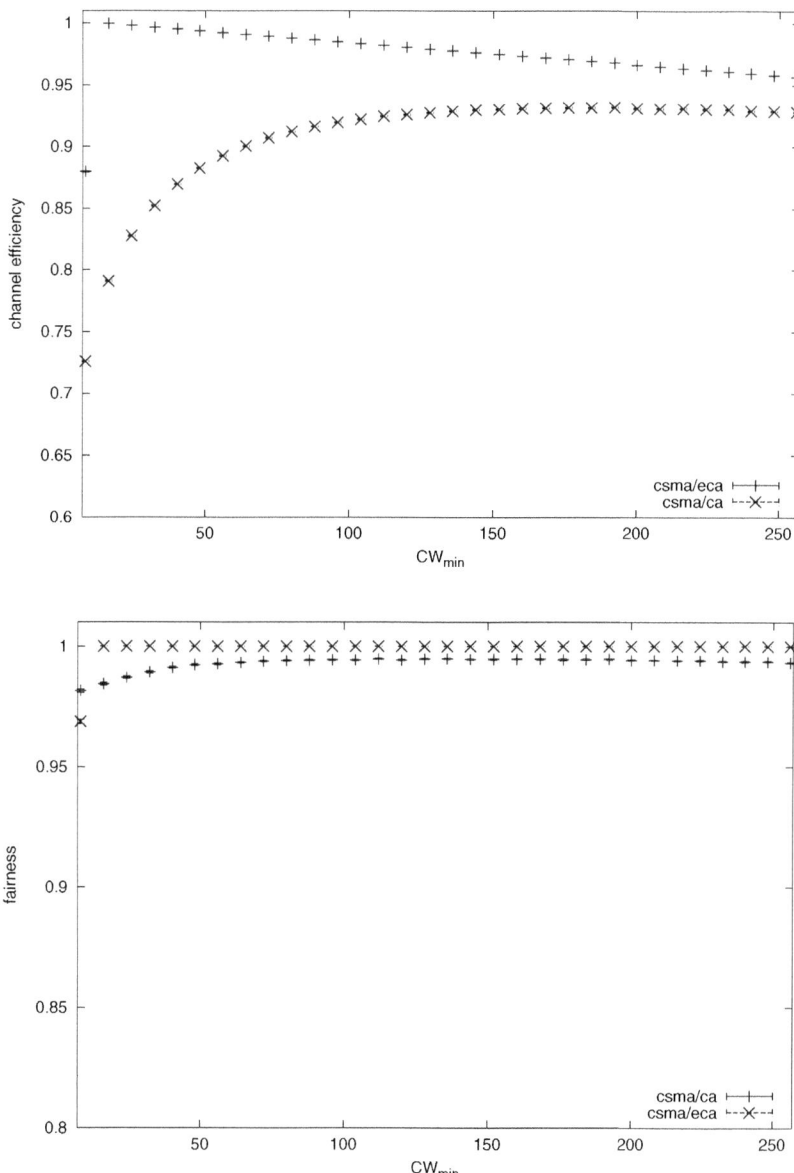

Fig. 5. Long term (steady-state) efficiency and fairness as a function of CW_{min} when the number of contenders equals 8

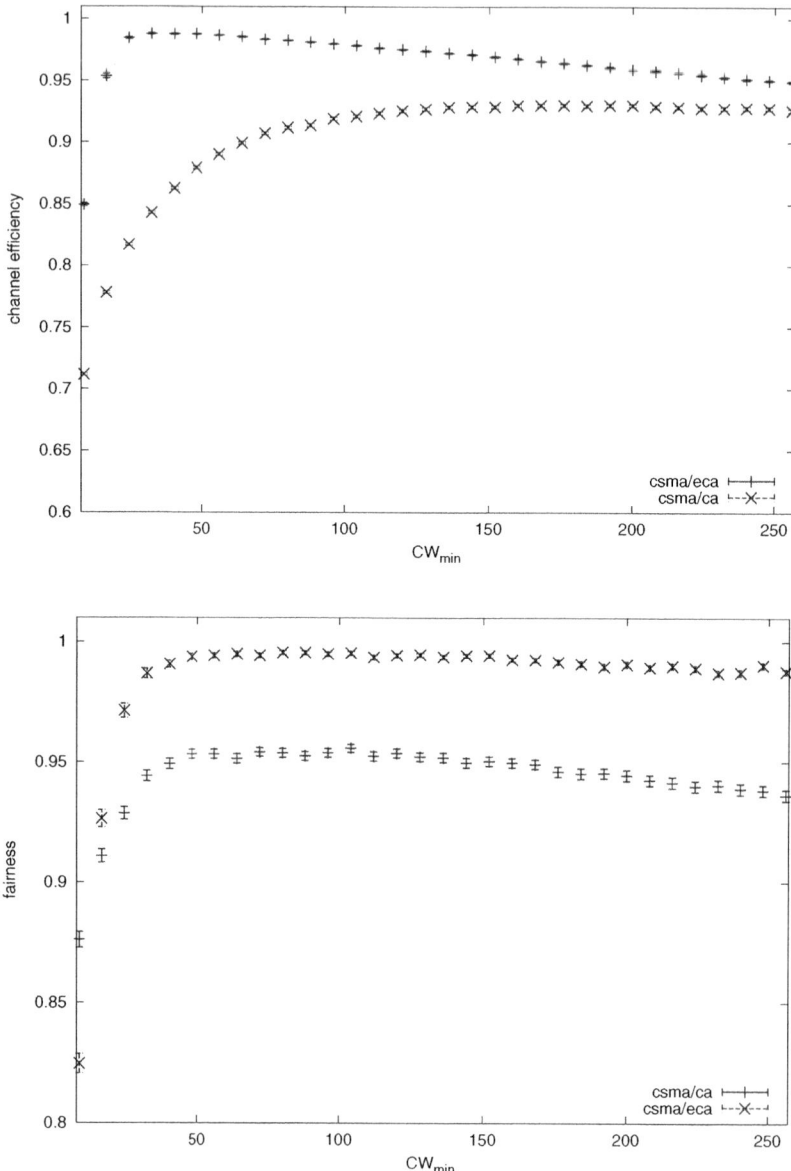

Fig. 6. Short term (transient) efficiency and fairness as a function of CW_{min} when the number of contenders equals 8

that only implements the contention mechanism. In all the plots, the values obtained using CSMA/CA are also provided for comparison purposes. All the simulations in this section are repeated 1000 times and average values and 95% confidence intervals have been computed. To assess short-term performance, we perform simulations that are 1000 slots long and include the transient-state. On the other hand, to assess long-term performance, the simulations are 10,000 slots long and the transient-state is not taken into account.

Fairness will be computed using Jain's fairness index [13] on the number of successful slots by the end of the simulation. Since we will perform simulations of different lengths, we will obtain results for short-term and long-term fairness. All of them are interesting properties of a medium access mechanism.

From the theoretical standpoint, and under the ideal channel and steady-state assumptions, CSMA/ECA delivers its maximum performance when the minimum contention window doubles the number of contending stations ($CW_{min} = 2\varsigma$), i.e., when the deterministic backoff after successes is equal to the number of contending stations ($V = \varsigma$). In this particular case all the slots are filled with successful transmissions and the stations transmit in a round-robin fashion. In fact, both the efficiency and the fairness are equal to one in these ideal conditions.

Fig. 5 plots the efficiency and fairness of CSMA/ECA in long simulations where the steady-state assumption is fulfilled. Since the number of contenders is eight, the maximum efficiency (one) is obtained when $V = \varsigma = 8$ and $CW_{min} = 2\varsigma = 16$. The maximum fairness one is obtained for any $V \geq \varsigma = 8$ and $CW_{min} \geq 2\varsigma = 16$.

Unfortunately, this simple approach to performance optimization may not be useful for real network configuration. The reason is that in CSMA/ECA the steady-state assumption may be unrealistic. Notice that in shorter simulations where the transient-state has been taken into consideration (See Fig. 6^2), the maximum of the performance curves differ from those obtained in the steady-state.

5   Dynamic Parameter Adjustment

An interesting observation is that the curves for CSMA/ECA are relatively flat when $CW_{min} \geq 2\varsigma = 16$. In this region, CSMA/ECA delivers high performance results across all the different plots. The ability of CSMA/ECA to deliver high performance for a wide range of values of CW_{min} will be very useful in real deployments. Our goal is to propose an approach that dynamically adjusts the configuration parameter CW_{min} to guarantee that the network stays in this favorable region of operation.

---

[2] There are small oscillations in the fairness curves presented in Fig. 6. They are a measure error related to the use of a measurement window (1000 slots) and they cannot be suppressed by increasing the number of simulations.

We want to guarantee that $CW_{min} \geq 2\varsigma$, and this can be achieved by setting[3]

$$CW_{min} \approx 8\varsigma. \tag{3}$$

Since CSMA/ECA is capable of preventing collisions and a deterministic backoff V is used after successes, each station successfully transmits in one out of V slots. Consequently, the ratio of busy slots in the overall system is:

$$\beta = \frac{\varsigma}{V} = \frac{2\varsigma}{CW_{min}}. \tag{4}$$

By substituting (3) into (4) we obtain that our target fraction of busy slots should be:

$$\beta_{target} = \frac{1}{4} \tag{5}$$

As in [7], we adopt a centralized approach to parameter adjustment. In particular, the access point takes measures of the channel, computes the configuration parameters and distributes this parameters to the stations using the beacon frames that are transmitted every 100ms. The information regarding CW_{min} is codified in a 4-bit field in the beacon frame. The standard restricts CW_{min} to be an integer power of two, and the information contained in the 4-bit field is, actually, the exponent.

Our proposed algorithm that adjusts CW_{min} to attain β_{target} while adhering to the standard restrictions is iterative. The access point measures the actual fraction of busy slots β during a 100ms interval, and then it uses that value and the previous minimum contention window $CW_{min,i}$ to compute the next contention window $CW_{min,i+1}$ as follows:

$$\begin{aligned} CW_{min,0} &= CW_{min}^{default}, & i = 0 \\ CW_{min,i+1} &= \max(CW_{min}^{default}, CW_{min,i} \cdot 2^{round(\log_2(\beta/\beta_{target}))}), & i > 0 \end{aligned} \tag{6}$$

where the $round()$ operator approximates a given value to the closest integer. Note that when $\beta = \beta_{target}$ the value of CW_{min} is maintained ($CW_{min,i+1} = CW_{min,i}$). Note also that we restrict CW_{min} to be larger or equal to the default value of the standard.

A common approach in the literature to evaluate the performance of an adaptation algorithm is to test its reaction when the number of contending stations suddenly changes. As an example, [6] studies the case in which the number of contenders jumps from 6 to 11. Similarly, [7] presents an example in which the number of stations instantaneously changes from 15 to 30.

In order to evaluate the performance of the proposed algorithm, we will stress it by simulating a scenario where twenty saturated stations simultaneously join an empty network. We will take measures such as the achieved efficiency, fairness

---

[3] The use of the value 8 is a design decision that ensures a fast convergence to the (collision-free) steady state and a good steady-state performance. See [11] for details on the duration of the transient state and the steady-state efficiency.

and number of successful transmissions. We also will see how the algorithm updates CW_{min} in oder to adapt to the new situation. Every 100ms, measures are taken and then the new value of the contention window is computed and sent using the beacon frame. The idealistic assumption that the access point transmits its frame in an empty slot has been used. It should not be difficult to prevent collision between data frames and the control beacon frame by using the *smart entry* approach presented in [11].

Table 2. Performance of the adaptive algorithm when the number of contending stations instantaneously changes from zero to twenty

time (s)	CW_{min}	Success	Collision	Empty	Efficiency	Fairness
0-0.1	32	33	27	30	0.55	0.45
0.1-0.2	64	53	5	31	0.91	0.62
0.2-0.3	128	56	2	90	0.95	0.72
0.3-0.4	256	56	1	208	0.94	0.80
0.4-0.5	256	57	0	210	0.96	0.87
0.5-0.6	256	56	0	202	0.96	0.93
0.6-0.7	256	57	0	203	0.96	0.93

The results of the simulation are presented in Table 2. It can be observed that the adaptation algorithm repeatedly doubles the contention window to make room for the new entrants. Meanwhile, CSMA/ECA reduces the collisions. As a result, the network is delivering high performance half a second after the sudden increase of the number of contending stations.

6   Conclusions

CSMA/ECA is a contention protocol that is identical to CSMA/CA, with the exception that it uses deterministic backoffs after successful transmission. In previous work, we have shown that CSMA/ECA significantly outperforms CSMA/CA. However, when used with static configuration parameters, CSMA/ECA is not suited for networks where the number of contenders is very large. We have addressed that particular problem in the present paper.

First, we have studied how the performance of CSMA/ECA varies as we change the configuration parameters. We have obtained results both for efficiency and fairness, since they are important measures to evaluate a random medium access control protocol.

We have seen that a theoretical approach that disregards the transient state is not useful to predict the performance of CSMA/ECA in a real network. Therefore, we have also used short simulations that reflect the behaviour of the protocol in a highly dynamic network.

We have used that information to propose an adaptation protocol to guarantee that CW_{min} is within the range of values that deliver high performance. We have verified that the proposed protocol works by means of simulation and we have included an example of the operation of the protocol.

Acknowledgements

This work was partially supported Spanish Government under project TEC2008-06055/TEC.

References

1. IEEE 802.11: Wireless LAN Medium Access Control (MAC) and Physical Layer (PHY) Specification (2007)
2. Cali, F., Conti, M., Gregori, E., Aleph, P.: Dynamic Tuning of the IEEE 802.11 Protocol to Achieve a Theoretical Throughput Limit. IEEE/ACM Trans. Netw. 8(6), 785–799 (2000)
3. Bianchi, G.: Performance Analysis of the IEEE 802.11 Distributed Coordination Function. IEEE J. Sel. Areas Commun. 18(3), 535–547 (2000)
4. Choi, J., Yoo, J., Choi, S., Kim, C.: EBA: An Enhancement of the IEEE 802. 11 DCF via Distributed Reservation. IEEE Trans. Mobile Comput. 4(4), 378–390 (2005)
5. Crowther, W., Rettberg, R., Walden, D., Ornstein, S., Heart, F.: A System for Broadcast Communication: Reservation-ALOHA. In: Hawaii Int. Conf. Syst. Sci., pp. 596–603 (1973)
6. Lopez-Toledo, A., Vercauteren, T., Wang, X.: Adaptive Optimization of IEEE 802.11 DCF Based on Bayesian Estimation of the Number of Competing Terminals. IEEE Trans. Mobile Comput. 5(9), 1283 (2006)
7. Patras, P., Banchs, A., Serrano, P.: A Control Theoretic Approach for Throughput Optimization in IEEE 802.11 e EDCA WLANs. In: Mobile Networks and Applications, pp. 697–708 (2009)
8. Barcelo, J., Bellalta, B., Sfairopoulou, A., Cano, C., Oliver, M.: CSMA with Enhanced Collision Avoidance: a Performance Assessment. In: IEEE VTC Spring (2009)
9. Barcelo, J., Bellalta, B., Cano, C., Sfairopoulou, A., Oliver, M.: Carrier Sense Multiple Access with Enhanced Collision Avoidance: a Performance Analysis. In: ACM IWCMC (2009)
10. Barcelo, J., Bellalta, B., Cano, C., Sfairopoulou, A., Oliver, M., Zuidweg, J.: Traffic Prioritization fo Carrience Sense Multiple Access with Enhanced Collision Avoidance. In: MACOM (2009)
11. Barcelo, J., Lopez-Toledo, A., Cano, C., Oliver, M.: Fairness and Convergence of CSMA with Enhanced Collision Avoidance. In: ICC (2010)
12. Eaton, J.W.: GNU Octave Manual. Network Theory Limited (2002)
13. Jain, R.: The Art of Computer Systems Performance Analysis. John Wiley & Sons, New York (1991)

A Test-Based Scheduling Protocol (TBSP) for Periodic Data Gathering in Wireless Sensor Networks*

Mario Orne Díaz-Anadón and Kin K. Leung

Electrical Engineering Department
Imperial College, London SW7 2BT, United Kingdom
{orne.diaz06,kin.leung}@imperial.ac.uk

Abstract. We propose TBSP, a TDMA protocol for gathering information periodically from multiple data sources to a central location across multiple hops. TBSP incurs a lower overhead than the existing protocols because it only requires the sensor nodes to keep track of their own schedule, whereas in the existing protocols each node needs to know the schedule of its neighbors. In order to gain a transmission slot, the sensor nodes transmit in different slots until they find one with sufficiently low interference. TBSP provides mechanisms to reduce the probability that the sensor nodes steal each other's slots or thwart each other's efforts. Our simulation study reveals that TBSP provides great energy savings if the network changes slowly and the sensor nodes can wait a dozen TDMA frames before obtaining a transmission slot. TBSP also provides the advantage of being more likely to keep the sensor network connected in sparse networks.

Keywords: adaptive; TDMA; wireless sensor networks.

1   Introduction

(WSNs) are used to collect signals such as temperature, acceleration or video from multiple locations within a certain geographic area. They can be deployed faster and at lower cost than their wired counterparts, which has led them to find applications in a range of environments, including water pipes, forests and battlefields [1]. A WSN consists of a large number of *sensor nodes*, which are devices equipped with a battery, a sensing module, a microcontroller and a radio transceiver. Usually, the sensor nodes have little energy and computational power. We propose a Test-Based Scheduling Protocol (TBSP) that obtains and maintains a TDMA schedule for periodically transmitting packets from a set of sensor nodes referred to as *data sources* to a special node referred to as *data sink* across multiple hops. TBSP is designed for applications in which the set

---

* This work is funded by UK EPSRC Research Grant EP/D076838/1, entitled: "Smart Infrastructure: Wireless Sensor Network System for Condition Assessment and Monitoring of Infrastructure".

A. Vinel et al. (Eds.): MACOM 2010, LNCS 6235, pp. 25–35, 2010.

of data sources, their data rate, or the properties of the wireless links change infrequently. The main goal of TBSP is to adapt the transmission schedule to those infrequent changes in an energy-efficient way. The focus of TBSP is the adaptation of the schedule rather than its initial construction because the initial scheduling overhead is negligible in long-running monitoring applications. TBSP is scalable because it is distributed and its buffering requirements do not grow with the network size. Our simulation studies reveal that TBSP consumes less energy than the existing protocols in networks that change relatively slowly. In addition, TBSP is more likely to obtain a collision-free schedule than the existing protocols because the existing protocols assign a given transmission slot to a set of nodes if the hop distance between them exceeds a certain number, whereas TBSP only assigns them the same slot if it has been proved empirically that they tolerate each other's interference. The rest of the paper is organized as follows. Section 2 reviews some related work, Section 3 describes our protocol, Section 4 presents our simulation methodology and results, and Section 5 concludes the paper.

2   Related Work

MAC protocols for wireless sensor networks can be classified as contention-based or TDMA-based [4]. The contention-based protocols respond quickly to traffic demand variations but waste bandwidth and energy due to back-off periods, packet collisions, idle listening and overhearing. The TDMA-based protocols avoid these problems during the data transmission phase but incur in scheduling and synchronization overhead. Overall, if the data are transmitted frequently and periodically, which is the case as in the applications that we are considering, the TDMA-based protocols outperform the contention-based protocols. Some hybrid protocols seeking to combine the benefits of contention and TDMA have been proposed [4], but for stationary networks they are less efficient than TDMA [5]. Therefore, we focus our attention in TDMA protocols, and more specifically in distributed TDMA protocols since centralized protocols such as [3] incur significant maintenance overhead in large networks.

Let us make some general definitions about distributed TDMA protocols. Since we are interested in periodic data-gathering in multihop networks, we assume that a *routing tree* rooted at the data sink has been established. In this tree, each node receives packets from its child nodes and relays these packets to its parent node. Time is divided in TDMA frames, and each TDMA frame consists of a number of transmission slots. The schedule is relatively periodic in the sense that, if a node is allocated a certain transmission slot, it is allocated that transmission slot in several consecutive TDMA frames. Let X be a sensor node and let Y be its parent node. We say that a transmission slot allocated to X is *unfeasible* if the joint interference from other nodes during that slot is so large that Y cannot receive packets from X or X cannot receive ACKs from Y. Feasible slots may become unfeasible, forcing their owners to seek new transmission slots. If a node's slot becomes unfeasible because of the interference of some nodes that recently started using that slot, we say that the node has been *expelled* from its slot or that the node has suffered an *expulsion*.

TRAMA [8] is a distributed TDMA scheduling protocol that adapts the schedule to traffic changes very quickly but also consumes much energy. FLAMA [7] greatly reduces the energy consumption by taking advantage of the fact that the adaptivity requirements in many monitoring applications are modest. However, the nodes in FLAMA incur significant overhead in either keeping track of the priorities of all their two-hop neighbors or listening in slots in which they do not receive packets. In addition, FLAMA does not minimize the buffering requirements or the *traversal time*, which we define as the maximum number of TDMA frames needed by a packet from the data sources to reach the data sink. FlexiTP [5] solves these problems by considering latency and memory requirements in its slot-selection algorithm. In Section 4, we present simulation results comparing the performance of FlexiTP with that of our protocol.

To our knowledge, every existing distributed TDMA protocol for periodic data gathering in WSNs uses a certain *interference model*, which is a model to decide when a node's interference is negligible. The most common interference model is to neglect the interference originated more than $k = 2$ hops away, which is a model referred to as the k-hop interference model. Unlike the existing algorithms, TBSP does not use an interference model. We say that TBSP is a *test-based* protocol because a node is allocated a given slot only if that slot has been proved feasible in an empirical *test* that consists in transmitting a dummy packet in that slot and checking whether the reception of this packet is acknowledged with an ACK packet.

The existing protocols obtain fewer feasible slots than TBSP does because their interference models may fail whereas TBSP bases its scheduling decisions in empirical tests. When the existing protocols assign an unfeasible slot to a certain node X, X incorrectly assumes that its parent node is unreachable, which makes X disconnected from the network if X does not have other neighbors. TBSP is less likely to suffer this connectivity problem because, up to a maximum number of times, if X cannot communicate with its parent node during a slot, X tries to change its slot, not its parent. Another disadvantage of the existing protocols is that they require that every node listens to its neighbors in case it receives messages from them indicating changes in their schedules. In contrast, TBSP does not incur this idle-listening overhead because all scheduling decisions are taken distributedly and without knowledge of other nodes' schedules.

3   TBSP

The purpose of the Test-Based Scheduling Protocol (TBSP) is to obtain and maintain a TDMA transmission schedule to transfer the data from the data sources to the data sink periodically. The minimum unit that can be scheduled to a node per TDMA frame is a Data Subframe (DS), which consists of two data transmission slots. Since each DS should be used during at least 20 consecutive TDMA frames, TBSP is only suitable for relatively static networks. Although we

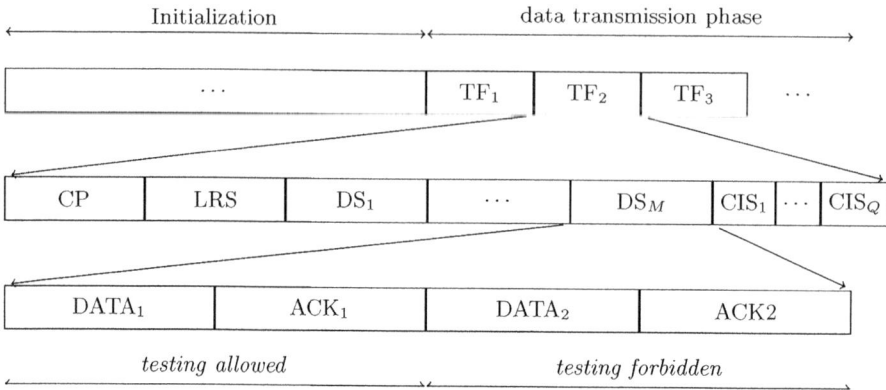

Fig. 1. Timing diagram of TBSP

assume infrequent network changes, we regard those changes as very important, and the goal of TBSP is to adapt the schedule to them in an energy-efficient way.

Every node seeking to obtain a DS is referred to as a *contender*, and the DS that it seeks to obtain is referred to as its *target DS*. A simplified description of the mechanism whereby a contender gains a DS is as follows. First, the contender listens during a TDMA frame in order to identify DSs with low interference. Then, the contender selects one of those DSs as its target DS. The target DS may be unfeasible due to the hidden terminal problem, and the contender determines whether that is the case by testing its target DS. The test consists in transmitting a packet referred to as *Testing Packet* during its target DS and checking whether it receives an ACK in response. The contender repeats the whole process with different target DSs until the test is positive, in which case it has gained its target DS and starts using it.

3.1   Timing Diagram

Figure 1 shows that TBSP consists of an initialization phase followed by a data transmission phase. The goals of the initialization phase are to obtain a routing tree and to synchronize the sensor nodes for the first time. These two goals are achieved by executing the distributed Bellman-Ford algorithm and the FTSP protocol [6], respectively. The data transmission phase extends during the rest of the network lifetime and consists of TDMA Frames (TFs), each of which is composed of a CSMA Period (CP), a Listening Request Slot (LRS), M Data Subframes (DSs) and Q Collision-Indication Slots (CISs). Every node keeps its transceiver active during the CP, the LRS, any DS during which it is scheduled to transmit or receive, and the CISs. We explain each of these slots as follows.

The CP is a CSMA period with two purposes. First, it is used to request and receive neighbor information by nodes that need to change their parent node in the routing tree. Second, it is used to request and receive synchronization information by nodes that are not engaged in periodic data transmissions. Other nodes do not use this synchronization method because the ACKs that they receive from their parents as a response to their periodic packet transmissions already contain synchronization information.

The Listening Request Slot (LRS) is a CSMA period used to transmit at most one Listening Request Packet (LRP), which is a packet whereby the contenders request their parents to listen during their target DSs.

The DSs are the Data Subframes that the sensor nodes use for their periodic packet. A DS is the minimum time unit that can be assigned to a node. It consists of two DATA slots with their respective ACK slots. We refer to these slots as $DATA_1$, ACK_1, $DATA_2$ and ACK_2. The ACK packets include synchronization information. During the $DATA_1$ slot the contenders can transmit Testing Packets, but not during the $DATA_2$ slot. If a node X receives a packet during the $DATA_2$ slot but not during the $DATA_1$ slot, the first packet was probably lost because another node, which we call Y, generated a collision by transmitting a Testing Packet. In this case, we say that Y is the *originator* of a *contender-induced collision* and X is its *victim*. If Y receives an ACK during ACK_1, Y seizes its target DS, thereby expelling X, which is undesirable.

The Collision-Indication Slots (CISs) are Q very short slots that are used to reduce the number of expelled nodes by propagating a *collision indication*, which is an indication that a contender-induced collision has occurred. We say that a node receives a collision indication if it senses the channel busy during any of the CISs. Collision indications are propagated as follows. In CIS_1, the victim transmits a dummy packet in order to make the channel busy. When the victim's one-hop neighbors sense the collision indication, they propagate it to their own one-hop neighbors during CIS_2. This process is repeated until CIS_Q so that the collision indication reaches all the nodes within Q hops of the victim. Hopefully, the originator of the collision also receives the indication, in which case it refrains from seizing its target slot in order to prevent an expulsion. This collision-indication mechanism is inappropriate in noisy environments because noise might be interpreted as a collision indication. In such noisy environments, we recommend not using the CISs at all (i.e. setting $Q = 0$), which according to our simulation results provides degraded but sufficient results. As an added benefit, the suppression of the CISs simplifies the protocol.

3.2   Algorithm to Gain a DS

Every contender uses a *DS-selection algorithm* to select its target DS. Then, each contender uses a *wait-selection algorithm* to decide the number of TDMA frames it waits before testing its target slot. This wait aims to address the hidden terminal problem by reducing the probability that two contenders' attempt to gain the same slot simultaneously, which would reduce the probability of obtaining a DS for both nodes. After the wait, the contender contends to transmit a Listening Request Packet in every TF until it succeeds. When it succeeds, the contender transmits a Testing Packet during the $DATA_1$ slot of its target DS. If it receives an ACK, it has obtained its target DS. Otherwise, the contender repeats the whole process until it obtains a DS. We detail the DS-selection algorithm and the backoff algorithm as follows.

The DS-selection algorithm. Each contender selects its target DS among the DSs that verify the following properties. First, the signal level detected during that DS by the contender must be low. Second, the contender should not have tested that DS yet. Third, if the contender seeks the DS in order to relay a packet from a child node, the number of the target DS must exceed the number of the DS in which it is scheduled to receive packets from its child. This third property is discussed in [5] and ensures small traversal time. We refer to the DS with the smallest number that verifies the previous properties as DS_{min}. If the contender has made fewer than a certain number of attempts (in our simulations this number is four), the contender selects DS_{min} as its target DS. Otherwise, the contender selects the first DS that verifies the above properties and whose number exceeds the number of DS_{min} by a small random integer.

The wait-selection algorithm. Every contender X keeps track of the number n_c of consecutive TFs without collision indications. Node X also selects a *wait parameter* n_b as a random integer between 0 and $N_b - 1$ (in our simulations, $N_b = 8$). In each TF, the algorithm executed by X depends on the value of $n_c - n_b$:

- If $n_c - n_b$ is negative, X does not contend during the LRS of the current TF. If X senses a collision indication, there is probably a contender nearby. In order to avoid ruining the efforts of that contender, X resets n_b as a random integer between N_b and $2N_b - 1$.
- If $n_c - n_b$ is non-negative, X contends to transmit a Listening Request Packet (LRP) in the LRS using CSMA. The backoff period to transmit the LPR is a random quantity that is shorter for contenders closer to the data sink or with full buffers. After X transmits the LRP, it transmits its Testing Packet in the $DATA_1$ slot of its target DS. If X receives an ACK from its parent Y during slot ACK_1, X has obtained its target DS, unless the ACK packet from Y indicates that Y's buffer is full, in which case X resets its backoff parameter n_b as a random integer between $3N_b$ and $4N_b - 1$. If, on the other hand, X receives no response from Y, X sets $n_c = 0$ and n_b with a random integer between 0 and $N_b - 1$, and reruns the DS-selection algorithm.

4   Performance Evaluation

In order to evaluate the performance of TBSP, we develop a simulator using the Python programming language. The code of our simulator and all our simulation parameters are available in [2]. We compare our protocol with FlexiTP [5] because FlexiTP is the only existing distributed and adaptive TDMA protocol for periodic data gathering that shares our goal of minimizing the buffering requirements and the traversal time. The major difference between TBSP and FlexiTP is that FlexiTP uses the traditional 2-hop interference model whereas TBSP follows our test-based approach of not keeping track of other nodes' schedules and assigning a given DS to a set of nodes if they have been proved empirically to be able to tolerate each other's interference.

4.1   Simulation Scenario

The transmit and receive power are 63 mW, which are common in current hardware. The noise figure of the receiver is 4.8 dB. The wireless channel has attenuation exponent of 3.5, path loss at 100 m of 80 dB, and standard deviation of the log-normal fading of $\sigma_f = 8$ dB. A transmission succeeds if the SINR at the receiver exceeds 20 dB. Although there is no common transmission range for all the network, we define t as a node's hypothetical transmission range if there were no fading and interference; in our case, t is 48 m.

The sensor nodes are randomly deployed within a square of side $3t$ and the data sink lies in the middle of one of the sides of the square. We define the *node density* ρ as $N(\pi t^2)/A$, where N is the number of nodes and A is the area of the square. Therefore, ρ is the average number of one-hop neighbors per node. We discard any deployments where more than 10 % cannot reach the data sink, which is rare for $\rho \geq 7$. All the nodes are data sources and the routing tree is the shortest path tree. Our results are computed as the average of 700 simulation runs.

In order to evaluate the adaptivity of the two protocols, we simulate a network that changes periodically at discrete points in time. We refer to each interval during which no changes occur as a *network cycle*. At the beginning of each network cycle, we remove a number $S = 3$ of nodes from the network and add an identical number of new nodes at random locations within the monitored area. These removals and additions force some sensor nodes to seek new parents and transmission slots. We compare the way in which FlexiTP and TBSP enable those nodes to obtain new slots as follows.

4.2   Total Energy Consumption E_t

We now define three energy metrics, all of which exclude the energy consumed in data transmissions, time synchronization and parent node selection because these components are the same for TBSP and FlexiTP. First, we define the *fixed energy consumption* E_f as the energy consumed per TDMA frame and per node when no contenders are present in the network. Second, we define the *variable energy consumption* E_v as the sum of extra energy consumed by all the nodes each time that a transmission slot is acquired. Third and most importantly, E_t is the total energy consumed per node and per TDMA frame and is given by

$$E_t = E_f + \frac{E_v f_s}{N}, \tag{1}$$

where N is the total number of nodes in the network and f_s is the average number of slots that have to be gained in the network per TDMA frame.

We also define n_n as the number of contenders forced to seek a slot by a topology change, and n_e as the number of contenders forced to seek a slot by an expulsion. We define the *adaptation latency* as $l = n_w/n_n$, where n_w is the total number of TDMA frames waited by all the contenders before obtaining a slot. In other words, the adaptation latency is the average number of TDMA frames to required to gain a slot multiplied by $1 + \epsilon$, where $\epsilon = n_0/n_n$ is the expected number of expelled nodes each time that a node obtains a slot. We make this definition of adaptation latency in order to consider the negative influence of expulsions.

In FlexiTP, E_t is equal to E_v and proportional to T_f, which is the duration of the FTS. The FTS is a CSMA slot used by the FlexiTP nodes to notify their 2-hop neighbors of the slots that they are going to use. The adaptation latency l decreases linearly with T_f until a certain point. We refer to the l, T_f and E_t at this point as l_0, T_f^0 and E_t^0 respectively. In our simulations, we use set $T_f = T_f^0$ because it allows us to compute the energy consumption for any adaptation latency l_1 with the expression $E_t = E_t^0 l_0/l_1$.

In order to compare the energy consumption of FlexiTP with that of TBSP in a fair way, we have to consider the adaptation latency l simultaneously. Figure 2 presents the adaption latency as a function of the node density. FlexiTP's latency is much smaller than that of TBSP for the simulated parameters, and we take this into account when we compare E_t for the two protocols. Figure 2 shows that the use of CISs (which are the slots shown in Figure 1) in TBSP reduces the latency, but only for high node densities. For high node densities the CISs reduce the latency because they reduce the number n_e of expulsions. However, for low densities the CISs do not reduce the latency because n_e is already very small without using any CISs.

Table 1 compares FlexiTP and TBSP in terms of E_f, E_v and E_t for different node densities. Let us first examine E_f and E_v for the two protocols independently. The two protocols suffer an increase in the energy consumption with the node density. In the case of FlexiTP, this is because a larger T_f^0 is needed for

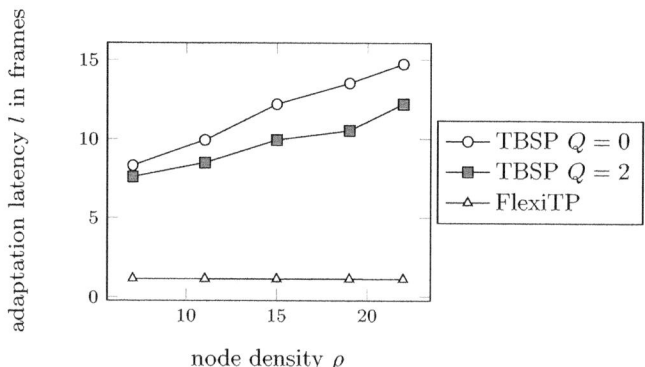

Fig. 2. Adaptation latency l. Note that although the latency of FlexiTP is up to 15 times smaller than that of TBSP, FlexiTP consumes at least 47 times more energy than TBSP does according to Table 1.

Table 1. Comparison of energy metrics. The last column shows $\bar{G}$, which is the number of times that TBSP outperforms FlexiTP in terms of E_t when we set the same maximum adaptation latency for the two protocols.

Simulation parameters			Energy metrics in mJ			
Protocol	ρ	E_f	E_v	E_t	G	$\bar{G}$
FlexiTP	7	18.425	0.0	18.43		
	14	31.017	0.0	31.02		
	21	38.916	0.0	38.92		
TBSP with $Q = 0$	7	0.031	65.9	0.36	51.2	6.2
	14	0.031	193.1	0.51	60.8	5.0
	21	0.031	335.6	0.59	66.0	4.5
TBSP with $Q = 2$	7	0.094	59.4	0.39	47.3	6.2
	14	0.094	157.0	0.49	63.3	6.4
	21	0.094	276.0	0.55	70.8	5.8

a node to communicate its list of transmission slots to its increased number of neighbors. In the case of TBSP, this is because more DSs are used for periodic transmissions around each contender, which increases the number of failed attempts to gain a DS and the number of expulsions. The node density is also important because it determines the usefulness of the CIS mechanism. The CISs are only beneficial for high node densities, since for low node densities the slight reduction of the number of expelled nodes does not warrant the overhead it incurs.

Now let us examine in Table 1 the value of the total energy consumption E_t. This is the most important energy metric that we consider since it considers both E_f and E_v using (1). Equation (1) shows that the speed of change of the wireless links, which is controlled by f_s, greatly affects E_t. Table 1 uses $f_s = 1/20$, which implies that the network varies relatively slowly. For this value, the quotient between E_t for FlexiTP and TBSP, which we define as G, ranges between 47.3 and 70.8. However, in order to provide a fair comparison, we have to compare the two protocols for the same adaptation latency. We provide this fair comparison by defining $\bar{G}$. We define $\bar{G}$ as the number of times that TBSP outperforms FlexiTP when FlexiTP uses the shortest FTS that achieves the same adaptation latency as TBSP does. Table 1 shows that TBSP is approximately 6 times more energy efficient than FlexiTP, but the results would vary for other values of f_s. If we reduce f_s, the advantage of TBSP over FlexiTP grows, and if we increase f_s over $1/3$, FlexiTP becomes more efficient than TBSP. However, in environments with a large f_s, both TBSP and FlexiTP are outperformed by contention-based MAC protocols and TRAMA [8].

4.3   Probability That a Node Is Assigned an Unfeasible Slot

Figure 3 shows the probability p_u that a node is assigned an unfeasible slot. For FlexiTP, this probability decreases with the node density because a greater

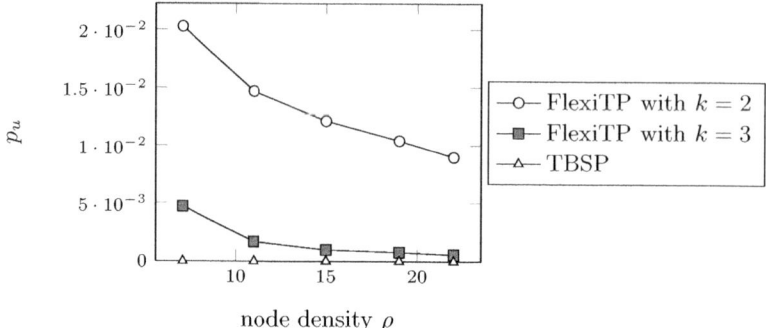

Fig. 3. Probability p_u that a node is allocated an unfeasible slot

node density increases the number of conditions that are required for a pair of nodes to be assigned the same transmission slot. Figure 3 presents two versions of FlexiTP. The first one is the original version, which neglects the interference from a node if it lies at least $k = 2$ hops away, and the second one is a modified version that uses $k = 3$. The modified version consumes twice as much energy as the original version because it requires that each node reports its schedule to a greater number of neighbors. Although this modified version of FlexiTP reduces p_u, the obtained p_u is non-negligible. In FlexiTP, when a node obtains an unfeasible slot, it unduly concludes that its parent node is unreachable since according to its schedule information there is no interference in that slot. Therefore, the node seeks a new parent, but in sparse networks, the node may not have any other neighbor, in which case the node cannot reach the data sink. In contrast, TBSP guarantees $p_u = 0$ because it only assigns a slot to a node if the node's ability to communicate during that slot has been proved with a Testing Packet.

4.4   Other Properties of the Computed Schedule

Our DS-selection algorithm seeks to obtain a schedule with the following properties. First, the schedule is short, which means that many nodes share each DS. A short schedule requires a smaller number M of DSs per TF (shown in Figure 1), thereby decreasing the frame rate and increasing the throughput. Second, the schedule enjoys a small traversal time (i.e. the packets from the data sources reach the data sink within a small number of TDMA frames). Third, the schedule imposes buffering requirements that do not grow with the network size. FlexiTP also seeks a schedule with these properties. Our simulation results show that TBSP performs similarly in these three aspects but we do not show these results due to space limitations.

5   Conclusions

The energy consumption of different scheduling protocols varies widely with the maximum tolerable value of the adaptation latency, which is approximately the average number of TDMA frames until a node obtains a transmission slot. If the network changes slowly and an adaptation latency of 15 frames is tolerable, for our simulation parameters, TBSP consumes six times less energy than Flexi-iTP. TBSP is also likely to outperform the other existing distributed TDMA protocols because it spares the sensor nodes from the burden of keeping track of their neighbors' schedules. Additionally, since TBSP does not use an interference model, the sensor nodes are less likely to incorrectly assume that their parent nodes are unreachable, and thus they are also less likely to become disconnected.

References

1. Arampatzis, T., Lygeros, J., Manesis, S.: A survey of applications of wireless sensors and wireless sensor networks. In: Proc. IEEE Mediterranean Conf. Control and Automation, pp. 719–724 (2005)
2. Díaz-Anadón, M.O.: Randsched implementation (2009),
   http://github.com/ornediaz/wsnpy.git
3. Gandham, S., Ying, Z., Qingfeng, H.: Distributed minimal time convergecast scheduling in wireless sensor networks. In: Proc. IEEE Int'l Conf. Distributed Computing Systems, p. 50 (2006)
4. Langendoen, K.G.: Medium access control in wireless sensor networks. In: Wu, H., Pan, Y. (eds.) Medium Access Control in Wireless Networks. Practice and Standards, vol. II, pp. 535–560. Nova Science Publishers, Hauppage (May 2008)
5. Lee, W.L., Datta, A., Cardell-Oliver, R.: FlexiTP: A flexible-schedule-based TDMA protocol for fault-tolerant and energy-efficient wireless sensor networks. IEEE Trans. Parallel Distrib. Syst. 19(6), 851–864 (2008)
6. Maróti, M., Kusy, B., Simon, G., Lédeczi, Á.: Robust multi-hop time synchronization in wireless sensor networks. In: Proc. Int'l Conf. Wireless Networks, ICWN (June 2004)
7. Rajendran, V., García-Luna-Aceves, J., Obraczka, K.: Energy-efficient, application-aware medium access for sensor networks. In: Proc. IEEE Intl. Conf. Mobile Ad-Hoc and Sensor Systems (MASS), pp. 630–637 (November 2005)
8. Rajendran, V., Obraczka, K., García-Luna-Aceves, J.: Energy-efficient, collision-free medium access control for wireless sensor networks. Springer Journal on Wireless Networks 12(1), 63–78 (2006)

Adaptive Minimum Conditional Bit-Error-Rate Linear Multiuser Detection for STBC-MC-CDMA Systems Transmitting over Mobile Radio Channels

Leandro D'Orazio[1], Claudio Sacchi[2], Jérôme Louveaux[3], and Luc Vandendorpe[3]

[1] SIEMENS S.p.A. Industry Sector
Viale Piero e Alberto Pirelli 10
I-20128 Milano (Italy)
leandro.dorazio.ext@siemens.com
[2] University of Trento
Department of Information Engineering and Computer Science (DISI)
Via Sommarive 14, I-38050, Povo (Trento), Italy
sacchi@disi.unitn.it
[3] Université Catholique de Louvain
Communications and Remote Sensing Laboratory
Place du Levant 2, B-1348 Louvain-la-Neuve (Belgium)
{louveaux,vandendorpe}@uclouvain.be

Abstract. This paper proposes an adaptive Minimum Conditional Bit-Error-Rate (MCBER) Multi-User Detector (MUD) for a Space-Time Block Coded Multi-Carrier Code Division Multiple Access (STBC MC-CDMA) system that is based on Least-Mean-Square (LMS) optimization algorithm. Theoretical basics and simulation results have been presented and discussed for the proposed detector. Results shown in terms of BER evidenced that the MCBER detector outperforms the adaptive LMS-based implementation of MMSE MUD and also the ideal MMSE MUD exploiting the perfect knowledge of channel state-information. The moderate computational effort and the reduced sensitivity to parameterization seem to represent other relevant strengths of the proposed approach.

Keywords: Minimum BER, Multi-User Detection, STBC, MIMO, MC-CDMA.

1 Introduction

Future mobile broadband communications working over wireless channels are required to provide high performance services in terms of speed, capacity and quality. Several non-conventional transmission solutions based on single-carrier and multi-carrier modulations (DS-CDMA, OFDM, MC-CDMA, etc.) have been proposed in order to obtain such ambitious targets. In such a perspective, Space-Time-Block-Coding (STBC) MIMO MC-CDMA techniques have been regarded by recent literature among the most promising technologies for future 4G cellular systems [1]. In STBC systems, the diversity in space and frequency domain can be jointly

A. Vinel et al. (Eds.): MACOM 2010, LNCS 6235, pp. 36–46, 2010.

exploited in order to provide performances very close to the single-user bound in multipath fading channels [1]. STBC MC-CDMA systems can profit by multi-user detection (MUD) in order to suppress the interference among the users. MMSE-MUD techniques have been proposed in [2]. It is interesting to note that MMSE combining is adopted in MC-CDMA (SISO and MIMO) as the classical linear multi-user detection criterion. Other MC-CDMA linear combining receivers, like Maximal-Ratio Combining (MRC), Orthogonality Restoring Combining (ORC) and Equal Gain Combining (EGC) are best-suited for the single-user case and may exhibit irreducible error floors when adopted in the multi-user transmission case [3][4]. Although MMSE technique is a good choice in order to provide a simple implementation of adaptive receivers (Least-Mean-Square – LMS – and Recursive-Least-Squared – RLS – implementations are allowed), it is intrinsically suboptimal because it minimizes the Mean Squared Error between the received signal and the noiseless signal pattern. This results in a maximization of the Signal-to-Noise plus Interference Ratio (SINR), rather than in the minimization of the Bit Error Rate (BER) that is the expected target of an efficient receiver. The Minimum Bit Error Rate (MBER) criterion for multi-user detection [5] has been successfully applied to DS-CDMA [6], MC-CDMA [7] and SDMA [8], but, at our best knowledge, never to STBC MC-CDMA systems.

In this paper, we are proposing a multi-user detector for STBC MC-CDMA transmission systems based on the Minimum Conditional Bit Error Rate (MCBER). Such a criterion has been already applied in [9] to SISO MC-CDMA systems. MCBER is a slight modification of the original MBER criterion that allows reducing the computational complexity of the detection algorithm to a linear order with the user number without significant loss in the BER performances. The practical implementation of the proposed MCBER detector relies on an adaptive optimization strategy based on the concept of deterministic gradient. In particular, we considered a Least Mean Square (LMS) algorithm [10]. Such a choice has been motivated by the necessity of testing a computationally-tractable algorithm like gradient descent, widely employed in practical MUD implementations, e.g. in DS/CDMA MMSE-MUD [11], MC-CDMA MMSE-MUD [12], and MBER-MUD [7]. In order to complete the state-of-the-art overview and to better highlight the contribution of this paper, we can mention related some related approaches shown in [3] and [4], where the Signal-To-Noise plus Interference Ratio (SINR) of SISO MC-CDMA systems is maximized with respect to some combining parameters, i.e.: the partial equalization combining parameter [3] and the orthogonality restoring threshold [4]. Our approach is different, as it considers the explicit optimization of linear STBC MC-CDMA combining with respect to a quasi-optimal analytical criterion directly involving system BER. Performance achieved by methodology shown in [3] and [4] are very close to those ones yielded by MMSE-MUD criterion that, by definition, maximizes SINR. We shall show that the adaptive LMS-based MCBER receiver outperforms both LMS-based MMSE receiver and even ideal MMSE receiver based on the direct inversion of the channel matrices. Such a relevant result has been obtained by spending and affordable computational effort.

The paper is organized as follows: Section II will contain the description of the system model. Section III will detail the proposed MCBER detector. Section IV will show some selected simulation results. Finally, paper conclusions are drawn in Section V.

2 System Model

A synchronous multiuser MIMO MC-CDMA system based on Alamouti's Space Time Block Coding [13] supporting K users is considered in this paper. Two transmit antennas at the transmitter side and one receive antenna at the receiver side are employed. The extension to scenarios characterized by an increased number of transmitting and receiving antennas is straightforward. A block diagram of the considered STBC MC-CDMA transmitter and receiver is shown in Fig.1.

Two consecutive data bits are BPSK modulated and encoded by using state-of-the-art Alamouti's scheme: two consecutive BPSK symbols of the generic user k ($k=1...K$), ($\left[a_k^1, a_k^2\right]$), are mapped to the transmitting antennas according to the code matrix Φ_k given by:

$$\Phi_k = \frac{1}{\sqrt{2}}\begin{bmatrix} a_k^1 & -a_k^{2^*} \\ a_k^2 & a_k^{1^*} \end{bmatrix} \tag{1}$$

This matrix represents the so-called STBC block. The encoder outputs are transmitted in two consecutive transmission periods T_1 and T_2. During the first transmission period T_1, the two BPSK symbols a_k^1 and a_k^2 are sent to two separate I-FFT-based multicarrier spreading blocks using a unique Hadamard-Walsh sequence c_k of length N ($c_k \triangleq \left[c_k(0),...,c_k(N-1)\right]^T$, being $c_k(n)$ the n-th chip of the spreading code, $[\bullet]^T$ the matrix transposition operator, and N the number of subcarriers). Finally, the RF converted MC-SS signals are simultaneously transmitted by antenna A and antenna B, respectively. In the same way, during the successive transmission period T_2, the symbol $-a_k^{2^*}$ is transmitted by antenna A and the symbol $a_k^{1^*}$ is transmitted by antenna B, respectively. The superscript operator * means here the complex conjugate operation. In this work, we considered the use of BPSK modulation, therefore $a_k^j \in \{-1,1\}$ (with $j \in \{1,2\}$).

In order to make the notation more compact, in the multi-user case, we define two vectors of bits: $\mathbf{A}^1 \triangleq \left[a_0^1, a_1^1, ..., a_{K-1}^1\right]^T$ and $\mathbf{A}^2 \triangleq \left[a_0^2, a_1^2, ..., a_{K-1}^2\right]^T$ and the orthonormal code matrix C as:

$$\mathbf{C} = \begin{bmatrix} c_0(0) & c_1(0) & \cdots & c_{K-1}(0) \\ c_0(1) & c_1(1) & \cdots & c_{K-1}(1) \\ \vdots & \vdots & \cdots & \vdots \\ c_0(N-1) & c_1(N-1) & \cdots & c_{K-1}(N-1) \end{bmatrix} \tag{2}$$

At the receive antenna, the received signals over two consecutive symbol periods, denoted by $\mathbf{R}^1$ and $\mathbf{R}^2$ at the time T_1 and T_2, respectively, can be expressed (after the FFT operation required for MC-SS de-multiplexing) as follows:

$$\begin{cases} \mathbf{R}^1 = \mathbf{H}_A\mathbf{C}\mathbf{A}^1 + \mathbf{H}_B\mathbf{C}\mathbf{A}^2 + \mathbf{N}^1 \\ \mathbf{R}^2 = -\mathbf{H}_A\mathbf{C}(\mathbf{A}^2)^* + \mathbf{H}_B\mathbf{C}(\mathbf{A}^1)^* + \mathbf{N}^2 \end{cases} \tag{3}$$

where $N^i = [\eta_0, \eta_1, ..., \eta_{N-1}]^T$ (with $i \in \{1,2\}$) is the Additive White Gaussian Noise (AWGN) vector (all vector components are independent and identically-distributed), and $H_{ant} = diag\{h_0^{ant}, h_1^{ant}, ..., h_{N-1}^{ant}\}$ (with $ant \in \{A,B\}$) is the $N \times N$ diagonal channel matrix, being h_n^{ant} the complex channel coefficient related to subcarrier n and to the transmit antenna ant. We reasonably assumed that fading is flat over each subcarrier (in order to avoid ISI), and almost time-invariant during two consecutive transmission periods (i.e.: the coherence time is much greater than symbol period).

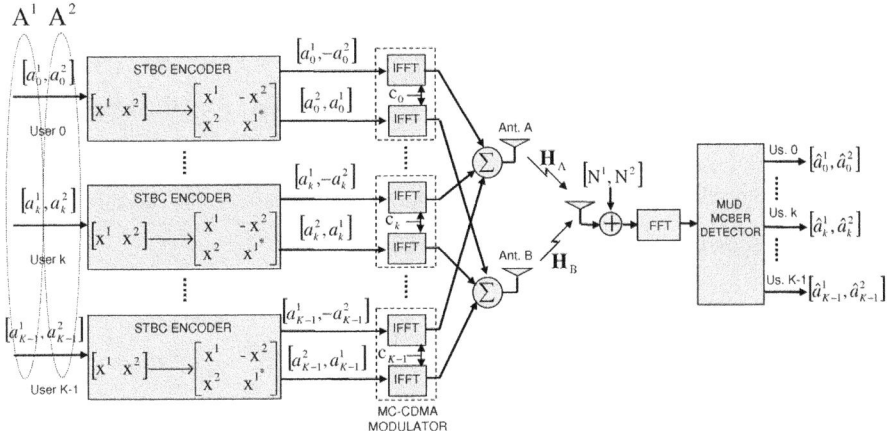

Fig. 1. The considered STBC MC-CDMA system

3   Minimum Conditional BER Decision Criterion

In this section, we shall derive an analytical expression for the error probability of a STBC MC-CDMA system using generic linear combining at the receiver side. Then, this expression will be used to formulate a LMS implementation of the MCBER detector, starting from the formulation of the ideal M-BER criterion.

3.1   Analytical Computation of the Probability of Error

A generic baseband linear multiuser detector generates two decision variables based on the linear combination of the received signal samples. The decision variables for the user k, denoted by $x_{k,1}$ and $x_{k,2}$, can be expressed as follows:

$$\begin{cases} x_{k,1} = w_k^{1*} R^1 + w_k^2 R^{2*} \\ x_{k,2} = w_k^{2*} R^1 - w_k^1 R^{2*} \end{cases} \quad (4)$$

being $w_k^1 = \left[w_k^1(0), w_k^1(1), ..., w_k^1(N-1)\right]^T$ and $w_k^2 = \left[w_k^2(0), w_k^2(1), ..., w_k^2(N-1)\right]^T$ the vectors of receiver gains used by the generic k-th user in order to recombine the baseband output of the FFT-based demodulation stage. The two bits contained in a single STBC block (corresponding to the BPSK symbols $\left[a_k^1, a_k^2\right]$) can be estimated by observing the real part of $x_{k,1}$ and $x_{k,2}$, respectively.

In our dealing, we are considering the employment of a digital BPSK modulation with real antipodal symbols ($A^i = \left(A^i\right)^* i = 1,2$). Under such hypothesis and substituting (3) in (4), we can write the real parts of $x_{k,1}$ and $x_{k,2}$ as follows:

$$\begin{cases} \Re(x_{k,1}) = \Re\left(w_k^{1*}H_A CA^1 + w_k^{1*}H_B CA^2 + w_k^{1*}N^1 - w_k^2 H_A^* CA^2 + w_k^2 H_B^* CA^1 + w_k^2 N^{2*}\right) \\ \Re(x_{k,2}) = \Re\left(w_k^{2*}H_A CA^1 + w_k^{2*}H_B CA^2 + w_k^{2*}N^1 + w_k^1 H_A^* CA^2 - w_k^1 H_B^* CA^1 - w_k^1 N^{2*}\right) \end{cases}$$

(5)

Conditioned on the transmitted bit vectors A^1 and A^2, the random variables $\Re(x_{k,1})$ and $\Re(x_{k,2})$ are Gaussian-distributed with mean values:

$$\begin{cases} \mu_{k,1} = \Re\left(w_k^{1*}H_A CA^1 + w_k^{1*}H_B CA^2 - w_k^2 H_A^* CA^2 + w_k^2 H_B^* CA^1\right) \\ \mu_{k,2} = \Re\left(w_k^{2*}H_A CA^1 + w_k^{2*}H_B CA^2 + w_k^1 H_A^* CA^2 - w_k^1 H_B^* CA^1\right) \end{cases}$$

(6)

and variances:

$$\begin{cases} \sigma_{k,1}^2 = \sigma^2\left(\left\|w_k^1\right\|^2 + \left\|w_k^2\right\|^2\right) \\ \sigma_{k,2}^2 = \sigma^2\left(\left\|w_k^1\right\|^2 + \left\|w_k^2\right\|^2\right) \end{cases}$$

(7)

respectively. To make easier the derivation of the probability of error, we can refer to the so-called sign-adjusted decision variables $x_{k,1}^S$ and $x_{k,2}^S$, defined as follows: $x_{k,1}^S = a_k^1 \Re(x_{k,1})$ and $x_{k,2}^S = a_k^2 \Re(x_{k,2})$. These two sign-adjusted decision variables, conditioned on the transmitted bit vectors A^1 and A^2, are Gaussian-distributed as well with mean values $a_k^1 \mu_{k,1}$ and $a_k^2 \mu_{k,2}$ and variances $\sigma_{k,1}^2$ and $\sigma_{k,2}^2$, respectively.

Thus, the probability to have an error in a STBC block, conditioned on A^1 and A^2, can be written as follows:

$$P_{E/A^1,A^2} = \Pr\left(x_{k,1}^S < 0, x_{k,2}^S > 0\right) + \Pr\left(x_{k,1}^S > 0, x_{k,2}^S < 0\right) + \Pr\left(x_{k,1}^S < 0, x_{k,2}^S < 0\right) \qquad (8)$$

As $x_{k,1}^S$ and $x_{k,2}^S$ are independent random variables, equation (8) can be rewritten as:

$$P_{E/A^1,A^2} = \Pr\left(x_{k,1}^S < 0\right)\Pr\left(x_{k,2}^S > 0\right) + \Pr\left(x_{k,1}^S > 0\right)\Pr\left(x_{k,2}^S < 0\right) + \Pr\left(x_{k,1}^S < 0\right)\Pr\left(x_{k,2}^S < 0\right)$$

(9)

Using the former considerations and the standard Q function, it can be shown that:

$$\left\{ \begin{array}{ll} \Pr\left(x_{k,1}^S < 0\right) = Q\left(\dfrac{\mu_{k,1}^S}{\sigma_{k,1}}\right) & \Pr\left(x_{k,1}^S > 0\right) = 1 - Q\left(\dfrac{\mu_{k,1}^S}{\sigma_{k,1}}\right) \\[3mm] \Pr\left(x_{k,2}^S < 0\right) = Q\left(\dfrac{\mu_{k,2}^S}{\sigma_{k,2}}\right) & \Pr\left(x_{k,2}^S > 0\right) = 1 - Q\left(\dfrac{\mu_{k,2}^S}{\sigma_{k,2}}\right) \end{array} \right. \tag{10}$$

Using equation (10), we can express equation (9) as follows:

$$P_{E/A^1,A^2} = Q\left(\frac{\mu_{k,1}^S}{\sigma_{k,1}}\right) + Q\left(\frac{\mu_{k,2}^S}{\sigma_{k,2}}\right) - Q\left(\frac{\mu_{k,1}^S}{\sigma_{k,1}}\right) Q\left(\frac{\mu_{k,2}^S}{\sigma_{k,2}}\right) \tag{11}$$

Assuming that the two symbols contained in a STBC block have the same probability of error, the third term in (11) is negligible with respect to the first two ones. Thus, an approximation of (11) is given by:

$$P_{E/A^1,A^2} \cong \hat{P}_{E/A^1,A^2} = Q\left(\frac{\mu_{k,1}^S}{\sigma_{k,1}}\right) + Q\left(\frac{\mu_{k,2}^S}{\sigma_{k,2}}\right) \tag{12}$$

By considering that the 2^{2K} possible transmitted bit vectors are independent and equiprobable, the average probability of error in demodulation for the k-th user can be written as:

$$\hat{P}_E^{TOT}\left(w_k^1, w_k^2\right) = \frac{1}{2^{2K}} \sum_{\forall A^1} \sum_{\forall A^2} \hat{P}_{E/A^1,A^2} \tag{13}$$

3.2   LMS Implementation of the MCBER Detector

The couple of receiver gains vectors $\left(w_k^1, w_k^2\right)$ minimizing the average probability of error shown in equation (13) practically implement the ideal MBER detection criterion for the considered STBC MC-CDMA system. In [8], it is pointed out that no closed-form expression for MBER solution can be found. For this reason, a numerical solution has to be investigated. A possible solution can exploit the Least Mean Square (LMS) algorithm based on the concept of gradient descent [10]. LMS updating of the filter weights is done iteratively along the negative gradient of the error probability surface, along both directions w_k^1 and w_k^2. The updating rule at i-th iteration is given by:

$$\left\{ \begin{array}{l} w_k^{1^{i+1}} = w_k^{1^i} - \mu \cdot \nabla_1\left(\hat{P}_E^{TOT}\left(w_k^1, w_k^2\right)\right) \\[2mm] w_k^{2^{i+1}} = w_k^{2^i} - \mu \cdot \nabla_2\left(\hat{P}_E^{TOT}\left(w_k^1, w_k^2\right)\right) \end{array} \right. \tag{14}$$

where μ is the step-size parameter and ∇_1 and ∇_2 represent the gradient along the two directions w_k^1 and w_k^2, respectively. For the first iteration (namely: $i = 0$), we verified that the most efficient initialization is $w_k^{1\,0} = w_k^{2\,0} = c_k$. It can be shown that the two gradients involved in (14) can be expressed as follows (the mathematical detail is here omitted for sake of space):

$$
\begin{cases}
\nabla_1\left(\hat{P}_E^{TOT}\left(w_k^1, w_k^2\right)\right) = -\frac{2}{\sqrt{2\pi}}\frac{1}{2^{2K}}\sum_{\forall A^1}\sum_{\forall A^2}\left\{ e^{\frac{\mu_{k,1}^2}{2\sigma_{k,1}^2}}\frac{a_k^1}{\sigma_{k,1}}\left[\mathbf{H}_A\mathbf{CA}^1 + \mathbf{H}_B\mathbf{CA}^2 - \mu_{k,1}\frac{w_k^1}{\left\|w_k^1\right\|^2 + \left\|w_k^2\right\|^2}\right] + \right. \\
\left. + e^{\frac{\mu_{k,2}^2}{2\sigma_{k,2}^2}}\frac{a_k^2}{\sigma_{k,2}}\left[\mathbf{H}_A\mathbf{CA}^2 - \mathbf{H}_B\mathbf{CA}^1 - \mu_{k,2}\frac{w_k^1}{\left\|w_k^1\right\|^2 + \left\|w_k^2\right\|^2}\right]\right\} \\[2em]
\nabla_2\left(\hat{P}_E^{TOT}\left(w_k^1, w_k^2\right)\right) = -\frac{2}{\sqrt{2\pi}}\frac{1}{2^{2K}}\sum_{\forall A^1}\sum_{\forall A^2}\left\{ e^{\frac{\mu_{k,1}^2}{2\sigma_{k,1}^2}}\frac{a_k^1}{\sigma_{k,1}}\left[\mathbf{H}_B\mathbf{CA}^1 - \mathbf{H}_A\mathbf{CA}^2 - \mu_{k,1}\frac{w_k^2}{\left\|w_k^1\right\|^2 + \left\|w_k^2\right\|^2}\right] + \right. \\
\left. + e^{\frac{\mu_{k,2}^2}{2\sigma_{k,2}^2}}\frac{a_k^2}{\sigma_{k,2}}\left[\mathbf{H}_B\mathbf{CA}^2 + \mathbf{H}_A\mathbf{CA}^1 - \mu_{k,2}\frac{w_k^2}{\left\|w_k^1\right\|^2 + \left\|w_k^2\right\|^2}\right]\right\}
\end{cases}
$$

(15)

Thus, the LMS implementation of the MBER detector for the considered STBC MC-CDMA system is given by combining the updating rule (14) and the gradient expressions (15). The computational complexity of this detector is exponential in the number of users ($O(2^{2K})$), so its practical application becomes unfeasible as K increases. But, the computational burden of the MBER criterion can be reduced by minimizing the conditional probability of error instead of the average probability of error [9]. The resulting detector, referred to as MCBER detector, has a computational order $O(K)$ which is linear with respect to the user's number. The computational effort is therefore reduced with respect to the ideal MMSE detector (which is $O(K^3)$ [15-16]) but comparable with the LMS-based adaptive implementation of MMSE (which is again linear).

The LMS implementation of the MCBER detector can be obtained by using the same updating rule (14) combined with the following new gradient expressions:

$$
\begin{cases}
\nabla_1\left(\hat{P}_E^{TOT}\left(w_k^1, w_k^2\right)\right) = -\frac{2}{\sqrt{2\pi}}\frac{1}{2^2}\sum_{\forall a_k^1}\sum_{\forall a_k^2}\left\{ e^{\frac{\mu_{k,1}^2}{2\sigma_{k,1}^2}}\frac{a_k^1}{\sigma_{k,1}}\left[\mathbf{H}_A\mathbf{CA}^1 + \mathbf{H}_B\mathbf{CA}^2 - \mu_{k,1}\frac{w_k^1}{\left\|w_k^1\right\|^2 + \left\|w_k^2\right\|^2}\right] + \right. \\
\left. + e^{\frac{\mu_{k,2}^2}{2\sigma_{k,2}^2}}\frac{a_k^2}{\sigma_{k,2}}\left[\mathbf{H}_A\mathbf{CA}^2 - \mathbf{H}_B\mathbf{CA}^1 - \mu_{k,2}\frac{w_k^1}{\left\|w_k^1\right\|^2 + \left\|w_k^2\right\|^2}\right]\right\} \\[2em]
\nabla_2\left(\hat{P}_E^{TOT}\left(w_k^1, w_k^2\right)\right) = -\frac{2}{\sqrt{2\pi}}\frac{1}{2^2}\sum_{\forall a_k^1}\sum_{\forall a_k^2}\left\{ e^{\frac{\mu_{k,1}^2}{2\sigma_{k,1}^2}}\frac{a_k^1}{\sigma_{k,1}}\left[\mathbf{H}_B\mathbf{CA}^1 - \mathbf{H}_A\mathbf{CA}^2 - \mu_{k,1}\frac{w_k^2}{\left\|w_k^1\right\|^2 + \left\|w_k^2\right\|^2}\right] + \right. \\
\left. + e^{\frac{\mu_{k,2}^2}{2\sigma_{k,2}^2}}\frac{a_k^2}{\sigma_{k,2}}\left[\mathbf{H}_B\mathbf{CA}^2 + \mathbf{H}_A\mathbf{CA}^1 - \mu_{k,2}\frac{w_k^2}{\left\|w_k^1\right\|^2 + \left\|w_k^2\right\|^2}\right]\right\}
\end{cases}
$$

(16)

In this way, we can just compute the statistical mean over the data of each user. For the other data contained in A^1 and A^2, we can use the decisions made by the other users.

4 Experimental Results

The performances of the proposed LMS-based MCBER detector (hereinafter called: "LMS-based MCBER") are evaluated in a Rayleigh fading channel, fixing the following parameters: number of subcarriers N=8, transmission data rate r_b=1 Mb/s, coherence bandwidth of the channel 2.1 MHz, Doppler spread of the channel 100 Hz. In order to verify the effectiveness of the proposed approach, we considered two state-of-the-art receivers for comparison, i.e.: the ideal MMSE MUD receiver [2] (hereinafter called "Ideal MMSE") and the LMS adaptive implementation of MMSE receiver shown in [2] and [14] (hereinafter called "LMS-based MMSE"). Perfect knowledge of the channel matrices (H_A, H_B) and of the code matrix C is assumed at the receiver side. As lower bound, we considered the BER curve obtained by Maximal Ratio Combining (MRC) detection in the single-user case (i.e., the optimal single-user detection, supposing the absence of multi-user interference). In our simulations, we considered three different scenarios including K=2, K=4, and K=6 users. In each scenario, we measured the average BER versus SNR. The corresponding BER curves are shown for all the tested receivers in Fig. 2, Fig. 3, and Fig. 4, respectively. It can be seen that in all scenarios the proposed LMS-based MCBER detector clearly outperforms the LMS-based MMSE adaptive detector. Moreover, the proposed MCBER detector yields better performances than the ideal MMSE detector. Such a last improvement is clearly evident for K=2 and K=4 users; whereas it becomes slighter for K=6 users. In general, for an increasing user number, BER curves related to ideal MMSE and MCBER becomes closer and more distant with respect to the single-user bound. This is not a surprising fact. Indeed, such behaviour has been already noted by Chen and Hanzo in [6] for the MBER DS/CDMA case. The motivation is currently under investigation, but, probably, it could be related with the intrinsic sub-optimality of the minimum BER reception criteria. In fact, we are not considering the theoretically optimum Maximum-Likelihood detection, computing the optimal symbol vector on the basis of the ML metric maximization. Truly, we are proposing and discussing a diversity combining methodology based on the minimization of the probability of bit error that is conceptually different from ML detection. The step-size parameter μ of both LMS-based algorithms (MMSE and MCBER) has been chosen empirically for each scenario in order to minimize the overall BER over the various SNR values. From the parameters selection phase, we noted that LMS-based MCBER detector is characterized by a reduced sensitivity to parameterization with respect to state-of-the-art LMS-based MMSE implementation. Indeed, fixing the user number K, the step-size μ is substantially invariant with respect to SNR values. On the other hand, LMS-based MMSE multi-user detector would require at different value of μ for each SNR in order to provide satisfactory BER performances.

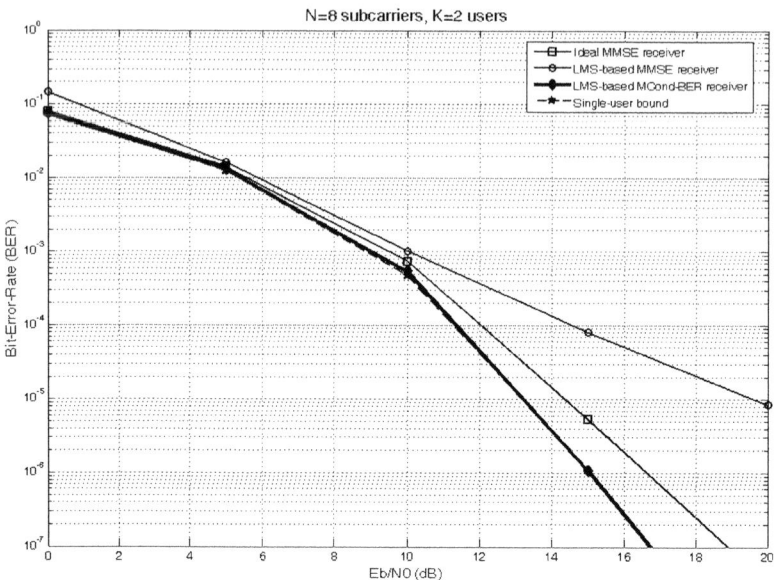

Fig. 2. BER performance yielded by the LMS-based MCBER, LMS-based MMSE and ideal MMSE multi-user detectors for N=8 subcarriers and K=2 users

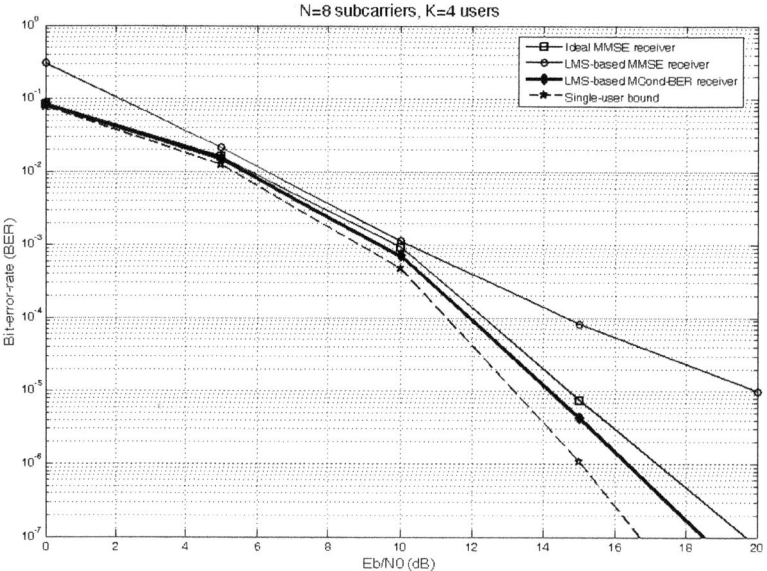

Fig. 3. BER performance yielded by the LMS-based MCBER, LMS-based MMSE and ideal MMSE multi-user detectors for N=8 subcarriers and K=4 users

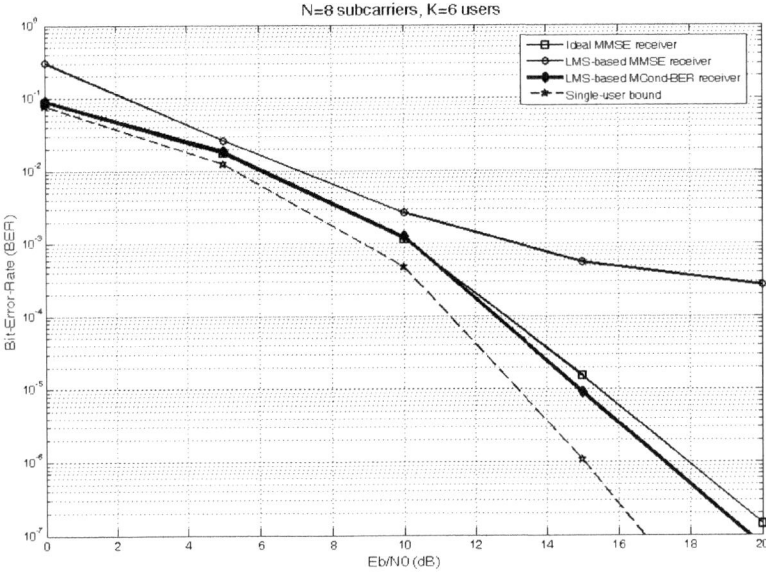

Fig. 4. BER performance yielded by the LMS-based MCBER, LMS-based MMSE and ideal MMSE multi-user detectors for N=8 subcarriers and K=6 users

5 Conclusion

In this paper, an adaptive LMS-based Minimum Conditional BER (MCBER) multi-user detector has been proposed for MIMO MC-CDMA systems using Alamouti's space-time block coding. The proposed approach allows to improve BER performances with respect to state-of-the-art detector (LMS-based MMSE, ideal MMSE), requiring a computational effort linearly increasing with users number. This is the main advantage with respect to state-of-the-art linear MUD approaches, generally based on SINR maximization. Future works might be related with the employment of Recursive-Least-Squared (RLS) optimization strategy, instead of LMS, in order to speed up convergence to optimal solution. Moreover, the impact of non ideal knowledge of channel state information on the MCBER performances should be studied as well.

Acknowledgements

This work has been partially supported by the Italian Ministry of University and Scientific Research, under the framework of SALICE (Satellite-Assisted Localization and Communication systems for Emergency services) research project (COFIN 2007RFTYY7_002).

References

1. Juntii, M., Vehkapera, M., Hara, S.: MIMO MC-CDMA Communications for Future Cellular Systems. IEEE Communications Magazine, 118–124 (February 2005)
2. Auffray, J.M., Helard, J.F.: Performance of multicarrier CDMA technique combined with space-time block coding over Rayleigh channel. In: IEEE Seventh International Symposium on Spread Spectrum Techniques and Applications, September 2-5, vol. 2, pp. 348–352 (2002)
3. Conti, A., Masini, B., Zabini, F., Andrisano, O.: On the Down-Link Performance of Multi-Carrier CDMA Systems with Partial Equalization. IEEE Trans. on Wireless Comm. 6(1), 230–239 (2007)
4. Masini, B.M., Conti, A.: Adaptive TORC Detection for MC-CDMA Systems. IEEE Trans. on Comm. 57(11), 3460–3471 (2009)
5. Chen, S.: Adaptive minimum bit-error-rate filtering. IEE Proceedings – Vision, Image and Signal Processing, Special Issue on Non-linear and Non-Gaussian Signal Processing 151(1), 76–85 (2004)
6. Chen, S., Samingan, A.K., Mulgrew, B., Hanzo, L.: Adaptive minimum-BER linear multiuser detection for DS-CDMA signals in multipath channels. IEEE Trans. on Signal Processing 49(6), 1240–1247 (2001)
7. Yi, S.J., Tsimenidis, C.C., Hinton, O.R., Sharif, B.S.: Adaptive minimum bit error rate multiuser detection for asynchronous MC-CDMA systems in frequency selective Rayleigh fading channels. In: Proc. of 14th Personal, Indoor and Mobile Radio Communications Conference (PIMRC 2003), September 7-10, vol. 2, pp. 1269–1273 (2003)
8. Sheng, S., Livingstone, A., Hanzo, L.: Minimum bit-error rate design for space-time equalization-based multiuser detection. IEEE Trans. on Comm. 54(5), 824–832 (2006)
9. Dayal, P., Desai, U.B., Mahanta, A.: Minimum conditional probability of error detection for MC-CDMA. In: Proc. of IEEE Eighth International Symposium on Spread Spectrum Techniques and Applications, August 30-September 2, pp. 51–55 (2004)
10. Widrow, B., Kamenetsky, M.: On the statistical efficiency of the LMS family of adaptive algorithms. In: Proc. of 2003 IEEE Internat. Joint Conf. on Neural Networks (IJCNN 2003), Portland (OR), July 20-24, vol. 4, pp. 2872–2880 (2003)
11. Castoldi, P.: Multiuser detection in CDMA mobile terminals. Artech-House, Boston (2002)
12. Kalofonos, D.N., Stojanovic, M., Proakis, J.G.: On the performance of adaptive MMSE detectors for a MC-CDMA system in fast Fading Rayleigh channel. IEEE Trans. on Wireless Communications 2(2), 229–239 (2003)
13. Alamouti, S.M.: A Simple Transmit Diversity Technique for Wireless Communications. IEEE J. Selec. Areas in Comm. 16(8), 1451–1458 (1998)
14. Hu, X., Chew, Y.H.: Performance of Space-Time Block Coded MC-CDMA System over Frequency Selective Fading Channel using Semi-blind Channel Estimation Technique. In: Proc. of the 2003 IEEE Wireless Communications and Networking Conf. (WCNC 2003), New Orleans (LA), March 16-20, vol. 1, pp. 414–419 (2003)
15. Sun, W., Li, H., Amin, M.: MMSE detection for space-time coded MC-CDMA. In: Proc. of the 2003 IEEE International Conference on Communications (ICC 2003), May 11-15, vol. 5, pp. 3452–3456 (2003)
16. Tran, X.N., Le, A.T., Fujino, T.: Performance Comparison of MMSE-SIC and MMSE-ML Multiuser Detectors in a STBC-OFDM System. In: Proc. of 16th IEEE International Symposium on Personal Indoor and Mobile Radio Communications (PIMRC 2005) (September 2005)

On the Performance of Single LDGM Codes for Iterative Data Fusion over the Multiple Access Channel

Javier Del Ser[1], Javier Garcia-Frias[2], Pedro M. Crespo[3],
Diana Manjarres[1], and Ignacio (Iñaki) Olabarrieta[1]

[1] TECNALIA-TELECOM, 48170 Zamudio, Spain
{jdelser,dmanjarres,iolabarrieta}@robotiker.es
[2] University of Delaware, Newark, DE 19716, USA
jgarcia@ee.udel.edu
[3] CEIT and TECNUN (University of Navarra), 20009 San Sebastian, Spain
pcrespo@ceit.es

Abstract. One of the applications of wireless sensor networks currently undergoing active research focuses on the scenario where the information generated by a data source S is simultaneously sensed by N nodes and therefrom transmitted to a common receiver. Based on the received information from such N nodes, such receiver infers the original information from S potentially more accurately than in the case of a single sensor. Often referred to as the CEO (*Central Estimating Officer*) problem [1], in this scenario we propose the use of single Low Density Generator Matrix (LDGM) codes for the transmission of the information registered by the nodes over the Multiple Access Channel (MAC). The corresponding receiver iterates between a soft demodulator, the set of N LDGM decoders and a soft-information fusion stage. Simulation results for the AWGN MAC channel show that 1) the proposed coding scheme outperforms the suboptimum limit assuming separated Slepian-Wolf distributed coding and capacity-approaching codes; and 2) the end-to-end Bit Error Rate (BER) performance is lower bounded, for increasing N, by the error floor due to the inherent ambiguity of the MAC channel when dealing with correlated sources. This paves the way for future research aimed at applying concatenated coding schemes to this setup.

Keywords: LDGM codes, data fusion, multiple access channel.

1 Introduction

Wireless Sensor Networks (WSN) have emerged during the last decade as a means to cooperatively monitor physical or environmental conditions in a distributed and cost-efficient manner. In general such networks consist of a large number of nodes – densely deployed over a wide geographical area – with reduced sensing and processing capabilities operating under a high degree of autonomy. The exigent transmission and reception requirements of sensor networks have

A. Vinel et al. (Eds.): MACOM 2010, LNCS 6235, pp. 47–57, 2010.

unchained an upsurge of challenging paradigms within the scientific community, mainly aimed at minimizing the power consumption and improving the battery lifetime of the constituent nodes, e.g. distributed compression, scheduling or multihop transmission techniques. Among such paradigms we concentrate on the centralized data fusion scenario where N nodes sense a common parameter of interest $\mathcal{P}$ (possibly subject to a non-zero probability of sensing error) and forward their sensed data to a common receiver. In this setup, information fusion stands for the combination of the sensors' data to improve the reliability of the information estimated by the central receiver (see Figure 1.a).

In this context, since the landmark paper by Chair and Varshney [2] several contributions have considered the data fusion problem in diverse uncoded communication scenarios, e.g. multihop networks subject to fading [3,4] and delays [5], and asynchronous multiple access channels [6,7]. On the other hand, the research activity on coded scenarios has gravitated on the performance of different Turbo-like codes for the simplistic case of parallel AWGN channels, e.g. LDGM [8], Irregular Repeat-Accumulate (IRA) [9], and concatenated Zigzag [10] codes. In such references it was shown that by iteratively exchanging soft information between the sensors' decoders and the data fusion stage, a significant error rate improvement is attained with respect to schemes where decoding and data fusion are separately and sequentially executed.

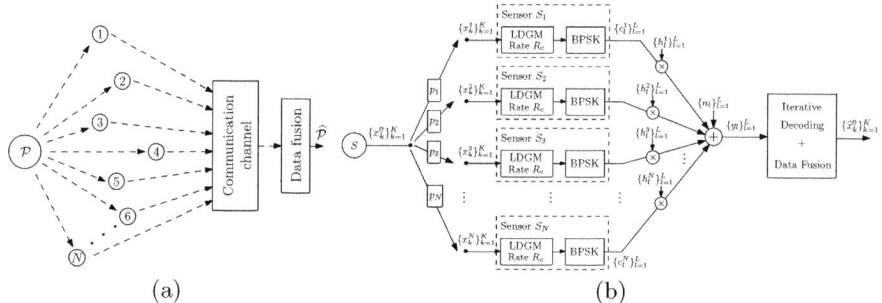

Fig. 1. (a) Generic data fusion setup where N nodes sense a certain physical parameter $\mathcal{P}$; (b) Block diagram of the considered scenario

Following this research trend, this paper focuses on the iterative data fusion scenario where the data sensed by the N nodes is transmitted to the common receiver over a Multiple Access Channel (MAC). This problem is similar to that posed in [11,12,13,14], where the authors proposed the use of correlation-preserving LDGM codes for the transmission of correlated sources over the AWGN and flat fading multiple access channel. Due to the structure of their generator matrices, LDGM codes keep a high degree of correlation between codewords corresponding to different correlated sensors and consequently, the sum of transmitted signals in the MAC is mostly constructive[1]. However, such

---

[1] This particular property of LDGM codes has been also exploited for the transmission of correlated information sources over the Gaussian broadcast channel [15,16].

references focused on the reliable transmission of the information generated by a pair of correlated sensors, whereas here we deal with the reliable communication of an information source read by a set of N sensors subject to a nonzero probability of sensing error. Extensive simulation results for the AWGN MAC will verify that single LDGM codes suffice for outperforming the suboptimum limit assuming ideal Slepian-Wolf distributed source coding and capacity-achieving channel codes. On the other hand, it is shown that the end-to-end error rate performance is limited by the ambiguity of the received signal about the symbols transmitted by the sensors. This last conclusion lays the basis for future research aimed at applying concatenated coding schemes.

The rest of this paper is organized as follows: Section 2 details the system model of the data fusion scenario, whereas Section 3 elaborates on the design of the iterative decoding and data fusion procedure. Next, Section 4 discusses Monte Carlo simulation results and finally, Section 5 ends the paper by drawing some concluding remarks.

2   System Model

Let us consider the system model depicted in Figure 1.b, where N sensors $\{S_n\}_{n=1}^N$ obtain blocks of K measures $\{x_k^n\}_{k=1}^K$ ($n = 1, \ldots, N$) from a physical parameter, which is modeled as a sequence of K i.i.d binary random variables $\{x_k^0\}_{k=1}^K$ with $P_{x_k}(0) = P_{x_k}(1) = 0.5\ \forall k$. It is further assumed that every sensor S_n is subject to a nonzero probability of sensing error $p_n < 0.5$, i.e. $Pr\{x_k^n \neq x_k^0\} = p_n\ \forall k \in \{1, \ldots, K\}$. Each sensor sequence $\{x_k^n\}_{k=1}^K$ is encoded by means of a LDGM code, i.e. a linear code with low density generator matrix $G = [\mathbf{I}\ \mathbf{P}]$ and parity check matrix $\mathbf{H} = [\mathbf{P}^T\ \mathbf{I}]$, where $\mathbf{I}$ denotes the identity matrix and $\mathbf{P}$ is a $K \times (L - K)$ sparse binary matrix. The coding rate is therefore given by $R_c = K/L$ (sensed symbols per coded symbol). In order to exploit the correlation as much as possible, the generator matrices are set equal for all the sensors. The output encoded sequence $\{c_l^n\}_{l=1}^L$ includes both systematic bits ($c_l^n = x_l^n$, $l \in \{1, \ldots, K\}$) and parity bits ($c_l^n = p_l^n$, $l \in \{K + 1, \ldots, L\}$).

The codewords $\{c_l^n\}_{l=1}^L$ ($n \in \{1, \ldots, N\}$) are next BPSK (*Binary Phase Shift Keying*) modulated and transmitted to a common receiver through a MAC. Referring to Figure 1.b, the received sequence will be given by

$$y_l = \sum_{n=1}^N h_l^n \phi(c_l^n) + n_l = b_l + n_l, \tag{1}$$

where $\phi : \{0, 1\} \rightarrow \{-\sqrt{E_c}, +\sqrt{E_c}\}$ denotes the modulation mapping, E_c stands for the average energy per channel symbol and sensor, and $\{n_l\}_{l=1}^L$ are i.i.d. circularly symmetric complex Gaussian random variables with zero mean and variance per dimension σ^2. In this paper the AWGN MAC channel will be considered, i.e. $h_l^n = 1\ \forall l \in \{1, \ldots, L\}$ and $\forall n \in \{1, \ldots, N\}$. Once $\{y_l\}_{l=1}^L$ is received at destination, the receiver demodulates, decodes and fuses the data from the sensors to obtain an estimate $\{\hat{x}_k^0\}_{k=1}^K$ of the original information sequence $\{x_k^0\}_{k=1}^K$. This is done by applying the message-passing Sum-Product Algorithm (SPA, see

[17] and references therein) over the factor graph that jointly describes the MAC channel, the whole set of N LDGM codes and the statistical relation between the sensor sequences and x_k^0, as will be explained in the next section.

3   Iterative Joint Decoding and Data Fusion

The optimum joint receiver would symbolwise estimate the original information $\{x_k^0\}_{k=1}^K$ from $\{y_l\}_{l=1}^L$ as $\{\widehat{x}_k^0\}_{k=1}^K$ by applying the Maximum A Posteriori (MAP) decision criterium, i.e.

$$\widehat{x}_k^0 = \underset{x_k^0 \in \{0,1\}}{\arg\max} \, P(x_k^0|\{y_l\}_{l=1}^L), \tag{2}$$

where $P(\cdot|\cdot)$ denotes conditional probability. A practical suboptimum scheme capable of obtaining $\{\widehat{x}_k^0\}_{k=1}^K$ from the received signal would operate under a sequential approach. First, the conditional probabilities of the encoded symbol c_l^n given the received sequence would be obtained, for $l \in \{1,\ldots,L\}$ and $n \in \{1,\ldots,N\}$, as

$$P(c_l^n|y_l) = \sum_{\sim c_l^n} P(c_l^1,\ldots,c_l^N|y_l) \propto \sum_{\sim c_l^n} \exp\left(-\frac{|y_l - \phi(c_l^1)h_l^1 - \ldots - \phi(c_l^N)h_l^N|^2}{2\sigma^2}\right),$$

where $\sim c_l^n$ denotes that all binary variables are summed over except c_l^n. Once the above L conditional probabilities for the n-th sensor codeword are computed, an estimation $\{\widehat{x}_k^n\}_{k=1}^K$ of the original sensor sequence $\{x_k^n\}_{k=1}^K$ would be produced by iterative LDGM decoding based on $\{P(c_l^n|y_l)\}_{l=1}^L$, independently from the other $N-1$ LDGM decoding procedures. Finally, the N recovered sensor sequences $\{\widehat{x}_k^n\}_{k=1}^K$ would be fused, for $k \in \{1,\ldots,K\}$, as

$$\widehat{x}_k^0 = \begin{cases} 1 & \text{if } \sum_{n=1}^N \widehat{x}_k^n \geq \lceil N/2 \rceil, \\ 0 & \text{if } \sum_{n=1}^N \widehat{x}_k^n < \lceil N/2 \rceil, \end{cases} \tag{3}$$

i.e. by symbolwise majority voting among the estimated N sensor sequences.

However, notice that since we assume $p_n < 0.5 \; \forall n \in \{1,\ldots,N\}$, the sensor sequences are spatially correlated, i.e. it can be easily derived that $Pr\{x_k^m = x_k^n\} = (p_m p_n + (1-p_m)(1-p_n))$ for $n \neq m$. This correlation should be exploited at the receiver in order to enhance the reliability of the fused sequence $\{\widehat{x}_k^0\}_{k=1}^K$. As opposed to the aforementioned sequential receiver, a practical design that efficiently exploits the correlation between sensor sequences hinges on describing the joint probability distribution involving all the variables of the system by means of factor graphs, as well as on the marginalization for $\widehat{x}_k^0$ via the message-passing Sum-Product Algorithm (SPA). This design methodology permits to reduce the computational complexity with respect to a direct marginalization based on exhaustive evaluation of the entire joint probability distribution.

Let us elaborate further on this design[2] by considering Figure 2.a, where the overall factor graph structure of the joint receiver is depicted for $N = 4$ sensors. This graph is built by properly interconnecting different compounding factor subgraphs: the graph modeling the statistical relationship between x_k^0 and $\{x_k^n\}_{n=1}^N \forall k \in \{1, \ldots, K\}$ (labeled as SENSING), the factor graph mapping sensor sequence $\{x_k^n\}_{k=1}^K$ to codeword $\{c_l^n\}_{l=1}^L$ through the LDGM parity check matrix $\mathbf{H}$, and the statistical relationship between the received sequence $\{y_l\}_{l=1}^L$ and the N codewords $\{c_l^n\}_{l=1}^L$, with $n \in \{1, \ldots, N\}$ (labeled as MAC). Please observe that the interconnection between such factor subgraphs is done via variable nodes corresponding to c_l^n and x_k^n. This interconnected set of subgraphs embodies a overall cyclic factor graph over which the SPA algorithm iterates in the order MAC $\mapsto$ LDGM$_1 \mapsto \ldots \mapsto$ LDGM$_n \mapsto$ SENSING until a fixed number of iterations $\mathcal{I}$ is reached. It is also important to notice that since the LDGM code is systematic, variable nodes $\{c_l^n\}_{l=1}^L$ and $\{x_k^n\}_{k=1}^K$ collapse into a single node $\forall n \in \{1, \ldots, N\}$, not shown in the plots for the sake of clarity.

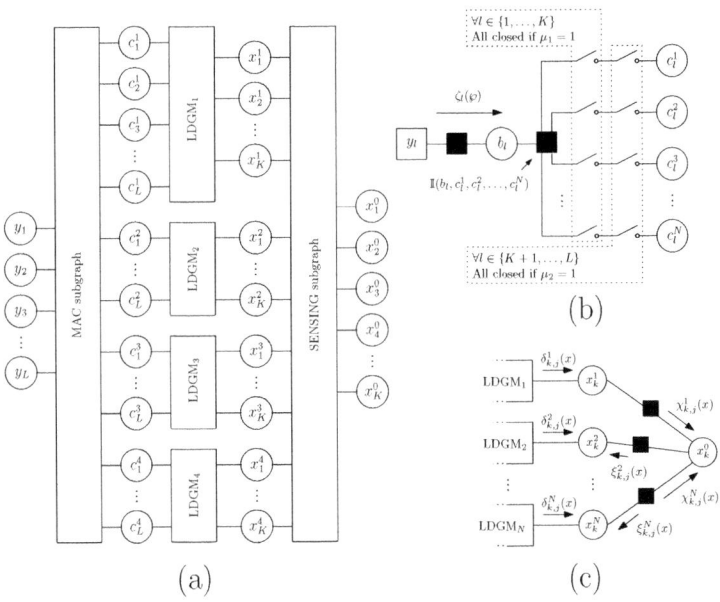

Fig. 2. (a) Overall factor graph of the proposed receiver; (b) MAC factor subgraph; (c) SENSING factor subgraph

The MAC factor subgraph is represented in Figure 2.b. First note that variable nodes $\{c_l^n\}_{n=1}^N$ are linked to the received symbol y_l through the auxiliary variable node b_l, i.e. the noiseless version of the MAC output y_l defined in expression (1). Let $\mathcal{B}$ denote the set of 2^N possible values of b_l imposed by the

---

[2] In what follows it is assumed that the reader has a tacit knowledge of factor graphs and the SPA (see [17] for a complete tutorial).

2^N possible combinations of $\{\phi(c_l^n)\}_{n=1}^N$ and the N MAC coefficients $\{h_l^n\}_{n=1}^N$. The message $\zeta_l(\wp)$ corresponding to $b_l = \wp \in \mathcal{B}$ is given by the conditional probability distribution imposed by the AWGN channel, i.e.

$$\zeta_l(\wp) = \Theta_l \exp(-|y_l - \wp|^2/2\sigma^2), \tag{4}$$

where the value of the constant Θ_l is selected so as to satisfy $\sum_{\wp \in \mathcal{B}} \zeta_l(\wp) = 1$ $\forall l \in \{1, \ldots, L\}$. On the other hand, the function associated to the check node relating $\{c_l^n\}_{n=1}^N$ to b_l is an indicator function given by

$$\mathbb{I}(b_l, c_l^1, c_l^2, \ldots, c_l^N) = \begin{cases} 1 & \text{if } \sum_{n=1}^N h_l^n \phi(c_l^n) = b_l, \\ 0 & \text{otherwise.} \end{cases} \tag{5}$$

With respect to Figure 2.b, it is also important to note the set of switches controlled by binary variables μ_1 and μ_2, which permit to disconnect systematic $(l \in \{1, \ldots, K\})$ and parity $(l \in \{K+1, \ldots, L\})$ variable nodes from the MAC factor subgraph. The reason being that, as detailed in Section 4, by properly setting these switches the degradation of the iterative SPA due to short-length cycles in the underlying factor graph is minimized.

Finally let us consider Figure 2.c corresponding to the SENSING subgraph where the soft information from each the sensors is fused to provide an estimation of x_k^0. First we introduce the following notation:

- $\delta_{k,j}^n(x)$: a *posteriori* soft information for the value $x \in \{0,1\}$ of the node x_k^n, which is computed, at iteration j, as the product of the *a posteriori information* generated by the SPA when applied to MAC and LDGM subgraphs.
- $\chi_{k,j}^n(x)$: soft information on x_k^0 (for the value $x \in \{0,1\}$) contributed by sensor S_n at iteration j.
- $\xi_{k,j}^n(x)$: *extrinsic* soft information for $x_k^n = x \in \{0,1\}$ built from the information provided by the rest of sensors at iteration j.

Under the above definitions, the SPA applied to this subgraph renders (see [17, equations (5), (6)])

$$\chi_{k,j}^n(x) = \Gamma_{k,j}^n \left[(1 - p_n)\delta_{k,j}^n(x) + p_n\delta_{k,j}^n(1-x) \right], \tag{6}$$

$$\xi_{k,j}^n(x) = \Upsilon_{k,j}^n \left[\sum_{\sim x_k^n} (1 - p_n) \prod_{m \neq n} \chi_{k,j}^m(x) + p_n \prod_{m \neq n} \chi_{k,j}^m(1-x) \right], \tag{7}$$

where p_n denotes the sensing error probability which in turn establishes the amount of correlation between sensors. Factors $\Gamma_{k,j}^n$ and $\Upsilon_{k,j}^n$ account for the normalization of each pair of messages, e.g. $\xi_{k,j}^n(0) + \xi_{k,j}^n(1) = 1$ $\forall k, n, j$. Finally, the estimation of x_k^0 at iteration j, denoted as $\hat{x}_k^0(j)$, is given by

$$\hat{x}_k^0(j) = \arg\max_{x \in \{0,1\}} \prod_{n=1}^N \chi_{k,j}^n(x), \tag{8}$$

i.e. by the product of all messages arriving to variable node x_k^0 at iteration j.

4   Simulation Results

In order to verify the performance of the proposed system, extensive Monte Carlo simulations have been carried out for the AWGN MAC channel. In all the simulations, the number of sensors is $N \in \{2, 4, 6\}$ whereas, with no loss of generality, the sensing error probability is set equal for all the sensors, i.e. $p_n = p = 5.025 \cdot 10^{-3}$. Identical randomly-generated LDGM codes with rate $R_c = 1/3$, variable and check degree distributions[3] $[d_v; d_c] \in \{[8\ 4], [10\ 5], [12\ 6]\}$, and input blocklength $K = 10000$ have been used at the sensors. The number of iterations for the SPA has been set to $\mathcal{I} = 50$. Performance plots will focus on the end-to-end Bit Error Rate (BER) versus E_b/N_0 per sensor (energy per bit to noise power spectral density ratio) averaged over 2000 different information sequences per simulated point.

Two different performance limits are included for each simulated scenario. The horizontal line stands for the residual probability of estimating x_k^0 in error provided that all sensor symbols $\{x_k^n\}_{n=1}^N$ are perfectly recovered, which can be computed, for even N, as

$$P_e \geq 0.5 \binom{N}{N/2} p^{N/2}(1-p)^{N/2} + \sum_{n=N/2+1}^{N} \binom{N}{n} p^n (1-p)^{N-n}, \qquad (9)$$

i.e. as the probability of decoding more than $N/2$ sensors in error. On the other hand, the minimum E_b/N_0 per sensor required for reliable transmission of all sensors can be obtained by combining Slepian-Wolf [18] and Shannon's Separation theorems. The Separation theorem is known not to hold for the MAC channel; however, its application provides an upper bound of the aforementioned minimum E_b/N_0 per sensor. Under the assumption of equal $p_n = p$, encoding rate R_c and transmit energy per sensor E_c, this suboptimum limit is given by the following inequality

$$R_c H(S_1, S_2, \ldots, S_N) \leq \mathbb{C}(E_c/N_0), \qquad (10)$$

where $\mathbb{C}(E_c/N_0)$ denotes the capacity of the MAC channel instance under consideration. The joint binary entropy of the sensors $H(S_1, \ldots, S_N)$ can be computed as

$$H(S_1, ..., S_N) = -\sum_{n=1}^{N} \binom{N}{n} Pr\{n\ 0\text{'s}\} \log_2 Pr\{n\ 0\text{'s}\}, \qquad (11)$$

with $Pr\{n\ 0\text{'s}\} = 0.5\left(p^n(1-p)^{N-n} + (1-p)^n p^{N-n}\right)$ denoting the probability of having a sequence with exactly n *zero* symbols. Under this definition observe that in expression (10), one obtains $E_c = R_c H(S_1, \ldots, S_N) E_b/N$.

---

[3] In other words, the parity matrix $\mathbf{P}$ of a (d_v, d_c) LDGM code has exactly d_v nonzero entries per row and d_c nonzero entries per column.

Having said this, Figure 4 depicts the end-to-end BER versus E_b/N_0 per sensor for the AWGN MAC channel, i.e. the MAC coefficients are set to $h_l^n = h^n = 1$ $\forall l, n$. In this case the minimum E_b/N_0 resulting from expression (10) is given by

$$\frac{E_b}{N_0} \geq \frac{2^{2R_c H(S_1,\ldots,S_N)} - 1}{2R_c H(S_1,\ldots,S_N)}, \tag{12}$$

which are included as vertical asymptotes in the aforementioned Figure. Also are depicted horizontal limits corresponding to the lowest achievable end-to-end error probability from expression (9). First observe that all the obtained BER curves are significantly below the E_b/N_0 limit (vertical dashed lines) from expression (12), which verify in practice the suboptimality of the computed separation-based bound. On the other hand, notice that the set of all BER curves for $N = 2$ coincide with the lowest achievable end-to-end error probability from expression (9) (horizontal dashed lines), while the waterfall region of such curves degrades as $[d_v\ d_c]$ increases. However, for $N \in \{4, 6\}$ the error floor due to the MAC ambiguity of the received sequence about which transmitted symbol corresponds to each sender is higher than the aforementioned minimum end-to-end error probability of error. By increasing $[d_v\ d_c]$ an error floor decrease is obtained at the cost of a degradation in the BER waterfall performance. To circumvent this issue, future research will be conducted towards applying concatenated encoding schemes to this data fusion setup.

It is important to remark that the results plotted in Figure 4 have been obtained by setting the variables controlling the switches from Figure 2.b to $\mu_1 = \mu_2 = 1$ during the first iteration, while for the remaining $\mathcal{I} - 1$ iterations $\mu_1 = \mu_2 = 0$ (i.e. the MAC subgraph is disconnected and does not participate in the message passing procedure). The rationale behind this setup lies on the

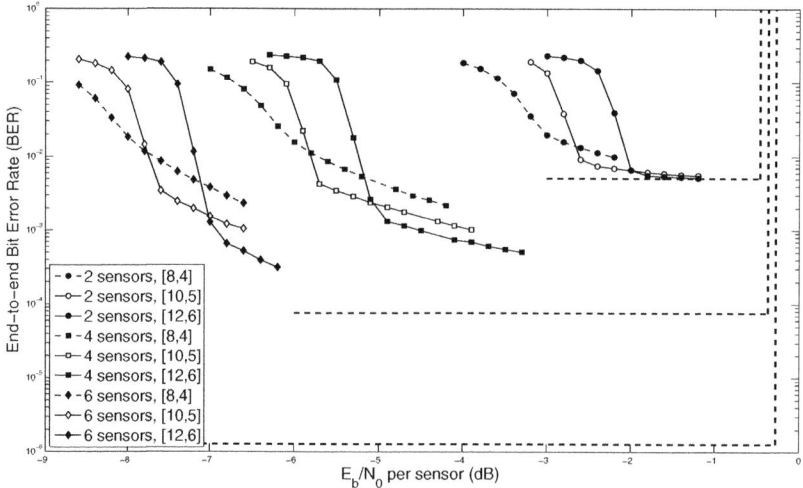

Fig. 3. BER versus E_b/N_0 per sensor for the AWGN MAC

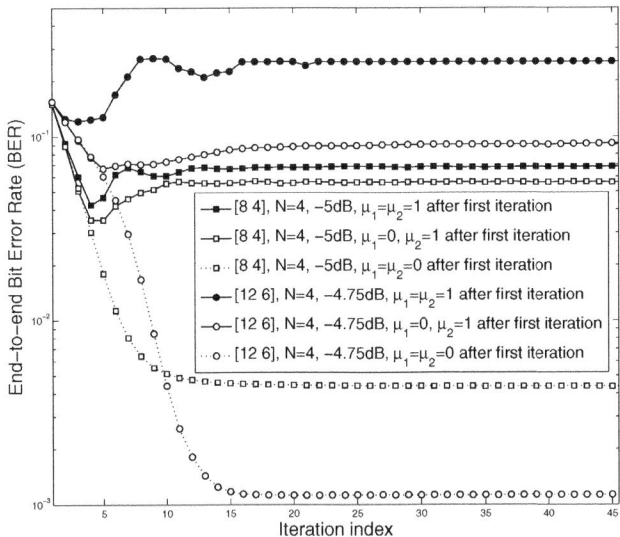

Fig. 4. End-to-end BER versus iteration index for the AWGN MAC and different combinations of $(\mu_1\ \mu_2)$

length-4 loop connecting variable nodes x_k^n, x_k^m ($m \neq n$), x_k^0 and b_k for $k \in \{1, \ldots, K\}$, which degrades significantly the performance of the message-passing SPA. This is graphically depicted in Figure 4.a for $N = 4$, $E_b/N_0 \in \{-5, -4.75\}$ (dB), $[d_v\ d_c] \in \{[8\ 4], [12\ 6]\}$ and different combinations of $(\mu_1\ \mu_2)$. Notice that when $(\mu_1\ \mu_2) = (0\ 1)$ only the switches corresponding to systematic encoded bits are disconnected from the first iteration onwards; in this case, short-length loops are also introduced in the graph through the parity symbol nodes c_l^n and the LDGM parity check matrix $\mathbf{H}$, and consequently the BER performance also degrades with the number of iterations.

5   Concluding Remarks

This paper has analyzed the performance of single LDGM codes for iterative data fusion over the AWGN multiple access channel. A detailed description of the proposed iterative joint decoding and data fusion procedure has been provided based on factor graphs and the Sum-Product Algorithm. Extensive simulations results have shown that our scheme clearly outperforms the suboptimum limit assuming separation between source and channel coding. Future research will aim at suppressing the error floor degradation due to the MAC ambiguity about the transmitted symbols by means of concatenated coding approaches.

Acknowledgments

This work was supported in part by the Spanish Ministry of Science and Innovation through the CONSOLIDER-INGENIO (CSD200800010) and the Torres-Quevedo

(PTQ-09-01-00740) funding programs, and by the Basque Government through the ETORTEK programme (*Future Internet* EI08-227 project). Javier Del Ser was on leave at VTT Research Center of Finland (Oulu) when this manuscript was submitted.

References

1. Berger, T., Zhang, Z., Viswanathan, H.: The CEO Problem. IEEE Transactions on Information Theory 42, 887–902 (1996)
2. Chair, Z., Varshney, P.K.: Optimal Data Fusion in Multiple Sensor Detection Systems. IEEE Transactions on Aerospace and Electronic Systems 22, 98–101 (1986)
3. Lin, Y., Chen, B., Varshney, P.K.: Decision Fusion Rules in Multi-hop Wireless Sensor Networks. IEEE Transactions on Aerospace and Electronic Systems 51, 475–488 (2005)
4. Del Ser, J., Olabarrieta, I., Gil-Lopez, S., Crespo, P.M.: On the Design of Frequency-Switching Patterns for Distributed Data Fusion over Relay Networks. In: International ITG Workshop on Smart Antennas (2010)
5. Thomopoulos, S.C.A., Zhang, L.: Distributed Decision Fusion with Networking Delays and Channel Errors. Information Science 66, 91–118 (1992)
6. Lin, Y., Chen, B., Tong, L.: Distributed Detection Over Multiple Access Channels. In: IEEE International Conference on Acoustics, Speech and Signal Processing, vol. 3, pp. 541–544 (2007)
7. Li, W., Dai, H.: Distributed Detection in Wireless Sensor Networks Using a Multiple Access Channel. IEEE Transactions on Signal Processing 55(3), 822–833 (2007)
8. Zhong, W., Garcia-Frias, J.: Combining Data Fusion with Joint Source-Channel Coding of Correlated Sensors. In: Information Theory Workshop (2004) (invited paper)
9. Zhong, W., Garcia-Frias, J.: Combining Data Fusion with Joint Source-Channel Coding of Correlated Sensors using IRA Codes. In: Conference on Information Sciences and Systems (2005)
10. Del Ser, J., Garcia-Frias, J., Crespo, P.M.: Iterative Concatenated Zigzag Decoding and Blind Data Fusion of Correlated Sensors. In: IEEE International Workshop on Scalable Ad Hoc and Sensor Networks, pp. 1–6 (2009)
11. Garcia-Frias, J., Zhao, Y., Zhong, W.: Turbo-like Codes for Transmission of Correlated Sources over Noisy Channels. IEEE Signal Processing Magazine 24(5), 58–66 (2007)
12. Zhao, Y., Zhong, W., Garcia-Frias, J.: Transmission of Correlated Senders over a Rayleigh Fading Multiple Access Channel. Signal Processing 86, 3150–3159 (2006)
13. Zhong, W., Chai, H., Garcia-Frias, J.: LDGM Codes for Transmission of Correlated Senders over MAC. In: Proceedings of the Allerton Conference on Communication, Control, and Computing (2005) (invited paper)
14. Zhong, W., Garcia-Frias, J.: Joint Source-Channel Coding of Correlated Senders over Multiple Access Channels. In: Proceedings of the Allerton Conference on Communication, Control, and Computing (2004)

15. Hernaez, M., Crespo, P.M., Del Ser, J., Garcia-Frias, J.: Serially-Concatenated LDGM Codes for Correlated Sources over Gaussian Broadcast Channels. IEEE Communications Letters 13(10), 788–790 (2009)
16. Hernaez, M., Crespo, P.M., Del Ser, J., Garcia-Frias, J.: Erratum to Serially-Concatenated LDGM Codes for Correlated Sources over Gaussian Broadcast Channels. IEEE Communications Letters 14(3), 235 (2010)
17. Kschischang, F.R., Frey, B.J., Loeliger, H.A.: Factor Graphs and the Sum-Product Algorithm. IEEE Transactions on Information Theory 47, 498–519 (2001)
18. Slepian, D., Wolf, J.K.: Noiseless Coding of Correlated Information Sources. IEEE Transactions on Information Theory 19, 471–480 (1973)

Preliminary Results on the Adoption of De Bruijn Binary Sequences in DS-CDMA Systems

Stefano Andrenacci, Ennio Gambi, and Susanna Spinsante

Università Politecnica delle Marche, D.I.B.E.T.,
Via Brecce Bianche 12, I-60131 Ancona, Italy
{s.andrenacci,e.gambi,s.spinsante}@univpm.it

Abstract. Code Division Multiple Access (CDMA) using Direct Sequence (DS) spread spectrum modulation provides multiple access capability essentially thanks to the adoption of proper sequences as spreading codes. The ability of a DS-CDMA receiver to detect the desired signal relies to a great extent on the auto-correlation properties of the spreading codes associated to the users; on the other hand, multi-user interference rejection depends on the cross-correlation properties of all the spreading codes in the set chosen. This paper provides preliminary results about the adoption of particular binary sequences, the De Bruijn ones, as spreading codes in DS-CDMA schemes. Though preliminary, the evaluations herein proposed encourage further investigation activities that are currently being developed.

Keywords: Spreading code, De Bruijn sequence, DS-CDMA.

1 Introduction

Code Division Multiple Access (CDMA) using Direct Sequence (DS) spread spectrum modulation is widely recognized as an efficient solution to resist against interference, to combat the effects of multipath fading, and to allow uncoordinated access by several users to a common radio network. In CDMA schemes, the multiple access capability is primarily due to coding, and there is no requirement for precise time or frequency coordination between the transmitters in the system. Each user in a DS-CDMA system is assigned a particular code sequence, which is modulated on the carrier along with the digital data: high-rate code sequences have the effect of spreading the bandwidth of the data signal [1].

As a consequence, the spreading sequence allocated to each user is an essential feature in any CDMA system, to provide the signal with a coded format and ensure a channel separation mechanism. In CDMA systems, all users transmit on the same frequency and at the same time, and the ability of a receiver to detect the desired signal relies to a great extent on the correlation properties of the spreading codes adopted [2]. Mutual interference among active users is inherent to a CDMA scheme, and, again, it strongly depends on the periodic and non-periodic cross-correlation properties of the spreading sequences [3]. The number of active users and their relative power levels affect the performance of a

A. Vinel et al. (Eds.): MACOM 2010, LNCS 6235, pp. 58–69, 2010.

CDMA system, besides the propagation channel conditions. But if the number of active users is fixed, and a specific channel condition is considered, it is possible to investigate the performance of a CDMA system as a function of the features and properties of the spreading codes chosen.

De Bruijn sequences have been studied for many years [4,5]. Their construction has been extensively investigated, and several different generation techniques proposed in the literature [6,7]. One of their most valued properties is the huge cardinality; on the other hand, not so much is known about their correlation features. If adequate, it would be possible to adopt de Bruijn sequences to implement the CDMA communication system, thanks to the huge number of different users that could share the radio channel.

In this paper we investigate the possibility of using binary De Bruijn sequences as spreading codes in DS-CDMA systems, by studying the correlation properties of such sequences. Given the huge amount of binary De Bruijn sequences obtainable, even for small values of the span parameter defined in the following, and considering the great complexity of the generation process [9], we limit this preliminary analysis to binary sequences of length 32, that form a set of 2048 different sequences. The paper is organized as follows: Section 2 provides a basic description of a DS-CDMA reference system; Section 3 introduces the main known properties of binary De Bruijn sequences. Section 4 evaluates the applicability of De Bruijn sequences in DS-CDMA; finally Section 5 concludes the paper.

2   Basic DS-CDMA System Model

The principle of DS-CDMA is to spread the user's information, i.e. data symbols, by a spreading sequence $c^{(k)}(t)$ of length L, defined as:

$$c^{(k)}(t) = \sum_{l=0}^{L-1} c_l^{(k)} p_{T_c}(t - lT_c) \tag{1}$$

where k denotes the k-th user, and $k = 0, \ldots, K-1$, being K the total number of active users. Function $p_{T_c}(t)$ represents the rectangular pulse which is equal to 1 for $0 \leq t < T_c$, and zero otherwise. Parameter T_c is called chip duration, and $c_l^{(k)}$ are the chips of the user's spreading sequence $c^{(k)}(t)$. The information signal of the k-th user, obtained through the spreading operation, is described by $x^{(k)}(t)$:

$$x^{(k)}(t) = d^{(k)} \sum_{l=0}^{L-1} c_l^{(k)} p_{T_c}(t - lT_c) \tag{2}$$

for $0 \leq t < T_d$, where T_d is the duration of one data symbol of the k-th user $d^{(k)}$, and $T_d = LT_c$.

The multiplication of information and spreading sequences is performed bit-synchronously, and the overall transmitted signal $x(t)$, due to all the K synchronous users (as in the case of downlink connection in a cellular system), is given by:

$$x(t) = \sum_{k=0}^{K-1} x^{(k)}(t) \tag{3}$$

In a DS-CDMA scheme, the proper choice of the spreading sequences associated to the users is a fundamental issue. The cross-correlation properties of the used spreading sequences affect the Multiple Access Interference (MAI): in order to reduce the system MAI, the cross-correlation function values should be as small as possible. Further, in order to ensure equal interference among all the transmitting users, the cross-correlation properties between different pairs of spreading sequences should be similar. Moreover, the auto-correlation function of the spreading sequences should have low sidelobes in order to achieve a reliable synchronization.

The received signal at the output of a radio channel with impulse response $h(t)$ can be expressed as:

$$y(t) = \sum_{k=0}^{K-1} r^{(k)}(t) + n(t) \tag{4}$$

where $r^{(k)}(t)$ is the noise-free received signal of user k, and $n(t)$ is the additive white Gaussian noise (AWGN). At the k-th receiver, the impulse response of the matched filter (MF) $h_{MF}^{(k)}(t)$ is adapted to both the transmitted waveform, including the spreading sequence $c^{(k)}(t)$, and to the channel impulse response $h(t)$. At the output of the matched filter of user k, the signal $z^{(k)}(t)$ may be written as:

$$z^{(k)}(t) = y(t) \otimes h_{MF}^{(k)}(t) = r^{(k)}(t) \otimes h_{MF}^{(k)}(t) + \sum_{g=0,g \neq k}^{K-1} r^{(g)}(t) \otimes h_{MF}^{(g)}(t) + n(t) \otimes h_{MF}^{(k)}(t) \tag{5}$$

After sampling at the time instant $t = 0$, the decision variable $\rho^{(k)}$ for user k is given by:

$$\rho^{(k)} = z^{(k)}(0) =$$
$$= \int_0^{T_d + \tau_{max}} r^{(k)}(\tau) h_{MF}^{(k)}(\tau) d\tau + \sum_{g=0,g \neq k}^{K-1} \int_0^{T_d + \tau_{max}} r^{(g)}(\tau) h_{MF}^{(g)}(\tau) d\tau +$$
$$+ \int_0^{T_d + \tau_{max}} n(\tau) h_{MF}^{(k)}(\tau) d\tau$$

being τ_{max} the maximum delay due to the radio channel. The information symbol $\hat{d}^{(k)}$ is finally estimated through threshold detection on $\rho^{(k)}$.

Three terms appear in the last equation: the first term is the desired signal of user k, the second term corresponds to the multiple access interference, and the third term is the additive noise. Due to the multiple access interference term, information bit estimation may be wrong with a certain probability, even at high Signal to Noise Ratio (SNR) values, leading to the well-known error-floor in the BER curves of DS-CDMA systems.

According to Pursley et al. [3], in asynchronous DS-CDMA systems, the average interference parameter may be expressed by:

$$r_{k,i} = 2L^2 + 4 \sum_{l=1}^{L-1} A_k(l)A_i(l) + \sum_{l=1-L}^{L-1} A_k(l)A_i(l+1) \tag{6}$$

where $A_k(l)$ denotes the aperiodic correlation function of the k-th user's spreading sequence $c^{(k)}(t)$ with period L. The aperiodic correlation function is, in its turn, defined as:

$$A_k(l) = \sum_{n=0}^{L-1-l} c_n^{(k)}c_{n+l}^{(k)}, \text{ for } 0 \leq l \leq L - 1 \tag{7}$$

$$A_k(l) = \sum_{n=0}^{L-1+l} c_{n-l}^{(k)}c_n^{(k)}, \text{ for } 1 - L \leq l < 0 \tag{8}$$

$$A_k(l) = 0, \text{ for } |l| > L \tag{9}$$

The average SNR at the output of a correlator receiver of the i-th user among the K users in the system, under AWGN environment, is given by:

$$SNR_i = \left\{ \frac{1}{6L^3} \sum_{k=1,k\neq i}^{K} r_{k,i} + \frac{N_0}{2E_b} \right\}^{-1/2} \tag{10}$$

and the average bit error probability for the i-th user is defined as

$$P_e^i = Q\left(\sqrt{SNR_i}\right), \tag{11}$$

provided a Gaussian distribution for the MAI term, and $Q(x) = \int_x^\infty e^{\frac{-u^2}{2}} du$. According to Eq. 10, the signal-to-noise ratio of user i-th in the system can be evaluated without knowledge of the cross-correlation functions of the spreading codes used, but by resorting to the proper aperiodic correlation definition. In the following, we will provide discussions about the correlation properties of binary De Bruijn sequences, that represent a specific set of maximal length sequences.

3 Binary De Bruijn Sequences and Their Correlation Properties

Binary De Bruijn sequences are a special class of nonlinear shift register sequences with maximal period $L = 2^n$: n is called the span of the sequence, i.e. the sequence may be generated by an n-stage shift register. In the binary case, the total number of distinct sequences of span n is $2^{2^{(n-1)}-n}$; in the more general case of span n sequences over an alphabet of cardinality α, the number of distinct sequences is $\frac{\alpha!^{\alpha^{(n-1)}}}{\alpha^n}$. In this paper we refer to binary De Bruijn sequences.

The states $S_0, S_1, \ldots, S_{N-1}$ of a span n De Bruijn sequence are exactly 2^n different binary n-tuples; when viewed cyclically, a De Bruijn sequence of length 2^n contains each binary n-tuple exactly once over a period.

Being maximal period binary sequences, the length of a De Bruijn sequence is always an even number. When comparing the total number of De Bruijn sequences of length L to the total number of available m-sequences or Gold sequences, similar but not identical length values shall be considered, as reported in Table 1. The Table confirms the exponential growth in the cardinality of De Bruijn sequences, at a parity of the span n, with respect to the other sequences.

Table 1. Length and total number of m-sequences, Gold, and De Bruijn sequences, for the same span n ($3 \leq n \leq 10$)

	m-sequences		Gold		De Bruijn	
n	length	# seq.	length	# seq.	length	# seq.
3	7	2	7	9	8	2
4	15	2	15	17	16	16
5	31	6	31	33	32	2048
6	63	6	63	65	64	2^{26}
7	127	18	127	129	128	2^{57}
8	255	16	255	257	256	2^{120}
9	511	48	511	513	512	2^{247}
10	1023	60	1023	1025	1024	2^{502}

About the auto-correlation values $\theta(k) = \sum_{i=0}^{L-1} c_i c_{i+k}$, $k = 0, 1, \ldots, L-1$, assumed by a De Bruijn sequence c of span n and period L, for a given shift k, the known results are as follows:

$$\theta(k) = 2^n \text{ for } k = 0$$
$$\theta(k) = 0 \text{ for } 1 \leq |k| \leq n-1$$
$$\theta(k) \neq 0 \text{ for } |k| = n$$

It is also known that $\theta(k) \equiv 0 \bmod 4$ for all k, for any binary sequence of period $L = 2^n$, with $n \geq 2$. As any binary De Bruijn sequence c comprises the same number of 1's and 0's, when converted into a bipolar form, the following holds:

$$\sum_{k=0}^{L-1} \theta_c(k) = \sum_{k=0}^{L-1}\sum_{i=0}^{L-1} c_i c_{i+k} = \sum_{i=0}^{L-1} c_i \sum_{k=0}^{L-1} c_{i+k} = 0 \tag{12}$$

A simple bound may be defined for the positive values of the correlation functions sidelobes for the De Bruijn sequences [8]:

$$0 \leq max\ \theta(k) \leq 2^n - 4 \left[\frac{2^n}{2n}\right]^+, \text{ for } 1 \leq k \leq L-1 \tag{13}$$

where $[x]^+$ denotes the smallest integer greater than or equal to x. In the case of binary De Bruijn sequences of span $n = 5$, the bound gives $0 \leq max\ \theta(k) \leq 16$.

It is easy to prove that for an arbitrary pair of De Bruijn sequences a and b, including $a = b$, with span n and period L, the cross-correlation function $r_{ab}(k) = \sum_{i=0}^{L-1} a_i b_{i+k}$, for $0 \leq k \leq L - 1$, exhibits the following properties:

$$r_{ab}(k) = r_{ba}(L - k), \text{ for } 0 \leq k \leq L - 1$$

$$\sum_{k=0}^{L-1} r_{ab}(k) = 0$$

$$r_{ab}(k) \equiv 0 \bmod 4, \text{ for } n \geq 2, \forall k$$

For the cross-correlation function of a pair of De Bruijn sequences a and b $(a \neq b)$ with span n, the following bound holds:

$$-2^n \leq r_{ab}(k) \leq 2^n - 4, \text{ for } 0 \leq k \leq L - 1 \tag{14}$$

It is worth noting that De Bruijn sequences may be piecewise orthogonal, meaning that it is possible to find two sequences having null cross-correlation for several values of the shift parameter k. On the other hand, it is also possible that two De Bruijn sequences have an absolute value of the cross-correlation equal to 2^n for some shift k (e.g. complementary sequences for $k = 0$), as stated by the bound equation above. This variability in the behaviour of the sequences' cross-correlation may affect the performance of the CDMA system, as it will be discussed in the following, when the spreading sequences associated to each user are chosen randomly from the whole set of span $n = 5$ sequences. This also motivates the need for a proper selection criterion to be applied on the whole set of sequences, in order to extract the spreading codes to use in the DS-CDMA system.

Adding a zero to the longest run of zeros in an m-sequence produces a so-called *primitive* De Bruijn sequence. Removing a zero from the longest run of zeros in each De Bruijn sequence produces a so-called *punctured* or *modified* De Bruijn sequence.

4   Applicability of Binary De Bruijn Sequences in DS-CDMA Systems

As previously stated in the Introduction, in this paper we limit our investigations to binary De Bruijn sequences of length 32, i.e. $n = 5$, which form a set of 2048 different sequences. Given the small value of the span parameter considered, it is possible to generate the set of binary De Bruijn sequences by means of an exhaustive approach, which may be intended as a brute force one: all the binary sequences of length 2^n are generated, then the ones satisfying the De Bruijn property are selected. For increasing values of n, the generation process becomes extremely complex, and more sophisticated techniques shall be applied [9]. A possible approach for sequence generation starts with n zeros and appends a one or a zero as the next bit of the sequence, as long as the n-tuple obtained has not appeared before, otherwise the generation path is discarded. This generation scheme, which we may call "tree approach" is fast to execute, but may suffer memory limitations, because all the sequences having the same span n must be generated at the same time.

The complete set of 2048 binary De Bruijn sequences of span $n = 5$ includes 1024 different sequences, and their corresponding complementary sequences, as it happens for any value of n. The cross-correlation function computed between two complementary De Bruijn sequences always shows a negative peak value of -2^n, corresponding to a shift $k = 0$. As a consequence, given the DS-CDMA context of application, it is necessary to avoid the presence of complementary sequences in the set from which spreading codes are chosen. This constraint will limit our analysis to 1024 sequences of span $n = 5$. Taking into account the requirement of low sidelobes for the auto-correlation functions of the selected CDMA spreading sequences, which allows for a better synchronization at the receiver, first of all we analyze the set of 2048 sequences, and select two subsets, including the sequences having a maximum absolute sidelobe level of the auto-correlation function equal to 4, and equal to 8, respectively. By this way, we get a subset named Φ_4 that contains 12 sequences, and a subset named Φ_8 that includes 772 sequences. Fig. 1 shows a sample auto-correlation function profile for a De Bruijn sequence belonging to Φ_4.

The exhaustive analysis of the statistical properties of the sequences included in the set Φ_4 shows, as expected, an average value of the auto-correlation functions equal to zero, and a standard deviation of 6.26. The same analysis performed on the 772 sequences in Φ_8 provides, again, an average value of the auto-correlation functions equal to zero, and a standard deviation equal to 6.34.

The same subsets of De Bruijn sequences have then been studied with regard to the statistical properties of the corresponding cross-correlation functions. The

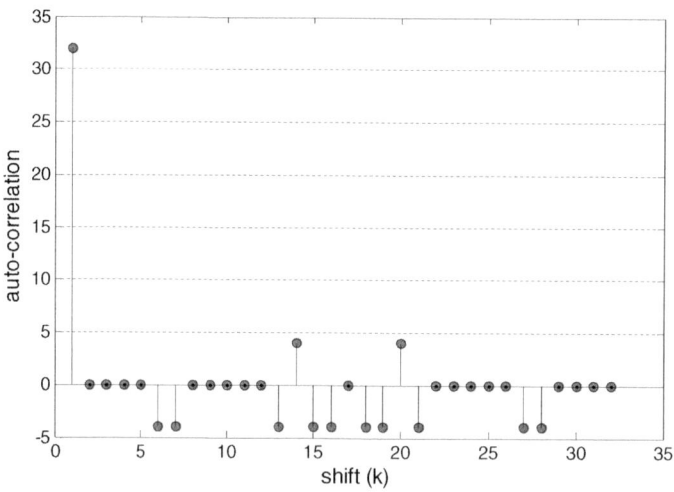

Fig. 1. Sample auto-correlation profile of a De Bruijn sequence in the subset Φ_4

different 66 cross-correlation functions computed over the subset Φ_4 provide a zero average value, as expected, a peak absolute value of 20, and a standard deviation of 5.74. Actually, high values of the cross-correlation functions (i.e. greater than 12) are sporadically obtained; however, when these values appear, and the cross-correlation between two sequences gets higher than 20, the effects on the DS-CDMA system performance are disruptive. Fig. 2 shows a sample cross-correlation profile computed between two sequences randomly chosen in the subset Φ_4.

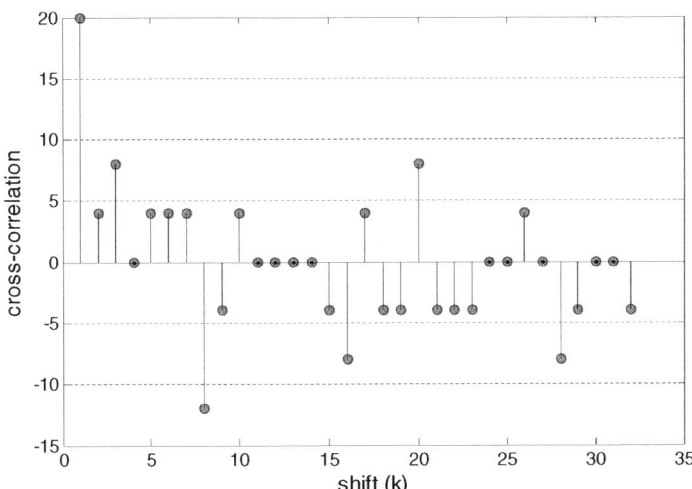

Fig. 2. Sample cross-correlation profile of a pair of De Bruijn sequences in the subset Φ_4

Once obtained the subset Φ_4 including sequences with favourable cross-correlation functions, we tested the possibility of adopting them as spreading codes in the downlink section of a DS-CDMA system, for different amounts of users. According to the average interference parameter definition provided in Eq. 6, we computed the average SNR at the output of a correlator receiver of the i-th user, under AWGN environment. The performance provided by the adoption of De Bruijn sequences are compared to those obtainable by the classical Gold sequences of length 31, and to the ideal behavior of the system. Simulation results are shown in Fig. 3, where the average Signal-to-Noise Ratio values are evaluated, for the E_b/N_0 parameter ranging from 0 to 15 dB, and for different numbers of users active in the system.

As previously discussed, the performance obtainable by the application of binary De Bruijn sequences as spreading codes in DS-CDMA systems are heavily affected by the proper selection of the sequences within a given subset. When the cross-correlation among the sequences selected takes low values, the De Bruijn sequences may perform better than classical Gold codes, even for increasing

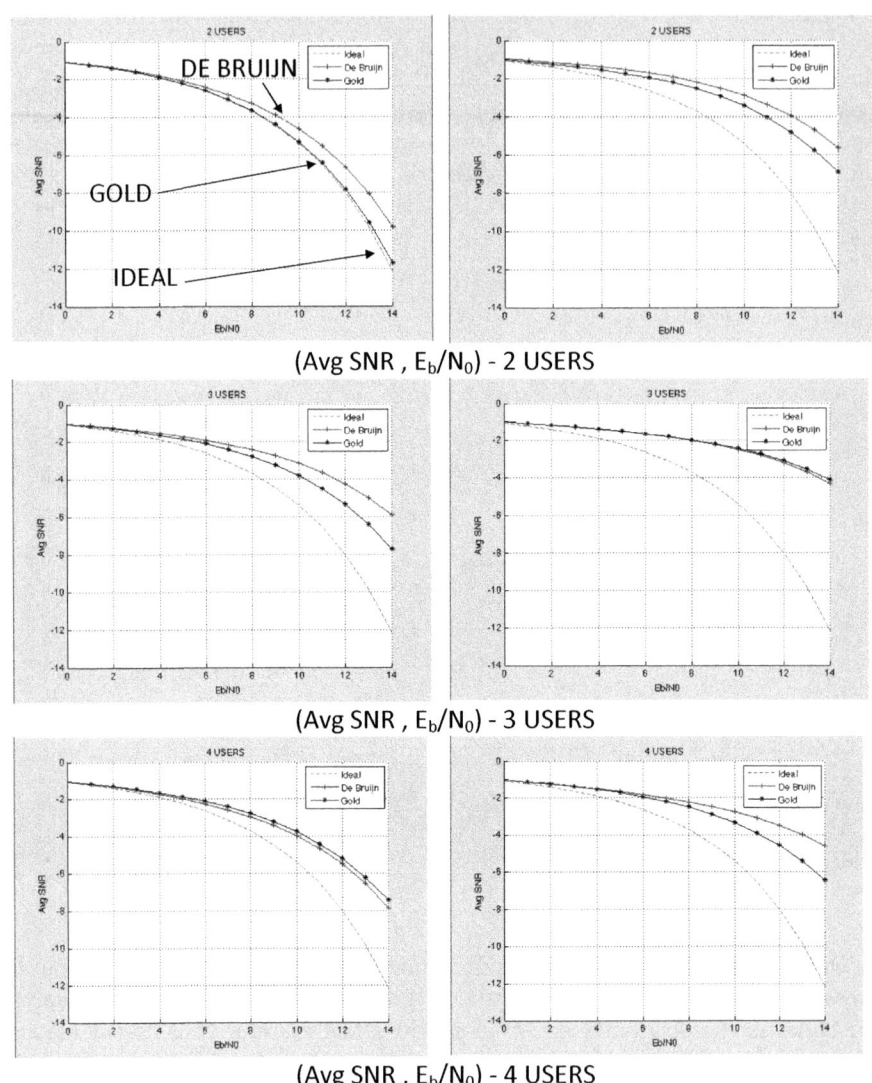

Fig. 3. Average SNR values for different numbers of users adopting De Bruijn spreading codes from the subset Φ_4, compared to Gold and ideal behaviour. Left and right graphs correspond to different maximum absolute cross-correlation values for the De Bruijn sequences used in simulations.

number of users, as shown in Fig. 3 in the case of 4 users, corresponding to a maximum absolute cross-correlation value of 7 for the Gold sequences, and 4 for the De Bruijn ones (in the left graph), and 9 for the Gold sequences, 8 for the De Bruijn ones (in the right graph). The advantage provided by the adoption of De Bruijn sequences, with respect to Gold sequences, is the huge cardinality

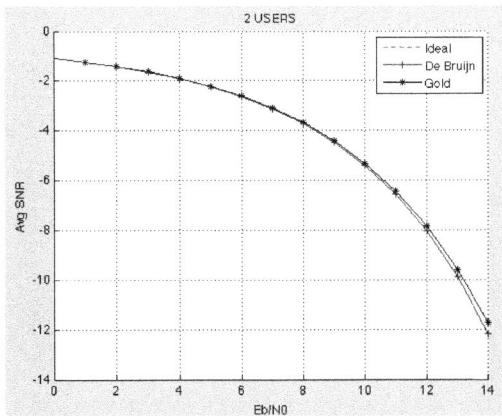

Fig. 4. Average SNR in the case of 2 users, when orthogonal De Bruijn spreading codes are chosen in the set Φ_4

of the initial set of available sequences, which allows to apply strict selection requirements, and to preserve enough cardinality at the same time. The results herein presented have been obtained by applying a correlation-based selection criterion on the set of span $n = 5$ sequences generated through an exhaustive approach. Research activities are currently ongoing in order to design a sequence generation algorithm which outputs the most suitable sequences according to precise constraints, defined on their auto- and cross- correlation profiles, without the need of generating all the possible sequences and then select the suitable ones.

In order to enforce the motivations for the study presented by this paper in its preliminary stage, we also show the performance provided by a two-users DS-CDMA system employing Gold sequences of length 31, and binary De Bruijn sequences of length 32, when the De Bruijn sequences chosen in Φ_4 are orthogonal, i.e. they have a null cross-correlation for any shift k. As expected, the corresponding curve is the same obtainable in the ideal case, as shown in Fig. 4. It is worthy to note that the number of orthogonal, or piecewise orthogonal, De Bruijn sequences increases as their length increases, thus allowing the extraction of a subset of sequences having the necessary cardinality to satisfy the system requirements. On the other hand, Fig. 5 shows that a random selection of the sequences in Φ_4 may also lead to negative performance, when the cross-correlation among the sequences gets a high absolute value, corresponding to 20 in the specific 4 users case presented.

As a final remark, it is possible to say that the results provided by the adoption of De Bruijn sequences deserve further investigations, especially for greater vaues of the span n, because when strict selection criteria are applied, the number of De Bruijn sequences suitable for CDMA applications decreases down to values similar to the number of Gold codes. On the other hand, the behaviour of modified De Bruijn sequences seems really interesting to study, in that lower values, almost equal to zero, are provided for the cross-correlation among

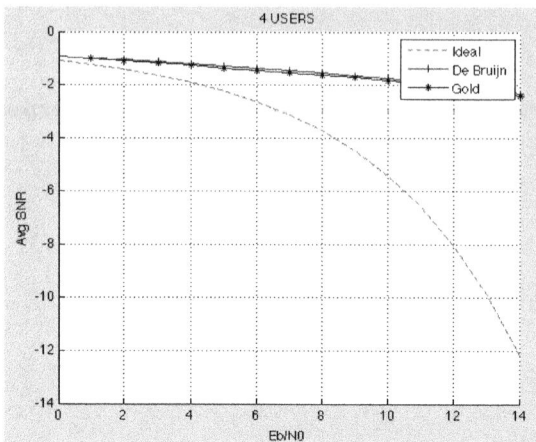

Fig. 5. Average SNR in the case of 4 users adopting De Bruijn spreading codes (blue curve) and Gold spreading codes (black curve) with the highest cross-correlation

sequences, that could make them really a good candidate for use as spreading codes in CDMA applications.

5   Conclusion

This paper presented some preliminary results about the application of binary De Bruijn sequences in DS-CDMA systems, as users' spreading codes. Binary De Bruijn sequences feature huge cardinality of the available sequence sets, even for small values of the span parameter, and may consequently allow the definition of proper selection criteria, based on thresholds applied to the auto- and cross-correlation profiles, though preserving a great amount of available codes. The results discussed in this paper are quite preliminary, however they encourage further studies and analyses, in order to extensively test the applicability of De Bruijn sequences in CDMA contexts, even by resorting to longer codes, that, however, require more sophisticated generation techniques.

References

1. Pursley, M.B.: Performance Evaluation for Phase-Coded Spread Spectrum Multiple-Access Communication - Part I: System Analysis. IEEE Trans. Commun. COM-25(8), 795–799 (1977)
2. Sarwate, D.V., Pursley, M.B.: Crosscorrelation Properties of Pseudorandom and Related Sequences. Proc. of the IEEE 68(5), 593–619 (1980)
3. Pursley, M.B., Sarwate, D.V.: Performance Evaluation for Phase-Coded Spread Spectrum Multiple-Access Communication – Part II: Code Sequence Analysis. IEEE Trans. Commun. COM-25(8), 800–802 (1977)

4. de Bruijn, N.: A combinatorial problem. Proc. Nederlandse Akademie van Weten-schappen 49, 758–764 (1946)
5. Fredricksen, H.: A survey of full length nonlinear shift register cycle algorithms. SIAM Rev. 24, 195–221 (1982)
6. Mitchell, C.J., Etzion, T., Paterson, K.G.: A Method for Constructing Decodable de Bruijn Sequences. IEEE Trans. Inf. Th. 42(5), 1472–1478 (1996)
7. Annexstein, F.S.: Generating De Bruijn Sequences: An Efficient Implementation. IEEE Trans. Comp. 46(2), 198–200 (1997)
8. Zhaozhi, Z., Wende, C.: Correlation properties of De Bruijn sequences. Systems Science and Mathematical Sciences, Academia Sinica 2(2), 170–183 (1989)
9. Etzion, T., Lempel, A.: Algorithms for the Generation of Full-Length Shift-Register Sequences. IEEE Trans. Inf. Th. IT-30(3), 480–484 (1984)

Analysis of Different Channel Sharing Strategies in Cognitive Radio Networks

Elena Bernal-Mor, Vicent Pla, and Jorge Martinez-Bauset

Dept. Comunicaciones, Universidad Politecnica de Valencia, UPV
Camino de Vera s/n, 46022, Valencia, Spain
elbermo@upvnet.upv.es, vpla@dcom.upv.es,
jmartinez@upvnet.upv.es

Abstract. The scarcity of the radio spectrum makes necessary to implement efficient methods to make use of it. In this paper we consider a primary network that rents a given number of its channels and a secondary network that has a number of dedicated channels and can use opportunistically the rented channels of the primary network. Considering this scenario different channel sharing strategies between primary and secondary users are analized and the optimal admission control policy respect to a cost function for secondary users is evaluated for each of the strategies.

Keywords: Cognitive radios, Call admission control, Quality of service.

1 Introduction

The scarcity of the radio spectrum makes necessary to implement efficient methods to make use of it. Today the most part of the radio spectrum is underutilized and the current research efforts in this field are devoted to the study of technlogies that enable dynamic spectrum access. Cognitive Radio (CR) concept has been proposed as a technology which provides the capacity to detect idle frequencies that are not occupied by licensed or primary users (PU) and enables secondary users (SU) to use these idle bands in an opportunistic manner [1].

In this paper we consider a primary network (PN) that rents a part of its channels and a secondary network (SN) that has a number of dedicated channels and can use opportunistically the rented channels of the PN if the PUs do not need them. That scenario has been proposed in [2] but the existence of channels that the PN does not rent were not considered. We propose different channel sharing strategies between PUs and SUs and we also obtain the optimal admission control (AC) policy for each strategy depending on how much harmful is considered the forced termination of an ongoing SU session compared with blocking a new SU session.

2 Model Description and Channel Sharing Strategies

We consider a CR network with a set of channels allocated to PUs and another set allocated to SUs. The PN has N_p channels that only PUs can access and N_r

A. Vinel et al. (Eds.): MACOM 2010, LNCS 6235, pp. 70–73, 2010.

rented channels that can be occupied by SUs when they are idle. Additionally, the SN has N_s dedicated channels that only can be occupied by SUs. If a SU is using a rented channel and a PU needs that channel, the SU vacates the channel and searches an idle channel among the channels that a SU can access, this is referred to as spectrum handover. If there are not idle channels the session of the SU is aborted.

We make the common assumptions of Poisson arrival processes and exponentially distributed random variables for service duration. The arrival rate for primary (secondary) users is λ_p (λ_s) and the service rate of primary (secondary) users is exponentially distributed with rate μ_p (μ_s).

Different channel sharing strategies between PUs and SUs are considered:

1. Any channel of the set of the PN can be rented to SUs with the restriction that the SUs cannot use more than N_r channels of the PN at the same time.
2. The rented channels are a fixed set of N_r primary network channels. There is no repacking for PUs, i.e. if a PU arrives at the network and the N_p channels are occupied, it occupies one of the rented channels. When one of the N_p channels becomes idle, the PU remains in the shared set of channels in spite of having idle channels in the dedicated set.
3. Is the same as strategy 2 but using repacking for PUs.

3   Analysis of the System

In this section we outline the analysis of the performance of strategies 1, 2 and 3 and a method to obtain the optimal AC policy for SUs respect to a cost function is presented.

For each strategy we use a finite quasy birth and death (QBD) Markov process to model the occupation of channels in the system.

To solve the QBD Markov processes and obtain the state stationary probabilities we use the linear level reduction (LLC) algorithm [3]. From the values of the state stationary probabilities, the blocking probabilites for SUs (P_s), PUs (P_p) and the dropping probability for SUs (P_{sh}) can be calculated.

Furthermore, for each strategy we determine the optimal AC policy for SUs modelling the system with Markov Decission Processes (MDP) [4]. The AC policy determines if a SU is accepted or not when arrives depending on the state of the system. In MDP when a decision is made (accepting or not a new SU) it is penalized with some inmediate cost.

In our system, the optimization problem is formulated as the minimization of the average cost rate per time unit. If π is the AC policy, we denote the average cost rate by γ^π and consider the problem of finding the policy π^* that minimizes γ^π, which we name the optimal policy. The cost structure has been chosen so that the cost rate represents a weighted sum of loss rates for new blocking SUs and for ongoing SUs whose service is aborted by the arrival of a PU that needs a rented channel. Let I denote the indicator function which takes the value 1 when all the rented channels are occupied with at least one SU and all the secondary dedicated channels are also occupied, and otherwise it takes the value 0. The average cost rate at state x under policy π is $\gamma_x(\pi) = (1 - \pi(x))\lambda_s + w\lambda_p I(x)$, where $\pi(x) = 1$ when the new SU session is accepted and 0 otherwise, and w is a weight that determines how much harmful is to abort a ongoing session of a SU compared with blocking a new session.

4   Numerical Results

For the numerical examples we consider, unless otherwise indicated, a system with parameters: $N_p = 3$, $N_r = 5$, $N_s = 2$, $\lambda_p = 3.2$, $\mu_p = 1$, $\lambda_s = 1.8$, $\mu_s = 1.25$ and $w = 1$. In Fig. 1 it is shown the optimal AC policy for strategy 1 for $w = 1$ and $w = 10$ and the complete sharing policy (CS) which accepts new requests of SUs as long as there are available channels for them. The higher w the more harmful is considered a forced termination of an ongoing SU session and thus, not accepting new SU is better. Therefore, the optimal AC policy is more restrictive with new SUs.

In Fig. 2, the variation of P_s and P_{sh} to λ_s are shown for each strategy for varying N_r, with $w = 10$ and using the optimal AC policy, note that the optimal policy may vary from one point to another as the load of SUs varies. Strategies 1 and 3 have the same results since the SUs find the same idle channels when arrive at the system. The difference between those strategies is in the interference that both PUs and SUs suffer when a PU occupies a channel that was being used by a SUs. Strategy 1 and 3 have smaller blocking probabilities and higher dropping probabilities than strategy 2. For both strategies, the higher λ_s the higher P_s and P_{hs}, which is logical as the system is more loaded. Regarding the behavior of the probabilities to the variation of N_r, the higher N_r the lower P_s and P_{sh}.

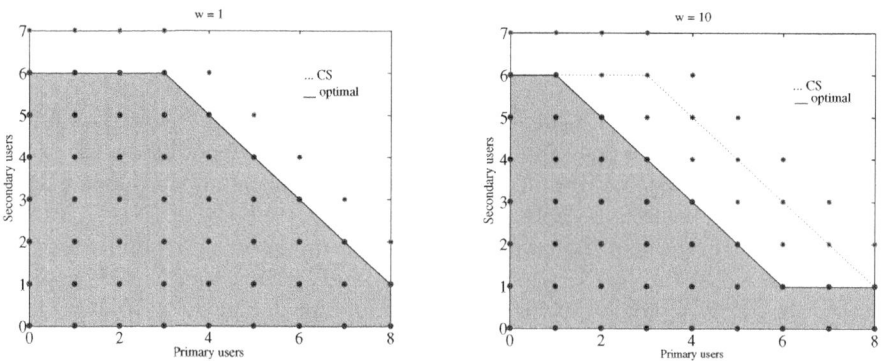

Fig. 1. Optimal admission control policy for strategy 1

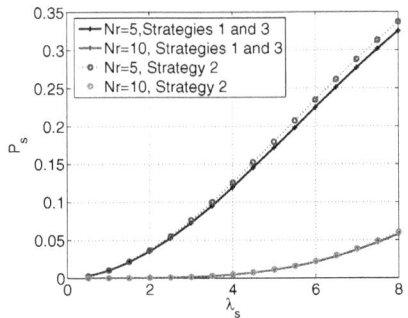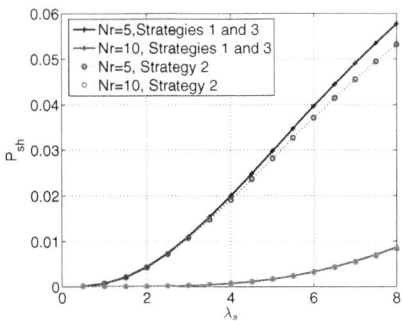

Fig. 2. Evolution of P_s and P_{sh} to λ_s and N_r

5   Conclusions

In this paper different channel sharing strategies between PUs and SUs for CR are studied. Furthermore, a method to obtain an optimal AC policy for each strategy respect a cost function is presented.

The results show that the more harmful is considered a forced termination of an ongoing SU session the more restricitive the optimal AC policy is. Regarding the blocking and forced termination probabilities, obviously a higher value of N_r yields lower values of P_s and P_{sh} but it will also entail a higher number of spectrum handovers for SUs, a higher operational cost at more channels are rented by the PN and thus it may mean a higher economic cost for SN. For future works, we plan to consider this fact, and study a trade-off between the values of blocking and forced termination probabilities and the operational cost in this type of networks.

Acknowledgments. This work was supported by the Spanish Governement and the EU through projects TSI2007-66869-C02-02 and TIN2008-06739-C04-02/TSI and by the UPV through the programm PAID-06-09. Elena Bernal-Mor was supported by the Spanish Governement under contract BES-2007-15030.

References

1. Akyildiz, I.F., Lee, W.Y., Vuran, M.C., Mohanty, S.: A Survey on Spectrum Management in Cognitive Radio Networks. IEEE Communications Communications 46(4), 40–48 (2008)
2. Tzeng, S.S.: Call Admission Control policies in Cellular Wireless Networks with Spectrum Renting. Computer Communications 32, 1905–1913 (2009)
3. Gaver, D.P., Jacobs, P.A., Latouche, G.: Finite Birth-and-Death Models in Randomly Changing Environments. Advances in Applied Probability 16, 715–731 (1984)
4. Ross, S.M.: Applied Probability Models with Optimization applications. Dover Publications, Inc., New York (1992)

A Queueing Model for SDMA Downlink Transmissions

Ruizhi Liao, Boris Bellalta, Miquel Oliver, and Núria Garcia

NETS Research Group - Dpt. of Information and Communication Technologies
Universitat Pompeu Fabra
Roc Boronat 138, 08018 Barcelona, Spain
Ph.: +34-935421498; Fax: +34-935422517
{ruizhi.liao,boris.bellalta,miquel.oliver,nuria.garcia}@upf.edu

Abstract. This paper presents a simple queuing model to explore the benefits of simultaneous downlink transmissions by utilizing the Multiple Input Multiple Output (MIMO) technology when a finite-buffer queue is considered. Results show that the system performance, in terms of frame losses and delay, can not be properly explored without considering the impact of the queueing process and its interactions with the number of available spatial streams.

1 Introduction

The spatial degrees of freedom provided by the MIMO technology have the potential to significantly improve the performance of wireless systems in both of the uplink and downlink. For instance, in Wireless Local Area Networks (WLANs), for the uplink, multiple stations can simultaneously communicate with the Access Point (AP) [1], and for the downlink, the bottleneck effect of AP can be mitigated by utilizing the spatial multiplexing feature to transmit frames to multiple stations in parallel [2]. Similar works can be found in [3,4]. Unfortunately, the analysis in [1,2,3,4] did not take the frame queueing analysis into account and thus, they are not able to provide performance measures such as the losses and/or the average transmission delay, which are key parameters for multimedia applications like the video conference and VoIP (Voice over Internet Protocol).

In this short-paper, a simple queueing model of a system using SDMA (Space Division Multiple Access) for downlink transmissions is developed. It allows to couple the parallel batch-service transmissions through the multiple antennas with considering the queue state, providing some first insights on the impact of queueing in such a system. A different approach for the same problem is presented in [6], although here a simpler alternative model is provided. The rest of the paper is organized as follows: Section 2 describes the multi-frame transmission queue. Section 3 presents the queueing model analysis using Markov Chains. Section 4 discusses the performance results and observations. Section 5 concludes the paper.

A. Vinel et al. (Eds.): MACOM 2010, LNCS 6235, pp. 74–78, 2010.

2   The Multi-frame Transmission Queue

We characterize the frame behavior at the link layer, where frames are queued and scheduled for transmission. Frames destined to different stations could be spatially assembled into a batch and sent out simultaneously via M multiple antennas. Thorough the paper this is called a Space-batch.

Consider a system with M antennas, a queue with K frames. Q_n denotes the number of frames in the queue at the instant that the n-th batch $\theta\{Q_n\}$ is sent out. We assume that there will be always $\theta\{Q_n\} = min(Q_n, M)$ frames available for the batch assembly, which is to say that in the Q_n frames there are at least $\theta\{Q_n\}$ frames destined for different stations. The queue evolving equation can be written as:

$$Q_n = \{Q_{n-1} - \theta\{Q_n\} + min(I_{n-1}, K - Q_{n-1})\} \tag{1}$$

where I_{n-1} denotes the number of incoming frames during the period between $n-1$ and n transmissions. An example of the modeled system is illustrated as Fig. 1.

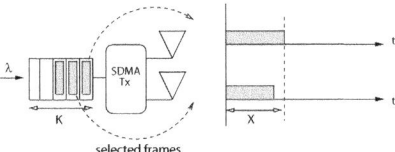

Fig. 1. SDMA Transmitter with $M = 2$ antennas and $K = 5$ frames

3   Queueing Analysis

A Markov Chain is utilized to model the queueing behavior. To understand how the system is modeled, a simple queueing system with $M = 2$ antennas and $K = 3$ frames queue length is considered. Its generalization to any M and K values is straightforward. A two-dimension state space is needed to model those new arriving frames that can not be included in the on-going Space-batch transmission and have to wait in the queue. Hence, the state space has to account for both frames in service and frames stored. Let (t, w) be a given system state with t denoting the number of frames in transmission and w denoting the number of frames waiting in the queue. The two-dimension state space $\mathcal{S}$ for this specific example can be written as $\mathcal{S} = \{(0,0), (1,0), (1,1), (1,2), (2,0), (2,1)\}$, whose components completely describe the state of the queue at any point of time. The state diagram for this system is illustrated in Fig. 2.

The desired performance metrics such as the blocking probability and average delay, as well as other system parameters such as the average Space-batch length and average queue occupation, can be computed from the steady-state probability distribution, π^s. This is obtained by solving the linear

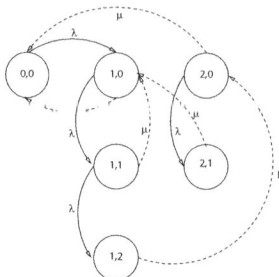

Fig. 2. Two-dimension state transition diagram for $M = 2$ antennas and $K = 3$ frames

system $\boldsymbol{\pi}^s \boldsymbol{Q} = 0$, where $\boldsymbol{Q}$ is the infinitesimal generator of the system [5,7]. And then we can calculate the blocking probability by $P_b = \sum_{m=1}^{M} \pi^s_{(m,K-m)}$, the average number of frames in the queue (including those in transmission) by $E[N] = \sum_{m=1}^{M} \sum_{i=0}^{K-m} (m+i) \cdot \pi^s_{(m,i)}$, and the average delay by applying the Little's Law $E[R] = E[N]/\lambda(1 - P_b)$. Finally, the average Space-batch length is computing by $E[\theta] = \frac{1}{1-\pi^s_{(0,0)}} \sum_{m=1}^{M} \sum_{i=0}^{K-m} m \cdot \pi^s_{(m,i)}$.

4 Performance Evaluation

The parameters considered to evaluate the proposed SDMA queuing model are:

1. Two queue length values: $K = 10$ and $K = 20$ frames.
2. Number of antennas: $M = 1$, 2, 4, and 8.
3. The service time, X, of a Space-batch is exponentially distributed with mean 0.01 seconds.
4. Poisson arrivals with rate λ. λ ranges from 10 to 1000 packets per second. The traffic load is $A = \lambda \cdot X$ Erlangs, independently of the number of frames transmitted simultaneously in a Space-batch.

The COST [8] libraries have been used to implement and simulate the considered system. As showed in the following plots, results from the queueing model and simulation are matched very well, which thus validates the correctness of the queueing model.

Figs. 3.a) and 3.b) show us the blocking probability. It increases as the traffic load gets heavier and decreases as we employ a longer queue or increase the number antennas. However, with respect to the number of antennas, an important result is that the blocking probability does not decrease proportionally to M. For example, with $K = 10$ and at a traffic load equal to $A = 2$ Erlangs, the P_b obtained with $M = 2$ equals to 0.15 and with $M = 4$ to 0.06, which means a considerable reduction. However, using $M = 8$ antennas, there is no gain compared with using $M = 4$ antennas. This can be explained by the second set of plots, which show the average number of frames included in a Space-batch against traffic load for different number of antennas.

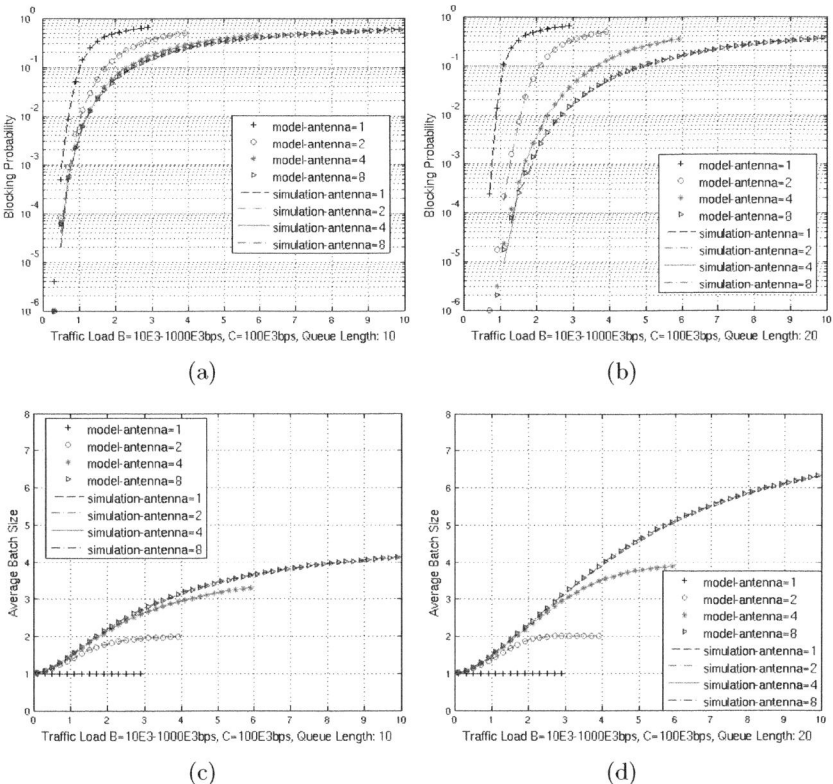

Fig. 3. Blocking probability & Average Space-batch against the Traffic load with different number of transmission antennas M

As we can see from Figs. 3.c) and 3.d), the average Space-batch length has no big difference when utilizing $M = 4$ or $M = 8$ antennas when $K = 10$ frames. This is because when using $M = 8$ antennas and a large Space-batch is sent out, the queue suddenly becomes empty. This implies that next Space-batch will contain very few frames as it is scheduled soon after the previous departure, showing a similar average Space-batch length than in the case of $M = 4$ antennas. In the second scenario, with $K = 20$ frames, we can easily distinguish the difference of the blocking probability and average Space-batch length between utilizing $M = 4$ and $M = 8$ antennas. This is because more frames are now allowed to be queued, and therefore we will have an increasing chance to transmit consistently large Space-batches as the traffic load gets higher.

5 Conclusions

A simple queuing model for SDMA downlink transmissions has been presented and validated. The experiments done show the potential of considering multiple antennas at the transmitter to improve the system performance. However, the results also show that the achieved gain is not linear with the number of antennas due to the presence of a finite transmission queue, which has to be dimensioned considering both the number of antennas and the desired system performance in terms of losses and delay. This work was partially supported by the Spanish Government under project TEC2008-06055/TEC.

References

1. Jin, H., Jung, B., Hwang, H., Sung, D.: A Throughput Balancing Problem between Uplink and Downlink in Multi-user MIMO-based WLAN Systems. In: Proceedings of IEEE Conference on Wireless Communications & Networking Conference, pp. 1795–1800 (2009)
2. Cai, L., Shan, H., Zhuang, W., Shen, X., Mark, J.W., Wang, Z.: A Distributed Multi-User MIMO MAC Protocol for Wireless Local Area Networks. In: Proceedings of IEEE Globecom, pp. 4976–4980 (2008)
3. Choi, Y., Lee, N., Bahk, S.: Exploiting Multiuser MIMO in the IEEE 802.11 Wireless LAN Systems. Wireless Personal Communications (2009)
4. Jin, H., Jung, B., Hwang, H., Sung, D.: A MIMO-based Collision Mitigation Scheme in Uplink WLANs. IEEE Communications Letters, 417–419 (2008)
5. Stewart, W.J.: Introduction to the Numerical Solution of Markov Chains. Princeton University Press, Princeton (1994)
6. Bellalta, B., Oliver, M.: A space-time batch-service queueing model for multi-user MIMO communication systems. In: Proceedings of the 12th ACM International MSWiM Conference, pp. 357–364 (2009)
7. Bellalta, B.: A Queuing Model for the Non-continuous Frame Assembly Scheme in Finite Buffers. In: Al-Begain, K., Fiems, D., Horváth, G. (eds.) ASMTA 2009. LNCS, vol. 5513, pp. 219–233. Springer, Heidelberg (2009)
8. Chen, G.: Component-oriented simulation toolkit (2004), http://www.ita.cs.rpi.edu/sense/cost.html

Queueing System with Alternating Service Rates for Free Space Optics-Radio Hybrid Channel

Vladimir Vishnevskiy and Olga Semenova

ZAO Research & Development Company "INSET"
Moscow, Russia
vishn@inbox.ru, olgasmnv@gmail.com

Abstract. We propose a queueing model for free space optics-radio hybrid channel. Hybrid channel is presented by a pair of alternating channels (free-space optical link and radio frequency channel). Time when optical link is/is not used mainly depends on the weather conditions and is approximated by the mixture of two exponential distributions. We consider a queueing system with alternating service rates and apply the matrix-analytical approach to obtain stationary distribution of system states and performance characteristics.

Keywords: hybrid channel, free-space optical communication, radio frequency channel, queueing system, alternating service, matrix-analytical approach.

1 Introduction

The objective of future wireless communication systems is to provide users with wide veriety of services such as internet access, file transfer, interactive data, voice and image transfer. The capacity of Radio Frequency (RF) wireless networks (e.g., 802.11x) is constrained by limits to link throughputs of tens of Mbps and distances of tens of meters. It does not scale well with increasing number of nodes in the system due to the interference between concurrent transmissions from neighboring nodes. Link availability can be maintained under most weather conditions with exception for heavy rain, that can take little effect. In order to improve the bandwidth and system capacity, free space optical (FSO) or optical wireless systems can be considered as an alternative communication media. With FSO, an optical transceiver is placed on each side of a transmission path to establish a network link. The transmitter is typically an infrared laser or LED that emits a modulated infrared laser signal. Commercial FSO currently provide several Gbts throughputs with several kilometers link distances. However, one of the major limitations of FSO technology is the need for optical links to maintain line-of-sight (LOS). The FSO link availability can be further limited by adverse weather conditions like fogs and heavy snowfalls.

The complementary property of RF and FSO motivates the engineers to design hybrid FSO/RF channels, in which the weaknesses of each link type are expected to be mutually mitigated. It is also natural to view the FSO/RF

A. Vinel et al. (Eds.): MACOM 2010, LNCS 6235, pp. 79–90, 2010.

combination as a way to solve the capacity scarcity problem in RF wireless networks, or at least turn it around to some extent. But it is obvious that an analysis prior to the equipment installation is essential, as a poorly designed path may result periods of system outages, increases system latency, decreases throughput or a complete failure of the communication across the link. Several practical designs and simulation models of the FSO/RF hybrid channel have been proposed and implemented in [1]–[4]. However, there is still no theoretical work that may give an insight how the hybrid channel with non-reliable links can be modeled by virtue of queueing systems.

To model the proposed hybrid channel with links of different rates, we propose a queueing system with two possible service rates: the first for the FSO link and the second for RF channel. Under the normal operating conditions, data is transmitted over the FSO link. When the FSO link fails and remains inaccessible over the certain time, packets are transmitted over the RF channel until the FSO link is available again. In the paper, the change of links is modelled as a queueing system with alternating service rates. In the model, we suppose that the periods of time when the FSO link is on/off are the mixture of two exponential distributions (partial case of hyperexponential distribution) approximating the experimental data presented below.

The paper is organized as follows. In Section 2, the results of the real experiment on the FSO link accessibility/inaccessibility time and their statistical analysis are presented. The queueing model for the hybrid channel is described in Section 3. Section 4 deals with the embedded Markov chain and stationary state probabilities which are obtained via matrix-analytical approach in Section 5. Numerical results are presented in Section 6.

2   Statistical Analysis of the Channel Accessibility/Inaccessibility Data

Recent papers studying the signal quality for free-space optical (FSO) communication deal with the outage probability that the meteorological optical range (MOR) is lower than a given threshold, see e.g. [5]. The outage probability is not sufficient to estimate the channel performance and one of the main factors is how often the weather conditions change. The optical channel accessibility is affected by other factors such as scintillation, adjustment, bearing fluctuations, effect of sun, etc. related to the atmosphere and weather conditions. Besides, they are accompanied by the transceiver construction features: optical wavelength, number of rays, ray divergence, receiver sensibility, transmitter power, etc. And it is more reasonable to estimate the optical channel accessibility/inaccessibility time rather than create a model taking into account those factors separately. For example, the experimental FSO accessibility time can be approximated by the mixture of two exponential distributions as shown on Fig. 1 and 2 (for 4.5 km channel).

To get the mixture distribution

$$F(t) = p\gamma_1 e^{-\gamma_1 t} + (1-p)\gamma_2 e^{-\gamma_2 t}$$

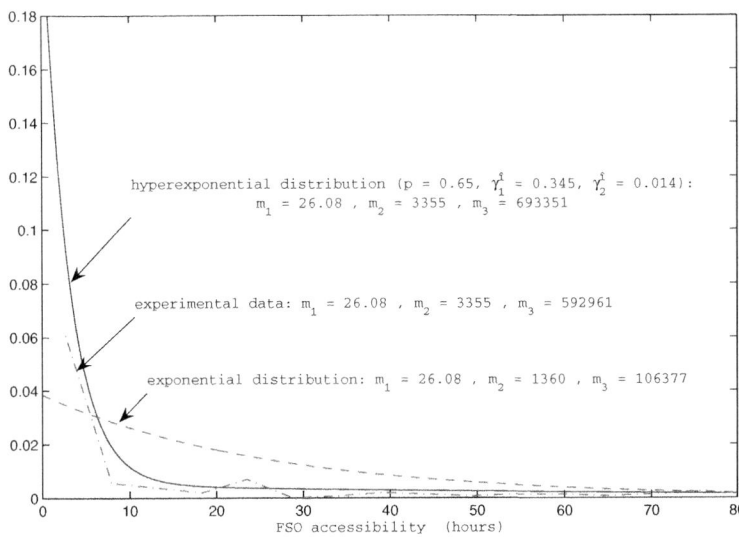

Fig. 1. FSO accessibility time obtained from the experiment and its approximation

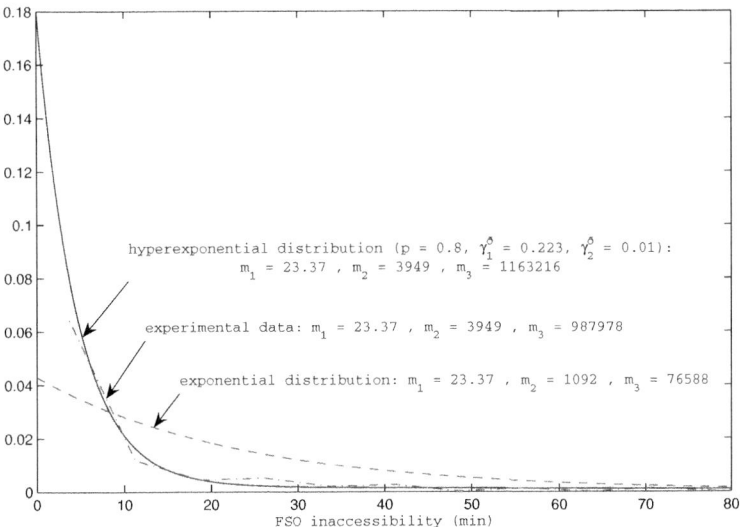

Fig. 2. FSO inaccessibility time obtained from the experiment and its approximation

which is the partial case of the hyperexponential one we use an approach from [6]. Let $\{X_1, ..., X_N\}$ be the data sample (e.g., accessibility period). First, calculate the first three moments m_1, m_2 and m_3 as

$$m_i = \frac{1}{i!N} \sum_{k=1}^{N} X_k^i, \quad i = 1, 2, 3$$

parameters

$$f_1 = \frac{m_3 - m_1 m_2}{m_2 - (m_1)^2}, \quad f_2 = \frac{m_1 m_3 - (m_2)^2}{m_2 - (m_1)^2}.$$

If the inequalities $m_2 > m_1^2$, $m_3 > m_1 m_2$, $m_1 m_3 > m_2^2$ and $f_1^2 > 4f_2$, are fulfilled we define the parameters of the mixture distribution as $\gamma_1 = \frac{2}{f_1 + \sqrt{f_1^2 - 4f_2}}$, $\gamma_2 = \frac{2}{f_1 - \sqrt{f_1^2 - 4f_2}}$, $p = \frac{\gamma_1(\gamma_2 m_1 - 1)}{\gamma_2 - \gamma_1}$. Otherwise, the values γ_1, γ_2 and p are chosen to satisfy the relations $p \neq \frac{(m_1)^2}{m_2}$, $\quad 0 \leq p \leq 1$, $\quad$ and $\quad m_1 \gamma_2 - 1 + p > 0$, where $\gamma_1 = \frac{p\gamma_2}{m_1\gamma_2 - 1 + p}$, $\gamma_2 = \frac{m_1(1-p) - \sqrt{(1-p)p_i(\gamma_2 - m_1^2)}}{(m_1)^2 - m_2 p}$. Note that the value p should be chosen to minimize $\left| \frac{p}{(m_1)^3} + \frac{1-p}{(m_2)^3} - m_3 \right|$ for the better approximation.

So, we assume that he probability density functions for periods of accessibility/inaccessibility are

$$F(t) = p\gamma_1 e^{-\gamma_1 t} + (1 - p)\gamma_2 e^{-\gamma_2 t}$$

with parameters γ_1, γ_2 and p defined as shown on Fig. 1-2.

3   Queueing Model for the Hybrid Channel

In the Section, we present queueing system with two alternating service rates for free space optics-radio hybrid channel.

Customers arrive to the system accordingly to the Poisson input with parameter λ. When the FSO (RF) link is on , the packet transmission times are exponentially distributed with the rate μ_1 (μ_2), $\mu_1 >> \mu_2$. Time when the system uses rate k ($k = 1, 2$) is limited by the random value ξ_k with probability density function

$$F_k(t) = p_k \gamma_k^{(1)} e^{-\gamma_k^{(1)} t} + (1 - p_k)\gamma_k^{(2)} e^{-\gamma_k^{(2)} t}, \quad k = 1, 2.$$

When time to use the rate 1 expires (FSO link fails), the current customer service (packet transmission) is interrupted and the system switches to rate 2 (to use RF channel) during the constant time Q_2. If time Q_2 expires and FSO channel is still inaccessible (e.g., the weather conditions are still bad) the customers start to be served at rate 2. Assume that the customer whose service was interrupted needs to be served anew. If the FSO link becomes accessible before time Q_2 expires, the customers start to be served at rate 1 immediately.

When FSO link becomes accessible, the system continues serving customers at rate 2 during the time having exponential distribution with parameter Q_1. And if FSO link remains accessible during this time period, the system starts using rate 1. Otherwise the system keeps using rate 2 until the FSO link becomes accessible again.

Using the results of [9], the sufficient stability condition can be obtained as

$$\rho_1 + \rho_1 V \left(\frac{p_1}{\gamma_1^{(1)}} + \frac{1 - p_1}{\gamma_1^{(2)}} \right)^{-1} < 1,$$

where $\rho_1 = \lambda/\mu_1$, V is the average time when the system does not use the FSO link

$$V = w_1 \left(\frac{p_2}{\gamma_2^{(1)}} + \frac{1 - p_2}{\gamma_2^{(2)}} \right) + (1 - w_1) \left(Q_2 + \frac{1}{\alpha Q_1} + \frac{V_2}{\alpha} \right),$$

with $\alpha = p_1 \frac{\gamma_1^{(1)}}{\gamma_1^{(1)}+Q_1} + (1-p_1) \frac{\gamma_1^{(2)}}{\gamma_1^{(2)}+Q_1}$ and $w_1 = 1 - p_2 e^{-\gamma_2^{(1)} Q_2} - (1-p_2) e^{-\gamma_2^{(2)} Q_2}$.

4 Embedded Markov Chain and Stationary State Probabilities

Below, we describe the embedded Markov chain to investigate the system behavior at service completion, service interruption moments and moments when the server changes its state. Let t_k be the k-th such moment, $k \geq 1$.

The system state at the moment t_k is presented by vector

$$X_k = (i_k, m_k, m'_k),$$

where i_k is the number of customers in the system at moment $t_k + 0$, m_k is the server's state at moment $t_k - 0$, m'_k is the server's state at moment $t_k + 0$. Here we assume the state (i, m, m) corresponds to the service completion, $m \in \{1, 2, 2'\}$. The server can be in one of four possible states:

1 – it works at rate 1 (FSO link is on),

1' – FSO link failed and the system waits for it accessibility (no longer than time Q_2),

2 – it works at rate 2 (packets are transmitted by RF channel),

2' – system works at rate 2, FSO link is accessible and the time to switch to FSO link is not expired.

Consider the stationary state probabilities

$$\pi(i, m, m') = \lim_{k \to \infty} P\{i_k = i, m_k = m, m'_k = m'\}, i \geq 0, m, m' \in \{1, 1', 2, 2'\}$$

and introduce the following probabilities:

$$f_i^{(k)} = \int_0^\infty \frac{(\lambda t)^i}{i!} e^{-\lambda t} e^{-\mu_k t} \left(p_k e^{-\gamma_k^{(1)} t} + (1 - p_k) e^{-\gamma_k^{(2)} t} \right) \mu_k dt,$$

is the probability that i customers arrive during the customer service time (when rate k is used),

$$y_i^{(k)} = \int_0^\infty \frac{(\lambda t)^i}{i!} e^{-\lambda t} e^{-\mu_k t} \left(p_k \gamma_k^{(1)} e^{-\gamma_k^{(1)} t} + (1 - p_k) \gamma_k^{(2)} e^{-\gamma_k^{(2)} t} \right) dt$$

is the probability that during the customer service time the rate k time expires (FSO channel fails ($k = 1$) or becomes accessible ($k = 2$)) and i customers enter the system during the uncompleted service time,

$$h_i^{(1)} = \int_0^{Q_2} \frac{(\lambda t)^i}{i!} e^{-\lambda t} \left(p_2 \gamma_2^{(1)} e^{-\gamma_2^{(1)} t} + (1 - p_2) \gamma_2^{(2)} e^{-\gamma_2^{(2)} t} \right) dt$$

is the probability that the switchover time from rate 1 to rate 2 is less than Q_2 and i customers arrive to the system,

$$h_i^{(2)} = \frac{(\lambda Q_2)^i}{i!} e^{-\lambda Q_2} \left(p_2 e^{-\gamma_2^{(1)} Q_2} + (1 - p_2) e^{-\gamma_2^{(2)} Q_2} \right)$$

is the probability that the switchover time from rate 1 to rate 2 is less than Q_2 and i customers arrive to the system during the switchover time,

$$r_i = \int_0^\infty \frac{(\lambda t)^i}{i!} e^{-\lambda t} e^{-Q_1 t} \mu_2 e^{-\mu_2 t} dt$$

is the probability that i customers enter the system during switchover time from rate 2 to rate 1,

$$m_i = \int_0^\infty \frac{(\lambda t)^i}{i!} e^{-\lambda t} e^{-Q_1 t} Q_1 e^{-\mu_2 t} dt$$

is the probability that during the customer service time when system changes from rate 2 to 1, the switchover time finishes and i customers enter the system.

The probabilities can be calculated as follows

$$f_i^{(k)} = p_k \frac{\mu_k \lambda^i}{\left(\gamma_k^{(1)} + \mu_k + \lambda \right)^{i+1}} + (1 - p_k) \frac{\mu_k \lambda^i}{\left(\gamma_k^{(2)} + \mu_k + \lambda \right)^{i+1}}, \quad k = 1, 2,$$

$$g_i^{(k)} = p_k \frac{\gamma_k^{(1)} \lambda^i}{\left(\gamma_k^{(1)} + \mu_k + \lambda \right)^{i+1}} + (1 - p_k) \frac{\gamma_k^{(2)} \lambda^i}{\left(\gamma_k^{(2)} + \mu_k + \lambda \right)^{i+1}}, \quad k = 1, 2,$$

$$h_i^{(1)} = \frac{\gamma_2^{(1)} \lambda^i p_2}{\gamma_2^{(1)} + \lambda} \left[\frac{1}{(\gamma_2^{(1)} + \lambda)^i} - e^{-(\gamma_2^{(1)} + \lambda) Q_2} \sum_{m=0}^{i} \frac{Q_2^m}{m!(\gamma_2^{(1)} + \lambda)^{i-m}} \right] +$$

$$+ \frac{\gamma_2^{(2)} \lambda^i (1 - p_2)}{\gamma_2^{(2)} + \lambda} \left[\frac{1}{(\gamma_2^{(2)} + \lambda)^i} - e^{-(\gamma_2^{(2)} + \lambda) Q_2} \sum_{m=0}^{i} \frac{Q_2^m}{m!(\gamma_2^{(2)} + \lambda)^{i-m}} \right],$$

$$r_i = \frac{\mu_2 \lambda^i}{(Q_1 + \mu_2 + \lambda)^{i+1}}, \quad m_i = \frac{Q_1 \lambda^i}{(Q_1 + \mu_2 + \lambda)^{i+1}}, \quad i \geq 0.$$

The stationary state probabilities $\pi(i, m, m')$, $i \geq 0$, satisfy the following balance equations

$$\pi(i, 1, 1) = \sum_{k=1}^{i+1} \left(\pi(k, 1, 1) + \pi(k, 2', 1) + \pi(k, 1', 1) \right) f_{i-k+1}^{(1)} + \tag{1}$$

$$+ \left(\pi(0, 1, 1) + \pi(0, 2', 1) + \pi(0, 1', 1) \right) s_1 f_i^{(1)},$$

$$\pi(i, 1, 1') = \sum_{k=1}^{i} \left(\pi(k, 1, 1) + \pi(k, 2', 1) + \pi(k, 1', 1) \right) g_{i-k}^{(1)} + \tag{2}$$

$$+ \left(\pi(0, 1, 1) + \pi(0, 2', 1) + \pi(0, 1', 1) \right) v_i^{(1)},$$

$$\pi(i, 1', r) = \sum_{k=0}^{i} \pi(k, 1, 1') h_{i-k}^{(r)}, \quad r = 1, 2, \tag{3}$$

$$\pi(i, 2, 2) = \sum_{k=1}^{i+1} \left(\pi(k, 2, 2) + \pi(k, 2', 2) + \pi(k, 1', 2) \right) f_{i-k+1}^{(2)} + \tag{4}$$

$$+ \left(\pi(0, 2, 2) + \pi(0, 2', 2) + \pi(0, 1', 2) \right) s_2 f_2^{(1)},$$

$$\pi(i, 2, 2') = \sum_{k=1}^{i} \left(\pi(k, 2, 2) + \pi(k, 2', 2) + \pi(k, 1', 2) \right) g_{i-k}^{(2)} + \tag{5}$$

$$+ \left(\pi(0, 2, 2) + \pi(0, 2', 2) + \pi(0, 1', 2) \right) v_i^{(2)},$$

$$\pi(i, 2', 2') = \sum_{k=1}^{i+1} \left(\pi(k, 2', 2') + \pi(k, 2, 2') \right) r_{i-k+1} + \tag{6}$$

$$+ \left(\pi(0, 2', 2') + \pi(0, 2, 2') \right) \frac{\lambda}{\lambda + Q_1} r_i,$$

$$\pi(i, 2', 1) = \sum_{k=1}^{i} \left(\pi(k, 2', 2') + \pi(k, 2, 2') \right) m_{i-k} \alpha + \tag{7}$$

$$+ \left(\pi(0, 2', 2') + \pi(0, 2, 2') \right) \left(\frac{\lambda \alpha}{\lambda + Q_1} m_{i-1} + \frac{Q_1 \alpha}{\lambda + Q_1} I_{\{i=0\}} \right),$$

$$\pi(i, 2', 2) = \sum_{k=1}^{i} \left(\pi(k, 2', 2') + \pi(k, 2, 2') \right) m_{i-k} (1 - \alpha) + \tag{8}$$

$$+ \left(\pi(0, 2', 2') + \pi(0, 2, 2') \right) \left(\frac{\lambda(1 - \alpha)}{\lambda + Q_1} m_{i-1} + \frac{Q_1(1 - \alpha)}{\lambda + Q_1} I_{\{i=0\}} \right),$$

where $I_{\{A\}}$ is the indicator function of the event A,

$$s_m = p_m \frac{\lambda}{\lambda + \gamma_m^{(1)}} + (1 - p_m) \frac{\lambda}{\lambda + \gamma_m^{(2)}}, \quad m = 1, 2,$$

$$v_i^{(m)} = s_m g_{i-1}^{(m)} I_{\{i>0\}} + (1 - s_m) I_{\{i=0\}}, \quad i \geq 0.$$

Note that equations (3) can be excluded from (1)–(5) and we have

$$\pi(i,1,1) = \sum_{k=1}^{i+1} \left(\pi(k,1,1) + \pi(k,2',1)\right) f^{(1)}_{i-k+1} + \sum_{k=1}^{i+1} \pi(k,1,1') \hat{f}^{(1)}_{i-k+1} + \quad (9)$$
$$+ \left(\pi(0,1,1) + \pi(0,2',1)\right) s_1 f^{(1)}_i + \pi(0,1,1') \left((s_1 f^{(1)}_i - f^{(1)}_{i+1}) h^{(1)}_0 + \hat{f}^{(1)}_{i+1}\right),$$

$$\pi(i,1,1') = \sum_{k=1}^{i} \left(\pi(k,1,1) + \pi(k,2',1)\right) g^{(1)}_{i-k} + \sum_{k=1}^{i} \pi(k,1,1') \hat{g}^{(1)}_{i-k} + \quad (10)$$
$$+ \left(\pi(0,1,1) + \pi(0,2',1)\right) v^{(1)}_i + \pi(0,1,1') \left(\hat{g}^{(1)}_i - h^{(1)}_0 g^{(1)}_i + v^{(1)}_i h^{(1)}_0\right),$$

$$\pi(i,2,2) = \sum_{k=1}^{i+1} \left(\pi(k,2,2) + \pi(k,2',2)\right) f^{(2)}_{i-k+1} + \sum_{k=1}^{i+1} \pi(k,1,1') \hat{f}^{(2)}_{i-k+1} + \quad (11)$$
$$+ \left(\pi(0,2,2) + \pi(0,2',2)\right) s_2 f^{(2)}_i + \pi(0,1,1') \left((s_2 f^{(2)}_i - f^{(2)}_{i+1}) h^{(2)}_0 + \hat{f}^{(2)}_{i+1}\right),$$

$$\pi(i,2,2') = \sum_{k=1}^{i} \left(\pi(k,2,2) + \pi(k,2',2)\right) g^{(2)}_{i-k} + \sum_{k=1}^{i} \pi(k,1,1') \hat{g}^{(2)}_{i-k} + \quad (12)$$
$$+ \left(\pi(0,2,2) + \pi(0,2',2)\right) v^{(2)}_i + \pi(0,1,1') \left(\hat{g}^{(2)}_i - h^{(2)}_0 g^{(2)}_i + v^{(2)}_i h^{(2)}_0\right),$$

where $\hat{f}^{(m)}_i = \sum_{k=0}^{i} h^{(m)}_k f^{(m)}_{i-k}$, $\quad \hat{g}^{(m)}_i = \sum_{k=0}^{i} h^{(m)}_k g^{(m)}_{i-k}$, $\quad i \geq 0, m = 1, 2.$

5   Matrix-Analytical Approach to Calculate the Stationary State Distribution

To calculate the stationary state probabilities $\pi(i,m,m')$, $i \geq 0$, $m, m' \in \{1, 1', 2, 2'\}$ (except $\pi(i,1',r)$ which can be easily calculated by formulas (3)) the matrix-analytical approach can be applied. We refer a reader to [7,8] for more details. Consider the vectors of stationary probabilities

$$\boldsymbol{\pi}_i = (\pi(i,1,1), \pi(i,1,1'), \pi(i,2,2), \pi(i,2,2'), \pi(i,2',2'), \pi(i,2',1), \pi(i,2',2)).$$

and rewrite the balance equations in the matrix form

$$\boldsymbol{\pi}_i = \sum_{k=0}^{i+1} \boldsymbol{\pi}_k P_{k,i}, \qquad i \geq 0, \quad (13)$$

where matrices $P_{i,l}$, $i \geq 0$, $l \geq min\{i-1, 0\}$, are defined as

$$P_{0,l} = \begin{bmatrix} A_l & O_{3\times 3} \\ O_{2\times 4} & B_l \\ C_l & O_{2\times 3} \end{bmatrix}, \quad (14)$$

$P_{i,l} = Y_{l-i+1}$ for $l \geq i-1$, where

$$
Y_l = \begin{bmatrix}
f_l^{(1)} & g_{l-1}^{(1)} & 0 & 0 & 0 & 0 & 0 \\
\hat{f}_l^{(1)} & \hat{g}_{l-1}^{(1)} & \hat{f}_l^{(2)} & \hat{g}_{l-1}^{(2)} & 0 & 0 & 0 \\
0 & 0 & f_l^{(2)} & g_{l-1}^{(2)} & 0 & 0 & 0 \\
0 & 0 & 0 & 0 & r_l\,\alpha m_{l-1} & (1-\alpha)m_{l-1} & 0 \\
0 & 0 & 0 & 0 & r_l\,\alpha m_{l-1} & (1-\alpha)m_{l-1} & \\
f_l^{(1)} & g_{l-1}^{(1)} & 0 & 0 & 0 & 0 & 0 \\
0 & 0 & f_l^{(2)} & g_{l-1}^{(2)} & 0 & 0 & 0
\end{bmatrix}, \quad l \geq 0, \tag{15}
$$

$$
A_l = \begin{bmatrix}
s_1 f_l^{(1)} & v_l^{(1)} & 0 & 0 \\
y_l^{(1)} & \hat{g}_l^{(1)} + h_0^{(1)}(v_l^{(1)} - g_l^{(1)}) & y_l^{(2)} & \hat{g}_l^{(2)} + h_0^{(2)}(v_l^{(2)} - g_l^{(2)}) \\
0 & 0 & s_2 f_l^{(2)} & v_l^{(2)}
\end{bmatrix},
$$

$$
B_l = \begin{bmatrix}
\frac{\lambda}{\lambda+Q_1}r_l & \frac{\lambda\alpha}{\lambda+Q_1}m_{l-1} + \frac{Q_1\alpha}{\lambda+Q_1}I_{\{l=0\}} & \frac{\lambda(1-\alpha)}{\lambda+Q_1}m_{l-1} + \frac{Q_1(1-\alpha)}{\lambda+Q_1}I_{\{l=0\}} \\
\frac{\lambda}{\lambda+Q_1}r_l & \frac{\lambda\alpha}{\lambda+Q_1}m_{l-1} + \frac{Q_1\alpha}{\lambda+Q_1}I_{\{l=0\}} & \frac{\lambda(1-\alpha)}{\lambda+Q_1}m_{l-1} + \frac{Q_1(1-\alpha)}{\lambda+Q_1}I_{\{l=0\}}
\end{bmatrix},
$$

$$
C_l = \begin{bmatrix}
s_1 f_l^{(1)} & v_l^{(1)} & 0 & 0 \\
0 & 0 & s_1 f_l^{(2)} & v_l^{(2)}
\end{bmatrix}, \quad l \geq 0,
$$

$$
y_i^{(m)} = (s_m f_i^{(m)} - f_{i+1}^{(m)})h_0^{(m)} + \hat{f}_{i+1}^{(m)}, \quad m = 1, 2, i \geq 0.
$$

In (15), $g_{-1}^{(1)} = g_{-1}^{(2)} = \hat{g}_{-1}^{(1)} = \hat{g}_{-1}^{(2)} = m_{-1} = 0$.

Effective and numerically stable algorithm for computing the vectors π_i, $i \geq 0$, which exploits a structure of generator Q consists of the following steps:

- Compute the matrices $P_{i,l}$, $i, l \geq 0$, from (14)–(15).
- Find the matrix G as solution of the matrix equation

$$
G = \sum_{l=i-1}^{\infty} P_{i,l} G^{l-i+1}.
$$

- Compute the matrices $\tilde{P}_{i,k}$, $k \geq 0$, $i = \overline{0,k}$, from

$$
\tilde{P}_{i,k} = P_{i,k} + \sum_{l=k+1}^{\infty} P_{i,l} \prod_{n=1}^{l-k} G^{(l-n)}, \quad i = \overline{0,k}, k \geq 0.
$$

- Compute the vector π_0 as the unique solution to the following system of linear algebraic equations:

$$
\pi_0(I - \tilde{P}_{0,0}) = \mathbf{0}, \quad \pi_0 \sum_{i=0}^{\infty} \Phi_i \mathbf{1} = 1,
$$

where $\Phi_0 = I$, $\Phi_k = \sum_{i=0}^{k-1} \Phi_i \tilde{P}_{i,k}(I - \tilde{P}_{k,k})^{-1}$, $k \geq 1$.
- Compute the vectors π_i, $i \geq 1$, as $\pi_k = \pi_0 \Phi_k$, $k \geq 1$.

Note that the stationary probabilities of the embedded Markov chain and the stationary distribution of the system states at arbitrary time coincide. Having the stationary distribution of the embedded Markov chain calculated we can compute the performance characteristics as shown in Section 6.

6   Numerical Results

Consider the system with $\lambda = 2000$, $\mu_1 = 15258$, $\mu_2 = 2441$. The parameters of the FSO link accessibility/inaccessibility distribution are estimated in Section 2 and defined as $\gamma_1^{(1)} = 9.57 \cdot 10^{-5}$, $\gamma_1^{(2)} = 4.019 \cdot 10^{-6}$, $\gamma_2^{(1)} = 3.72 \cdot 10^{-3}$, $\gamma_2^{(2)} = 1.684 \cdot 10^{-4}$ $(1/sec)$, $p_1 = 0.65$, $p_2 = 0.8$.

The main performance characteristic for the hybrid channel is the channel reliability which can be calculated as

$$R = \sum_{i=0}^{\infty} \pi(i, 1', 2).$$

Fig. 3 shows the dependence of the channel unreliability $(1 - R)$ on Q_2.

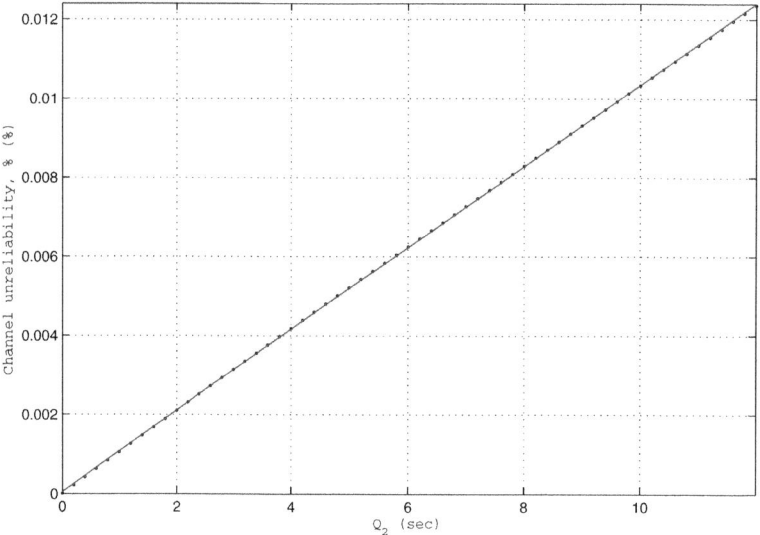

Fig. 3. Dependence of the channel unreliability on the switchover time from FSO to RF link

On Fig. 4, we present results obtained for the mean number of switches from FSO to RF link during a time unit which defined as

$$N_{OR} = \left(\lambda + \frac{2}{m_O + m_R}\right)\left[\sum_{i=0}^{\infty} \pi(i, 1, 1')\left(p_2 e^{-\gamma_2^{(1)}Q_2} + (1 - p_2)e^{-\gamma_2^{(2)}Q_2}\right) + \sum_{i=0}^{\infty} \pi(i, 2', 1)\right]$$

where $m_O = \frac{p_1}{\gamma_1^{(1)}} + \frac{1-p_1}{\gamma_1^{(2)}}$ and $m_R = \frac{p_2}{\gamma_2^{(1)}} + \frac{1-p_2}{\gamma_2^{(2)}}$ are the average periods when the FSO link is on/off (with $Q_1 = 0.0033$). The dependence of the queue length on the input intensity is shown on Fig. 5.

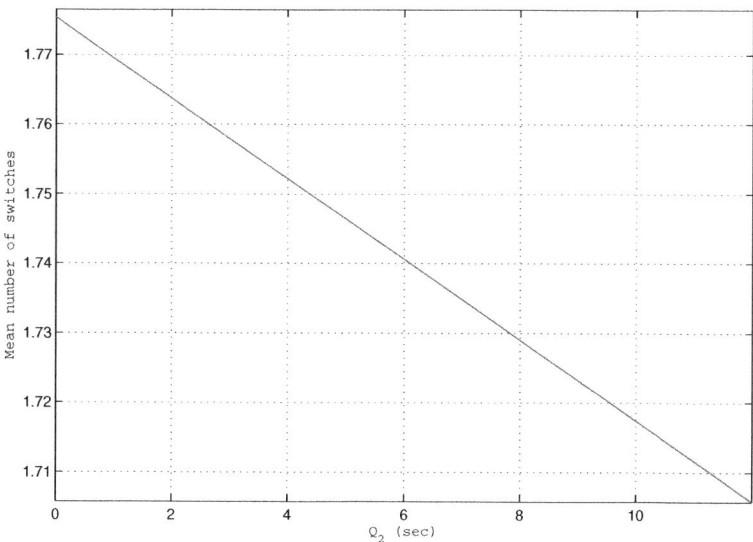

Fig. 4. The mean number of switches from FSO to RF link

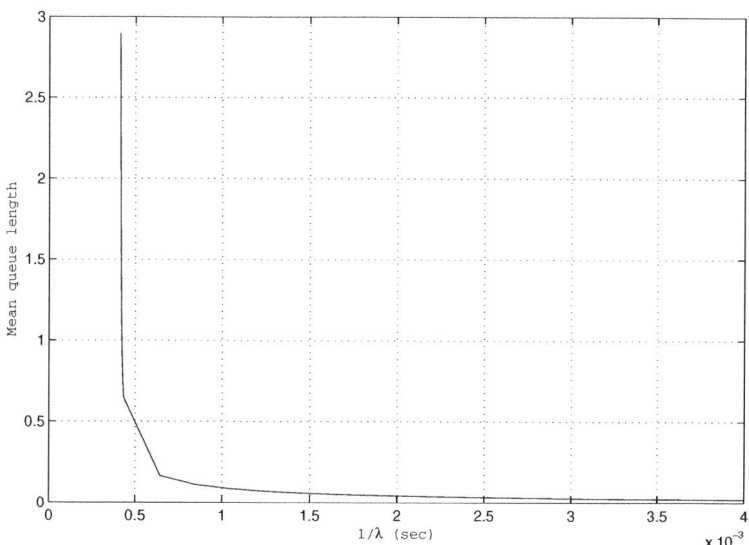

Fig. 5. Dependence of the queue length on the input intensity

References

1. Akbulut, A., Gokhan, H., Ari, F.: Design, availability and reliability analysis on an experimental outdoor FSO/RF communication system. In: International Conference ICTON, pp. 403–406 (2005)
2. Derenick, J., Thorne, C., Spletzer, J.: On the deployment of a hybrid free-space optic/ radio frequency (FSO/RF) mobile ad-hoc networks. In: Conference Proceeding, Intelligent Robots and Systems, pp. 3990–3996 (2005)
3. Wang, D., Abouzeid, A.A.: Throughput of hybrid radio-frequency and free-spaceoptical (RF/FSO) multi-hop networks. In: Information Theory and Applications Workshop, USA, pp. 1–8 (2007)
4. Nadeem, F., Leitgeb, E., Kvicera, V., Grabner, M., Awan, M.S., Kandus, G.: Simulation and analysis of FSO/RF switch over for different armospheric effects. In: International Conference ConTEL, pp. 39–43 (2009)
5. Letzepis, N., Nguyen, K.D., Guillen i Fabregas, A., Cowley, W.G.: Outage analysis of the hybrid free-space optical and radio-frequency channel. IEEE Journal on Selected Areas in Communications 27, 1709–1719 (2009)
6. Kazimirsky, A.: On approximation of arrival flows with MAPs of order two. In: Proceedings of IST 2002, Minsk, vol. 3, pp. 35–40 (2007)
7. Breuer, L., Dudin, A.N., Klimenok, V.I.: A retrial $BMAP/PH/1$ system. Queueing Systems 40, 433–457 (2002)
8. Ramaswami, V.A.: A stable recursion for the steady state vector in Markov chains of $M/G/1$ type. Commun. Statist.-Stochastic Models 4, 183–188 (1988)
9. Fricker, C., Jaibi, R.: Monotonicity and stability of periodic polling models. Queueing Systems 15, 211–238 (1994)

An Efficient Method for Proportional Differentiated Admission Control Implementation

Vladimir V. Shakhov

Institute of Computational Mathematics and Mathematical Geophysics of SB RAS
Novosibirsk 630090, Russia
shakhov@rav.sscc.ru

Abstract. In this paper, the admission control mechanism inspired in the framework of proportional differentiated services has been investigated. The mechanism provides a predictable and controllable network service for real-time traffic in terms of blocking probability. Implementation of proportional differentiated admission control is a complicated computational problem. Previously, asymptotic assumptions have been used to simplify the problem, but it is unpractical for real world applications. We improve previous solutions of the problem and offer an efficient non-asymptotic method for implementation of proportional differentiated admission control.

Keywords: QoS, resource management, admission control.

1   Introduction

Efficient implementation of admission control mechanisms is a key point for next generation wireless network development. Actually, over the last few years an interrelation between pricing and admission control in QoS-enabled networks has been intensively investigated. Call admission control can be utilized to derive optimal pricing for multiple service classes in wireless cellular networks [1]. Admission control policy inspired in the framework of proportional differentiated services [2] has been investigated in [3]. The proportional differentiated admission control (PDAC) provides a predictable and controllable network service for real-time traffic in terms of blocking probability. To define the mentioned service, proportional differentiated service equality has been considered and the PDAC problem has been formulated. The PDAC solution is defined by the inverse Erlang loss function. It requires complicated calculations. To reduce the complexity of the problem an asymptotic approximation of the Erlang B formula [4] has been applied. However, even in this case, the simplified PDAC problem remains unsolved.

In this paper we improve the previous results in [3] and withdraw the asymptotic assumptions of the used approximation. It means that for the desired accuracy of the approximate formula an offered load has to exceed a certain threshold.

A. Vinel et al. (Eds.): MACOM 2010, LNCS 6235, pp. 91–97, 2010.

The concrete value of the threshold has been derived. More over, an explicit solution for the considered problem has been provided. Thus, we propose a method for practical implementation of the PDAC mechanism.

2   Problem Statement

Let us consider the concept of admission control inspired in the framework of proportional differentiated services. In the above paper [3], whose notation we follow, PDAC problem is defined as

$$\delta_1 B_1(\rho_1, n_1) = \delta_2 B_2(\rho_2, n_2) = \cdots = \delta_K B_K(\rho_K, n_K). \tag{1}$$

Here

- K is a number of traffic classes. $K \geq 2$;
- δ_i - is the weight of class $i, i = 1, \ldots, K$. This parameter reflects the traffic priority. By increasing the weight, we also increase the admittance priority of corresponding traffic class;
- ρ_i - is the offered load of class i traffic;
- $n_i = \lfloor C_i/b_i \rfloor$, C_i is an allotted partition of the link capacity, b_i is a bandwidth requirement of class i connections;
- $B(\rho_i, n_i)$ - is the Erlang loss function (the blocking probability for traffic of class $i, i = 1, \ldots, K$).

It needs to find $C_1, C_2, \ldots, C_K$ taking into account known $\delta_i, \rho_i, b_i, i = 1, \ldots, K$ and the restriction imposed by given link capacity, C:

$$\sum_{i=1}^{K} C_i = C. \tag{2}$$

Let us remark that variations of C_i imply a discrete changing of the function $B(\rho_i, n_i)$. Hence, it is practicably impossible to provide the strict equality in (1). It is reasonable to replace (1) by an approximate equality as follows

$$\delta_1 B_1(\rho_1, n_1) \approx \delta_2 B_2(\rho_2, n_2) \approx \cdots \approx \delta_K B_K(\rho_K, n_K). \tag{3}$$

But, even in this case, the above problem is difficult and complex combinatorial problem. For its simplification, the following asymptotic approximation has been used [3]. If the capacity of link and the offered loads are increased together:

$$n \to \infty, \rho \to \infty \tag{4}$$

and $\rho > n$, then the Erlang loss function

$$B(\rho, n) = \frac{\frac{\rho^n}{n!}}{\sum_{i=0}^{n} \frac{\rho^i}{i!}} \tag{5}$$

can be approximated by

$$1 - \frac{n}{\rho}. \tag{6}$$

Taking into account the PDAC problem, the authors of [3] consider the limiting regime when

$$n_i \to \infty, C_i \to \infty, \tag{7}$$

and $\rho_i > C_i/b_i, i = 1, \ldots, K$. Under these conditions the asymptotic approximation of the Erlang B formula has been used and equations (1) have been replaced by simplified equations as follows

$$\delta_1 \left(1 - \frac{C_1}{b_1 \rho_1} \right) = \delta_2 \left(1 - \frac{C_2}{b_2 \rho_2} \right) = \ldots = \delta_K \left(1 - \frac{C_K}{b_K \rho_K} \right). \tag{8}$$

In practice, the limited regime (5) is not appropriate. But the simplification (6) can be used without the conditions (5). Actually, the approximation (4) can be applied without the condition (3). Let us prove it.

3   Approximate Erlang B Formula

We assert that for the desired accuracy of the approximation (4) an offered load has to exceed a certain threshold. The concrete value of the threshold is given by the following theorem.

Theorem 1. *For any small $\epsilon > 0$, if*

$$\rho \geq n + \frac{1}{\epsilon} \tag{9}$$

then

$$1 - \frac{n}{\rho} < B(\rho, n) < 1 - \frac{n}{\rho} + \epsilon. \tag{10}$$

Proof. Here and below we use the following designation:

$$\beta(\rho, n) = 1 - \frac{n}{\rho}.$$

Assume that $\rho > n$. First, we rewrite the Erlang B formula:

$$B(\rho, n) = \left(\sum_{i=0}^{n} \frac{n!}{i! \rho^{n-i}} \right)^{-1}.$$

Remark that

$$\sum_{i=0}^{n} \frac{n(n-1)\ldots(i+1)}{\rho^{n-i}} \leq \sum_{i=0}^{n} \left(\frac{n}{\rho} \right)^{n-i}.$$

Taking into account properties of geometrical progression we have

$$\frac{1}{B(\rho,n)} \leq \sum_{i=0}^{n} \left(\frac{n}{\rho}\right)^{n-i} < \frac{1}{\beta(\rho,n)}.$$

Hence

$$B(\rho,n) > 1 - \frac{n}{\rho}.$$

To prove the second inequality of the theorem we use the following upper bound of the Erlang loss function [5]

$$UB = \frac{n(1 - \frac{\rho}{n})^2 + 2\frac{\rho}{n} - 1}{2\frac{\rho}{n} - \rho(1 - \frac{\rho}{n})}.$$

Transform this as follows:

$$UB = \frac{\rho(\rho - n + 2) - n(\rho - n + 2) + n}{\rho(\rho - n + 2)}.$$

It implies

$$UB = 1 - \frac{n}{\rho} + \frac{n}{\rho(\rho - n + 2)}.$$

We have $n/\rho < 1$. Hence,

$$B(\rho,n) < UB < 1 - \frac{n}{\rho} + \frac{1}{\rho - n}.$$

Thus, for any ϵ such that

$$\epsilon > \frac{1}{\rho - n}, \tag{11}$$

it follows that

$$UP < 1 - \frac{n}{\rho} + \epsilon.$$

From the inequality (8) we obtain the condition (7).
    The proof is completed.

Note that the approximate formula (4) can provide the required accuracy ϵ in the case of $\rho < n - 2 + 1/\epsilon$. Actually, if $\epsilon = 0.01, n = 200$ then the required accuracy is reached for $\rho = 270 < 298$. Thus, the condition (7) is sufficient but not necessary. It guarantees the desired accuracy of the approximation for any small ϵ and n.

4   PDAC Solution

Assume that the solution $(C_1, C_2, \ldots, C_K)$ of the PDAC problem satisfies inequalities $\rho_i > C_i/b_i, i = 1, \ldots, K$. Let us derive an analytical solution for the PDAC problem under the condition (6). Without reducing generality, assume that $\max_i \delta_i = \delta_1 = 1, i = 1, \ldots, K$. Indeed, if $\delta_1 \neq 1$ then we define new weights $\hat{\delta}_i = \delta_i/\delta_1, i = 1, \ldots, K$. Thus, the condition (6) can be reformulated as follows

$$1 - \frac{C_1}{b_1\rho_1} = \delta_i \left(1 - \frac{C_i}{b_i\rho_i} \right), i = 2, \ldots, K. \tag{12}$$

According to the transitivity property, any solution of the PDAC problem under condition (6) is also a solution of the PDAC problem under condition (9). Therefore,

$$C_i = b_i\rho_i \left(1 + \frac{1}{\delta_i} \left(\frac{C_1}{b_1\rho_1} - 1 \right) \right), i = 2, \ldots, K. \tag{13}$$

Using the equality (2), we get

$$C_1 = \frac{C + S_2}{1 + S_1}, \tag{14}$$

where

$$S_1 = \frac{1}{b_1\rho_1} \sum_{j=2}^{K} \frac{b_j\rho_j}{\delta_j}; \; S_2 = \sum_{j=2}^{K} b_j\rho_j \left(\frac{1}{\delta_j} - 1 \right). \tag{15}$$

Thus, the formulas (10)-(12) provide the implementation of proportional differentiated admission control.

It is clear that for some values C, b_i, ρ_i, δ_i we can obtain $C_1 > C$ in (12) or $C_i < 0$ in (10). Therefore, the problem is unsolvable and PDAC implementation is impossible for the given parameters. It follows from the theorem that the approximation (4) is applicable even for $n = 1$ and any small $\epsilon > 0$ if $\rho > 1/\epsilon - 1$. In spite of this fact the solution above can not be useful for small values of the ratio C_i/b_i. In this case the loss function $B(\rho_i, n_i)$ is sensitive to fractional part dropping under calculation $n_i = \lfloor C_i/b_i \rfloor$. For example, if $b_i = 128$ kb/s,

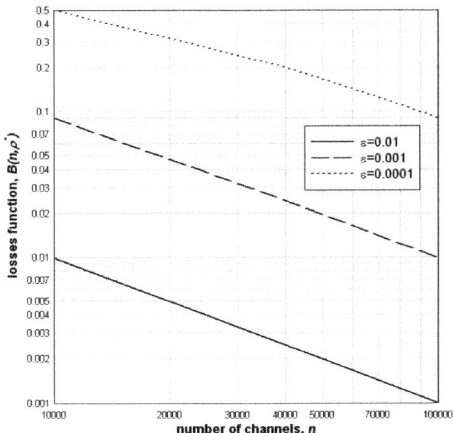

Fig. 1. The behavior of losses function $B(n, \rho^*)$ according to different values of ϵ

Table 1.

class	C_i, kb/s	n_i	$B(\rho_i, n_i)$	$\delta_i B(\rho_i, n_i)$
1	130887	1022	0.0803	0.0803
2	129786	1013	0.0877	0.079
3	128409	1003	0.0961	0.0769
4	126639	989	0.108	0.0756
5	124279	970	0.1243	0.0746

$\rho_i = 2$ and we obtain $C_i = 255$ kb/s, then the approximate value of the blocking probability is about 0.004. But $n_i = \lfloor C_i/b_i \rfloor = 1$ and $B(1,2) \approx 0.67$. Thus, the offered approximate formula is useful if the ratio C_i/b_i is relatively large.

From (10) it is clear that if the number of channel n is relatively small then high accuracy of approximation is reached for heavy offered load.

Let us remark that heavy offered load corresponds to high blocking probability. Generally, this situation is abnormal for general communication systems. But the blocking probability $B(n, \rho)$ decreases if the number of channels n increased relative accuracy ϵ. Let us designate $\rho^* = n + 1/\epsilon$. If the approximation (2) is admissible for ρ^* then it is also admissible for any $\rho > \rho^*$. In the Figure 1, the behavior of losses function $B(n, \rho^*)$ according to different ϵ is shown. Thus, the provided approximation is attractive for a performance measure of queuing systems with a large number of devices.

5   Example

Assume, $C = 640$ Mb/s, $K = 5, b_i = 128$ kb/s, $\rho_i = 1100, \delta_i = 1 - 0.1(i - 1), i = 1, \ldots, 5$. In average, there are 1000 channels per traffic class. Following the theorem above we conclude that the blocking probability can be replaced by the approximation (4) with accuracy about 0.01. Using (10)-(12), find a solution of the simplified PDAC problem and calculate the blocking probability for the obtained values. The results is shown in the Table 1.

Note that $\sum_{i=1}^{5} C_i = 640$ Mb/s and three channels per 128 kb/s have not been used. We get

$$\delta_i \left(1 - \frac{C_i}{b_i \rho_i}\right) = 0.0704, i = 1, \ldots, 5.$$

It is easy to see

$$\max_{i=1,\ldots,5} \left\{ B(\rho_i, n_i) - \left(1 - \frac{C_i}{b_i \rho_i}\right) \right\} < 0.01,$$

and

$$\max_{i,j} |B(\rho_i, n_i) - B(\rho_i, n_i)| < 0.01$$

If $K = 10, \delta_i = 1 - 0.05(i-1), i = 1, \ldots, 10$, and other parameters are the same then

$$\max_{i,j} |B(\rho_i, n_i) - B(\rho_i, n_i)| < 0.001.$$

If an obtained accuracy is not enough then the formulas (10)-(12) provide efficient first approximation for numerical methods.

6   Conclusion

In this paper a simple non-asymptotic approximation for the Erlang B formula is considered. We find the sufficient condition when the approximation is relevant. The proposed result allows rejecting the previously used limited regime and consider PDAC problem under finite network resources. Following by this way, we get explicit formulas for PDAC problem. The proposed formulas deliver high performance computing of network resources assignment under PDAC requirements. Thus, an efficient method for proportional differentiated admission control implementation has been provided.

References

1. Yilmaz, O., Chen, I.: Utilizing call admission control for pricing optimization of multiple service classes in wireless cellular networks. Computer Communications 32, 317–323 (2009)
2. Dovrolis, C., Stiliadis, D., Ramanathan, P.: Proportional differentiated services: Delay differentiation and packet scheduling. IEEE/ACM Trans. Networking 10, 12–26 (2002)
3. Salles, R.M., Barria, J.A.: Proportional Differentiated Admission Control. IEEE Commun. Lett. 8, 320–322 (2004)
4. Jagerman, D.: Some properties of the Erlang loss function. Bell Syst. Tech. J. 53(3), 525–551 (1974)
5. Harrel, A.: Sharp bounds and simple approximations for the Erlang delay and loss formulas. Management Sciences 34(8), 959–972 (1988)

A $Geo_m/G/1/n$ Queueing System with $LIFO$ Discipline, Service Interruptions and Repeat Again Service, and Restrictions on the Total Volume of Demands

Alexander Pechinkin and Sergey Shorgin

Institute of Informatics Problems, Russian Academy of Sciences,
Vavilova str. 44, building 2, 119333 Moscow, Russia
{apechinkin,sshorgin}@ipiran.ru

Abstract. Consideration is given to a discrete-time queueing system with inverse discipline, service interruption and repeat again service, second-order geometrical demand arrival, arbitrary (discrete) distribution of demand length and finite storage. Each demand entering the queue has random volume besides its length. The total volume of the demands in the queue is limited by a certain number. Formulae for the stationary probabilities of states and the stationary waiting time distribution in the queuing system are obtained.

Keywords: Queueing system, discrete time, finite buffer, the demand length and volume.

1 Introduction

The researchers in the field of queueing theory show a great interest in investigation of queueing systems (QS) where, along with the restriction on the total number of demands the restrictions on other parameters are assigned as well. Particularly, in modern information and computer systems each demand usually requires a certain memory size and the newly arriving demands can be accepted for serving only in case when their volume does not surpass resources available in the server.

Unfortunately, due to difficulties arising during construction of a Markovian process describing behavior of the system with restrictions on the total volume of demands, there are very few works devoted to development of analytical methods of calculation of such systems ([1], [2], [3]). But considering of the inversion service order ($LIFO$) in the systems with the restriction on the total volume of demands, as was shown in [4], [5], [6], could help to receive suitable-to-use algorithms for the calculation of stationary characteristics. It should be noted that all the quoted papers examined continuous-time QS while one of the distinctive features of modern information and communication systems is the universal implementation of digital technologies which generates a need for studying

A. Vinel et al. (Eds.): MACOM 2010, LNCS 6235, pp. 98–106, 2010.

discrete-time QS (and this studying in many cases meets additional difficulties related to the possibility of simultaneous occurrence of several events).

In the present article consideration is given to QS $Geo_m/G/1/n$ with $LIFO$ discipline functioning in discrete time with restriction on the total volume of demands, which is similar to QS $M_l/G/1/n$ functioning in the continuous time described in [4]. The relations are obtained allowing the calculation of the basic stationary characteristics of this system.

2 System Description

Let us consider a unilinear discrete-time QS $Geo_m/G/1/n$ with n, $0 \leq n < \infty$, waiting spaces, and with second kind Geometrical input flow, i.e. a demand flow with a_m, $0 \leq m \leq n+1$, the probability of a demand arrival during a time slot depends on the number m of demands residing in the QS just before the slot begins.

Each demand arriving in the system along with its length has random positive volume. Joint distribution of the demand length and volume is defined by the probability $B_k(x)$ of the event that the demand length (the number of service time slots) is equal to k, $k \geq 0$, and its volume does not exceed x. We shall assume the fulfillment of natural condition that the length of the demand cannot be equal to zero i.e. $B_0(x) = 0$ for any x. For the sake of simplicity we shall suppose that for each k, $k \geq 1$, there exists the derivative $b_k(x) = B'_k(x)$ (the density function of the volume).

Let:

$b(y) = \sum\limits_{k=0}^{\infty} b_k(y)$ be the density function of the demand volume distribution;

$B(y) = \int\limits_{0}^{y} b(u)\,du = \int\limits_{0}^{y} \sum\limits_{k=0}^{\infty} b_k(u)\,du = \sum\limits_{k=0}^{\infty} B_k(u)$ be the demand volume distribution function;

$b(k|y) = b_k(y)/b(y)$, $k \geq 1$, be the conditional probability that demand length is equal to k given that its volume is equal to y;

$B(k|y) = \sum\limits_{i=k}^{\infty} b(i|y)$, $k \geq 1$, be the conditional probability that the demand length is no less than k given that its volume is equal to y;

$\beta(s|y) = \sum\limits_{k=1}^{\infty} s^k b(k|y)$ be the generating function of the demand length given that its volume is equal to y;

$m_y = \sum\limits_{k=1}^{\infty} k b(k|y) = \sum\limits_{k=1}^{\infty} B(k|y)$ be the conditional mathematical expectation of the demand length given that its volume is equal to y.

Let also

$$\beta^*(s|y) = \sum\limits_{k=1}^{\infty} s^{k-1} b(k|y) = \frac{\beta(s|y)}{s}, \quad \tilde{\beta}(s|y) = \sum\limits_{k=1}^{\infty} s^{k-1} B(k+1|y) = \frac{s - \beta(s|y)}{s(1-s)}.$$

The total volume of demands in the QS is limited by a (nonrandom) number Y, $0 < Y < \infty$. If the aggregate volume of arriving demand and the demands in the system exceeds Y, then the arriving demand will be lost. For definiteness we shall assume that if at the moment of a new demand arrival the served demand leaves the queuing system, then its volume will be disregarded while defining the aggregate volume.

The inversion service order with service interruption and repeat again service is realized in the system whereby a demand accepted by the queuing system is directed to the server and drives out the demand being served to the beginning of the queue. In particular, if at the moment of new demand arrival the served demand leaves the system, then a new demand is directed to the server.

At the moment when a previously interrupted demand is directed from the queue to the server, the new service time of this demand will not depend on realized earlier service times of this demand during previous stages of its sojourn at the server. This new service time is distributed according to the law $\{b(k|y), \ k \geq 1\}$, where y is the (previous) volume of the demand.

Let us consider separately a service order of a demand arriving at a moment when there are no free places for waiting in the system, and at the moment any demands does not quit the system, and the volume of the arriving demand aggregated with total volume of other demands residing in the system is not greater than Y. This arriving demand will quit the system immediately, but a demand residing at the server (with volume y) will be served from this moment repeatedly (with the service time distribution $b(k|y)$).

We will suppose that the system has the stationary mode. Since for the QS under discussion the necessary and sufficient condition of the stationary mode existence has a complicated form, we will bring forward the following simple sufficient condition: $b(+0) > 0$.

3   Stationary Probabilities of States

Denote the stationary probability of absence of demands in the queuing system through p_0. Let $p_{k,1}(y), \ k \geq 0$, be the stationary density of probabilities (with argument y) of the event that there is 1 demand in the system (at the server), and the served length and the volume of the demand are equal to k and y respectively, and $p_{k,i}(y, z), \ i = \overline{2, n+1}, \ k \geq 0$, be the stationary density of probabilities (with arguments y and z) of the event that there are i demands in the system, and the (served) length and the volume of the serving demand are equal to k and y respectively, and the total volume of other demands situated in the system is equal to z.

As the total amount of demands situated in the system ranges between 0 and Y, then $p_{k,1}(y) = 0, \ k \geq 0$, at $y \notin D_1$ and $p_{k,i}(y, z) = 0, \ k \geq 0, \ i = \overline{2, n+1}$, at $(y, z) \notin D_2$, where D_1 is a segment $[0, Y]$, and D_2 is a triangle bounded by lines $y = 0, \ z = 0$ and $y + z = Y$. Therefore, without paying special attention to it we shall assume henceforth that $y \in D_1$ and $(y, z) \in D_2$.

By using the method of states elimination (see [8], p. 22), we obtain the system of equations

$$p_{k,1}(y) = (1 - a_1 B(Y - y)) \frac{B(k+1|y)}{B(k|y)} p_{k-1,1}(y), \quad k \geq 1, \tag{1}$$

$$p_{k,i}(y,z) = (1 - a_i B(Y-y-z))\frac{B(k+1|y)}{B(k|y)}\, p_{k-1,i}(y,z)\,, \quad i=\overline{3,n+1}\,, \quad k\geq 1\,, \quad (2)$$

with the initial condition

$$p_{0,1}(y) = a_0 b(y) p_0 + a_1 b(y) \int_0^Y \sum_{k=1}^{\infty} \frac{b(k|u)}{B(k|u)}\, p_{k-1,1}(u)\, du$$

$$+ a_1 B(Y-y) \sum_{k=1}^{\infty} \frac{B(k+1|y)}{B(k|y)}\, p_{k-1,1}(y)\,, \quad (3)$$

$$p_{0,2}(y,z) = a_1 b(y) \sum_{k=1}^{\infty} \frac{B(k+1|z)}{B(k|z)}\, p_{k-1,1}(z)$$

$$+ a_2 b(y) \int_0^{Y-z} \sum_{k=1}^{\infty} \frac{b(k|u)}{B(k|u)}\, p_{k-1,2}(u,z)\, du$$

$$+ a_2 B(Y-y-z) \sum_{k=1}^{\infty} \frac{B(k+1|y)}{B(k|y)}\, p_{k-1,2}(y,z)\,, \quad (4)$$

$$p_{0,i}(y,z) = a_{i-1} b(y) \int_0^z \sum_{k=1}^{\infty} \frac{B(k+1|u)}{B(k|u)}\, p_{k-1,i-1}(u,z-u)\, du$$

$$+ a_i b(y) \int_0^{Y-z} \sum_{k=1}^{\infty} \frac{b(k|u)}{B(k|u)}\, p_{k-1,i}(u,z)\, du$$

$$+ a_i B(Y-y-z) \sum_{k=1}^{\infty} \frac{B(k+1|y)}{B(k|y)}\, p_{k-1,i}(y,z)\,, \quad i=\overline{3,n+1}\,. \quad (5)$$

Solving equations (1) and (2) step by step from $k = 1$, we deduce:

$$p_{k,1}(y) = (1 - a_1 B(Y-y))^k B(k+1|y)\, p_{0,1}(y)\,, \quad k \geq 1\,, \quad (6)$$

$$p_{k,i}(y,z) = (1 - a_i B(Y-y-z))^k B(k+1|y)\, p_{0,i}(y,z)\,, \quad i=\overline{2,n+1}\,, \quad k\geq 1\,. \quad (7)$$

By plugging into the formulae (3), (4), and (5) expressions (6) and (7), we obtain:

$$p_{0,1}(y) = \frac{a_0 b(y)}{1 - a_1 B(Y-y)\tilde{\beta}(1 - a_1 B(Y-y)|y)}\, p_0$$

$$+ \frac{a_1 c_{11} b(y)}{1 - a_1 B(Y-y)\tilde{\beta}(1 - a_1 B(Y-y)|y)}\,, \quad (8)$$

$$p_{0,2}(y,z) = \frac{a_1 b(y)\tilde{\beta}(1 - a_1 B(Y - z)|z)p_{0,1}(z)}{1 - a_2 B(Y - y - z)\tilde{\beta}(1 - a_2 B(Y - y - z)|y)}$$
$$+ \frac{a_2 c_{12} b(y)}{1 - a_2 B(Y - y - z)\tilde{\beta}(1 - a_2 B(Y - y - z)|y)} , \qquad (9)$$

$$p_{0,i}(y,z) = \frac{a_{i-1} b(y) \int_0^z \tilde{\beta}(1 - a_{i-1} B(Y - z)|u)p_{0,i-1}(u, z - u)\, du}{1 - a_i B(Y - y - z)\tilde{\beta}(1 - a_i B(Y - y - z)|y)}$$
$$+ \frac{a_i c_{1i} b(y)}{1 - a_i B(Y - y - z)\tilde{\beta}(1 - a_i B(Y - y - z)|y)} , \quad i = \overline{3, n+1}, \quad (10)$$

where

$$c_{11} = \frac{a_0 c_{01}}{1 - a_1 c_{01}} p_0 , \qquad (11)$$

$$c_{01} = \int_0^Y \frac{b(u)\beta^*(1 - a_1 B(Y - u)|u)}{1 - a_1 B(Y - u)\tilde{\beta}(1 - a_1 B(Y - u)|u)}\, du , \qquad (12)$$

$$c_{12} = c_{12}(z) = \frac{a_1 c_{02}}{1 - a_2 c_{02}} \tilde{\beta}(1 - a_1 B(Y - z)|z)p_{0,1}(z) , \qquad (13)$$

$$c_{1i} = c_{1i}(z) = \frac{a_{i-1} c_{0i}}{1 - a_i c_{0i}} \int_0^z \tilde{\beta}(1 - a_{i-1} B(Y - z)|u)p_{0,i-1}(u, z - u)\, du ,$$
$$i = \overline{3, n+1}, \quad (14)$$

$$c_{0i} = c_{0i}(z) = \int_0^{Y-z} \frac{b(u)\beta^*(1 - a_i B(Y - u - z)|u)}{1 - a_i B(Y - u - z)\tilde{\beta}(1 - a_i B(Y - u - z)|u)}\, du ,$$
$$i = \overline{2, n+1}. \quad (15)$$

Formulae (6)–(15) allow to calculate step by step from $i = 1$ up to $n + 1$ the stationary probability densities of the investigated QS, accurate to the stationary probability p_0; implementation of normalization condition is required for defining p_0:

$$p_0 + \sum_{k=0}^{\infty} \int_0^Y p_{k,1}(y)\, dy + \sum_{i=2}^{n+1} \sum_{k=0}^{\infty} \iint_{z+u\leq Y} p_{k,i}(y,z)\, dy\, dz = 1 .$$

Let us also put down the expressions for some stationary characteristics related to the stationary state probabilities.

The stationary density of probabilities $p_i(y)$, $i = \overline{2, n+1}$, that there are i demands in the system having the total volume y is given by the formula

$$p_i(y) = \int_0^y \sum_{k=0}^\infty p_{k,i}(y - z, z)\, dz\,, \quad i = \overline{2, n+1}\,.$$

The stationary probability a of demand arrival at a time slot (the demand may be rejected by the system because of the restriction on the volume) takes the form

$$a = a_0 p_0 + \sum_{i=1}^{n+1} a_i \int_0^Y p_i(y)\, dy\,.$$

The stationary density of probabilities $p^*_{k,1}(y)$, $k \geq 1$, that the arriving demand (which won't be necessarily accepted by the system) will find in the system one demand of (served) length k and volume y, and the stationary density of the probabilities $p^*_{k,i}(y, z)$, $i = \overline{2, n+1}$, $k \geq 1$, that the arriving demand will find in the system i other demands, where the length and volume of the demand at the server are equal to k and y respectively, and the aggregate volume of other demands residing in the system is equal to z, are determined by the expressions:

$$p^*_{k,1}(y) = \frac{B(k+1|y)}{B(k|y)}\, p_{k-1,1}(y)\,, \quad k \geq 1\,,$$

$$p^*_{k,i}(y, z) = \frac{B(k+1|y)}{B(k|y)}\, p_{k-1,i}(y, z)\,, \quad i = \overline{2, n+1}\,, \quad k \geq 1\,.$$

The stationary probability p^*_0 that at the moment of a new demand arrival the server will finish serving a single demand situated in the system, and the stationary density of probabilities $p^*_i(y)$, $i = \overline{0, n}$, that at the moment of new demand arrival the server will finish serving a demand, and i other demands with the total volume y will remain in the system, are given by the formulae

$$p^*_0 = \sum_{k=1}^\infty \int_0^Y \frac{b(k|z)}{B(k|z)}\, p_{k-1,1}(z)\, dz\,,$$

$$p^*_i(y) = \sum_{k=1}^\infty \int_0^{Y-y} \frac{b(k|z)}{B(k|z)}\, p_{k-1,i+1}(z, y)\, dz\,, \quad i = \overline{1, n}\,.$$

The stationary probabilities π_y $(y < Y)$ and π that the arriving demand of volume y and the arriving demand of arbitrary volume will be admitted to the system, have the form

$$\pi_y = \frac{1}{a} \left(a_0 p_0 + a_1 p_0^* + \sum_{i=1}^{n} a_{i+1} \int_0^{Y-y} p_i^*(z)\, dz \right.$$

$$+ a_1 \int_0^{Y-y} \sum_{k=1}^{\infty} p_{k,1}^*(z)\, dz + \sum_{i=2}^{n} a_i \iint_{z+u \leq Y-y} \sum_{k=1}^{\infty} p_{k,i}^*(z,u)\, dz\, du \left. \right),$$

$$\pi = \frac{1}{a} \left((a_0 p_0 + a_1 p_0^*) B(Y) + \sum_{i=1}^{n} a_{i+1} \int_0^{Y} B(Y-y) p_i^*(y)\, dy \right.$$

$$+ a_1 \int_0^{Y} B(Y-y) \sum_{k=1}^{\infty} p_{k,1}^*(y)\, dy$$

$$+ \sum_{i=2}^{n} a_i \iint_{y+z \leq Y} B(Y-y-z) \sum_{k=1}^{\infty} p_{k,i}^*(y,z)\, dy\, dz \left. \right).$$

4 Stationary Distribution of Demand Sojourn Time in the System

In this section we shall present a calculation algorithm in terms of the generating function of the stationary distribution of the demand sojourn time in the system.

We shall call an (i, z)-system a system similar to the initial one, but with $n - i$, $0 \leq i \leq n$, waiting places, restriction z, $0 \leq z \leq Y$, on the total volume of demands and probability $\hat{a}_m = a_{m+i}$, $m = \overline{0, n-i+1}$, of demand arrival at time slot given there are m demands in (i, z) system. It is easy to see that the (i, z)-system represents the initial QS but with the requirement that it permanently contains i demands (in the queue) with the total volume $Y - z$; these demands never will be served.

It is convenient to consider that the busy period (BP) of an (i, z)-system ends at the moment of departure of the demand which had been the first at the server, even if at the same time a new demand is arriving in the system.

Let:

$\varphi(s|k, y; i, z)$, $i = \overline{0, n}$, $k \geq 1$, be the generating function of BP of an (i, z)-system opened by the demand of residual length k and volume y;

$\varphi(s|y; i, z)$, $i = \overline{0, n}$, $k \geq 1$, be the generating function of BP of an (i, z)-system opened by the demand of arbitrary length and volume y;

$\varphi(s|i, z)$, $i = \overline{0, n}$, be the generating function of BP of an (i, z)-system opened by the demand of arbitrary length and with volume not greater than z, multiplied by probability that the volume of the demand opening the BP is not greater than z.

Then for $\varphi(s|k,y;i,z)$, $\varphi(s|y;i,z)$, and $\varphi(s|i,z)$ the following recurrent relations are valid:

$$\varphi(s|k,y;n,z) = a_{n+1}B(z-y)\varphi(s|y;n,z)\sum_{m=1}^{k-1}s^m(1-a_{n+1}B(z-y))^{m-1}$$
$$+ s^k(1-a_{n+1}B(z-y))^{k-1}$$
$$= a_{n+1}B(z-y)\varphi(s|y;n,z)s\frac{1-(s[1-a_{n+1}B(z-y)])^{k-1}}{1-s(1-a_{n+1}B(z-y))}$$
$$+ s^k(1-a_{n+1}B(z-y))^{k-1}, \quad k\geq 1, \tag{16}$$

$$\varphi(s|k,y;i,z) = a_{i+1}\varphi(s|y;i,z)\varphi(s|i+1,z-y)\sum_{m=1}^{k-1}s^m(1-a_{i+1}B(z-y))^{m-1}$$
$$+ s^k(1-a_{i+1}B(z-y))^{k-1}$$
$$= a_{i+1}\varphi(s|y;i,z)\varphi(s|i+1,z-y)s\frac{1-(s[1-a_{n+1}B(z-y)])^{k-1}}{1-s(1-a_{n+1}B(z-y))}$$
$$+ s^k(1-a_{n+1}B(z-y))^{k-1}, \quad k\geq 1, \quad i=\overline{0,n-1}, \tag{17}$$

$$\varphi(s|y;i,z) = \sum_{k=1}^{\infty}b(k|y)\varphi(s|k,y;i,z), \quad i=\overline{0,n}, \tag{18}$$

$$\varphi(s|i,z) = \int_0^z b(y)\varphi(s|y;i,z),dy, \quad i=\overline{0,n}. \tag{19}$$

By plugging into the (18) the values of $\varphi(s|k,y;i,z)$ from formulae (16) and (17) and executing simple arithmetical calculation, we obtain:

$$\varphi(s|y;n,z) = s\beta^*(s[1-a_{n+1}B(z-y)]|y)$$
$$\times \left(1 - \frac{a_{n+1}B(z-y)s[1-\beta^*(s[1-a_{n+1}B(z-y)]|y)]}{1-s[1-a_{n+1}B(z-y)]}\right)^{-1}, \tag{20}$$

$$\varphi(s|y;i,z) = s\beta^*(s[1-a_{i+1}B(z-y)]|y)$$
$$\times \left(1 - \frac{a_{i+1}\varphi(s|i+1,z-y)s[1-\beta^*(s[1-a_{i+1}B(z-y)]|y)]}{1-s[1-a_{i+1}B(z-y)]}\right)^{-1},$$
$$i=\overline{0,n-1}. \tag{21}$$

Expressions (16) and (17) taken together with equations (20), (21), and (19) give a possibility to obtain $\varphi(s|k,y;i,z)$ step by step, from $i=n$ down to $i=0$.

Let $g(s|k,y)$ be the generating function of BP of sojourn time of demand accepted for serving and having length k and volume y, $g(s)$ be the generating function of BP of sojourn time of arbitrary demand accepted for serving. We have at last:

$$g(s|k,y) = \frac{1}{a\pi_y} \left((a_0 p_0 + a_1 p_0^*)\varphi(s|k,y;0,Y) \right.$$

$$+ \sum_{i=1}^{n} a_{i+1} \int_0^{Y-y} \varphi(s|k,y;i,z)p_i^*(z)\,dz + a_1 \int_0^{Y-y} \varphi(s|k,y;1,z)\sum_{m=1}^{\infty} p_{m,1}^*(z)\,dz$$

$$+ \left. \sum_{i=2}^{n} a_i \iint_{z+u\le Y-y} \varphi(s|k,y;i,z+u)\sum_{m=1}^{\infty} p_{m,i}^*(z,u)\,dz\,du \right), \quad k\ge 1, \quad y\le Y,$$

$$g(s) = \frac{1}{\pi} \sum_{k=1}^{\infty} \int_0^Y \pi_y b_k(y)g(s|k,y)\,dy.$$

It is obvious that in the QS under consideration the waiting time for the start of a demand serving is equal to zero, and the sojourn time of a demand at the server (taking into account possible interruptions of service) coincides with the total sojourn time of a demand in the system.

Acknowledgments. The work has been accomplished with financial support from the Russian Foundation for Fundamental Research (projects No. 08-07-00152 and No. 09-07-12032).

References

1. Romm, E.L., Skitovich, V.V.: On One Generalization of the Erlang Problem. Avtomatika i Telemekhanika 6, 164–167 (1971)
2. Alexandrov, A.M., Katz, B.A.: Serving heterogeneous customer flows. Izvestiya AN SSSR. Tekhnicheskaya Kibernetika 2, 47–53 (1973)
3. Tikhonenko, O.M.: Queueing models in data processing systems. Izdatel'stvo Universitetskoe, Minsk (1990)
4. Pechinkin, A.V., Pechinkina, O.A.: An $M_k/G/1/n$ system with $LIFO$ discipline and constraints on the total number of customers. Vestnik Rossiyskogo Universiteta Druzhby Narodov. Ser. Prikladnaya Matematika i Informatika 1, 86–93 (1996)
5. Pechinkin, A.V.: Queueing system with $LIFO$ discipline and constraints on the total number of customers. Vestnik Rossiyskogo Universiteta Druzhby Narodov, Ser. Prikladnaya Matematika i Informatika 2, 85–99 (1996)
6. Pechinkin, A.V.: $M_l/G/1/n$ system with $LIFO$ discipline and constrained total amount of items. Automation and Remote Control 4, 545–553 (1998)
7. Abramushkina, T.V., Aparina, S.V., Kuznetsova, E.N., Pechinkin, A.V.: Numerical techniques for calculating stationary probabilities of states of $M/G/1/n$ system with $LIFO\,PR$ discipline and constraints on the total number of customers. Vestnik Rossiyskogo Universiteta Druzhby Narodov, Ser. Prikladnaya Matematika i Informatika 1, 40–47 (1998)
8. Bocharov, P.P., D'Apice, C., Pechinkin, A.V., Salerno, S.: Queueing theory. Modern Probability and Statistics. VSP Publishing, Utrecht (2004)

Retrial Queueing Model $MMAP/M_2/1$ with Two Orbits

Konstantin Avrachenkov[1], Alexander Dudin[2], and Valentina Klimenok[2]

[1] INRIA Sophia Antipolis, France
K.Avrachenkov@sophia.inria.fr
[2] Department of Applied Mathematics and Computer Science
Belarusian State University
Minsk 220030, Belarus
dudin@bsu.by, klimenok@bsu.by

Abstract. A retrial single-server queueing model with two types of customers is considered. Arrivals occur according to the *Marked Markovian Arrival Process* ($MMAP$). In case of the server occupancy at the arrival epoch, the customer moves to the orbit depending on the type of the customer. One orbit is an infinite while the second one is a a finite. Service time distributions are exponential with the parameter depending on the type of a customer. Joint distribution of the number of customers in the orbits and some performance measures are computed. Numerical results are presented. Possible extensions of the model are outlined.

Keywords: $MMAP/M_2/1$ queueing model, retrials, stationary state distribution, asymptotically quasi-Toeplitz Markov chains.

1 Introduction

Retrial queues well describe behavior of many real world system, in particular, wire and wireless telecommunication networks, contact centers, etc. For references and examples see, e.g., the book [1]. Theory of retrial queues is developed in much less extent comparing to queues with losses or buffers. In particular, the overwhelming majority of research in retrial queues deals with the systems with homogeneous customers. This is evident shortcoming of the existing theory from the point of view of potential applications. This shortcoming can be overcome due to the following heuristic. If the have K different stationary Poisson arrival processes with intensities λ_k, $k = \overline{1, K}$ and service time of type-k customers is defined by the distribution function $B_k(t)$, then we can try to approximate characteristics of this system with heterogeneous customers by the respective characteristics of the system with homogeneous customers where the stationary Poisson arrival process has the rate $\Lambda = \sum_{k=1}^{K} \lambda_k$ and the service time distribution is defined by the function $B(t) = \sum_{k=1}^{K} \frac{\lambda_k}{\Lambda} B_k(t)$. However, this simple engineering

A. Vinel et al. (Eds.): MACOM 2010, LNCS 6235, pp. 107–118, 2010.

approximation can lead to significant errors in evaluation of the system performance measures. Situation becomes even worse if different types of customers have different impatience and, consequently, different retrial rates.

So, retrial queues with several types of customers deserve the special treatment. Recently, retrial models with two types of customers were investigated by a group of Spanish authors, see, e.g., [2] and references therein. The authors consider multi-server retrial queue with two stationary Poisson arrival processes and different retrial rates for customers of the different types. Restriction of the model in [2] is assumption that service times of both types of customers are identical. Motivation of consideration of retrial queues with two separate orbits presented in [2] is as follows. The requests may be generated by the different categories of users (e.g., by the clients of a contact center having different service level requirements, human generated retrials and automatic retrials generated by computer) having different inter-retrial times and impatience. The same motivation holds good for the model under study in this paper.

Here we consider the model with heterogeneous customers of two types arriving according to the Marked Markovian Arrival Process. Customers who meet the server busy move to the orbits. Each type of the customers has a separate orbit. Retrial rates from the orbits are different as well as the service rates of customers of different types. In [2], infinite capacity of both orbits are suggested and different approximate algorithms for computation of the stationary distribution of the system states are discussed. Because all these schemes somehow assume truncation of at least of one orbit, in our model we assume in advance that one orbit has the finite capacity. This assumption is not very restrictive in many applications (the number of users of a some system hardly can be infinite), moreover that the orbit capacity can be assumed to be pretty large if powerful computer can be employed for realization of our algorithm. This algorithm evidently follows from algorithm for computing the stationary distribution of so called asymptotically quasi-Toeplitz Markov chains ($AQTMC$), see [6].

The rest of the paper is organized as follows. In section 2, the mathematical model is described. The steady state distribution of the number of customers in the orbits is analyzed in section 3. Section 4 contains expression for some performance measures of the system. In Section 5, numerical results are presented. Section 6 concludes the paper, in particular, some possible extensions of the model are discussed.

2   Mathematical Model

We consider a retrial single server queueing system. Customers arrive according to the $MMAP$. This means the following, see, e.g., [5]. Arrival of customers in the $MMAP$ are directed by the underlying random process ν_t, $t \geq 0$. The process ν_t, $t \geq 0$, is an irreducible continuous time Markov chain with state space $\{0, 1, ..., W\}$. The sojourn time of this chain in state ν is exponentially distributed with parameter $\lambda_\nu > 0$. When the sojourn time in the state ν expires, with probability $p_{\nu,\nu'}^{(0)}$ the process ν_t jumps into the state ν' without generation of customers, $\nu, \nu' = \overline{0, W}, \nu \neq \nu'$, and with probability $p_{\nu,\nu'}^{(r)}$ the process ν_t jumps into the state ν' with generation of a customer of type r, $r = 1, 2,$ $\nu, \nu' = \overline{0, W}$.

The behavior of the $MMAP$ is completely characterized by the matrices D_0, D_r, $r = 1, 2$, defined by their entries $(D_r)_{\nu,\nu'} = \lambda_\nu p_{\nu,\nu'}^{(r)}$, $\nu, \nu' = \overline{0, W}, r = 1, 2$, and

$$(D_0)_{\nu,\nu} = -\lambda_\nu, \nu = \overline{0, W},$$

$$(D_0)_{\nu,\nu'} = \lambda_\nu p_{\nu,\nu'}^{(0)}, \ \nu, \nu' = \overline{0, W}, \nu \neq \nu'.$$

The matrix $D(1) = D_0 + D_1 + D_2$ represents the generator of the process $\nu_t, t \geq 0$. The average arrival rate λ is defined by

$$\lambda = \boldsymbol{\theta}(D_1 + D_2)\mathbf{e}$$

where $\boldsymbol{\theta}$ is the vector of a stationary distribution of the Markov chain $\nu_t, t \geq 0$. The vector $\boldsymbol{\theta}$ is the unique solution to the system

$$\boldsymbol{\theta}D(1) = \mathbf{0}, \ \boldsymbol{\theta}\mathbf{e} = 1.$$

Here $\mathbf{e}$ is a column-vector consisting of 1's and $\mathbf{0}$ is a row-vector consisting of zeroes.

The average arrival rate λ_r of the type r customers is defined by $\lambda_r = \boldsymbol{\theta} D_r \mathbf{e}$, $r = 1, 2$. The squared coefficient of variation $v^{(r)}$ of inter-arrival times for the type r customers is given by

$$v^{(r)} = \frac{2\boldsymbol{\theta}(-D_0 - D_{\bar{r}})^{-1}\mathbf{e}}{\lambda_r} - \left(\frac{1}{\lambda_r}\right)^2, \ \bar{r} \neq r, \ r\bar{r} = 1, 2.$$

The coefficient of correlation $C_{cor}^{(r)}$ of two successive intervals between type r customers arrival is computed by

$$C_{cor}^{(r)} = \left[\frac{\boldsymbol{\theta}(D_0 + D_{\bar{r}})^{-1}}{\lambda_r}D_r(D_0 + D_{\bar{r}})^{-1}\mathbf{e} - \left(\frac{1}{\lambda_r}\right)^2\right](v^{(r)})^{-1}, \ \bar{r} \neq r, \ r\bar{r} = 1, 2.$$

Assumption that the arrival process of customers of two types is defined by the $MMAP$ instead of two independent stationary Poisson arrival processes allows to effectively take into account different variance of inter-arrival times and correlation in arrival of customers of the same type and cross-correlation between two types. Note that the $MMAP$ is generalization or the well-known MAP (Markov Arrival Process) to the case of heterogeneous customers. Survey of papers devoted to the retrial queues with the MAP can be found in [4].

If the server is idle at a customer arrival epoch, the service is immediately started. Service time is assumed to be exponentially distributed with parameter μ_r depending on the type r of the customer. After the service completion, the customer leaves the system.

If the server is busy, the customer moves to the orbit depending on the type of the customer. The orbit for the type-1 customers has an infinite capacity. The customers from this orbit try to get service later on in exponentially distributed with parameter α times independently of other customers. The customer repeats attempts until it eventually succeeds to find the server being idle. The orbit for the type-2 customers has a finite capacity R. If the server is busy and the orbit is full at the type 2 customer arrival epoch, the customer is lost. The type-2 customers from this finite orbit try to get service later on in exponentially distributed with parameter γ intervals independently of the other customers. The customer repeats attempts until it succeeds to find the server being idle.

The structure of the system is presented in Figure 1.

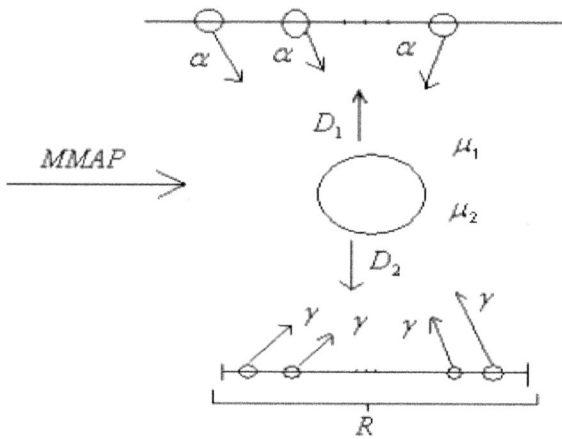

Fig. 1. The structure of the system

Our aim is to analyze the steady state behavior of this queueing model.

3   Stationary Distribution of the System States

Behavior of the described queueing model is defined by the four-dimensional continuous time Markov chain:

$$\xi_t = \{i_t, n_t, r_t, \nu_t, \}, \ t \geq 0,$$

where, at the moment t,

i_t, $i_t \geq 0$, is the number of customers presenting in the orbit 1,

r_t, $r_t = \overline{1, R}$, is the number of customers presenting in the orbit 2,

$$n_t = \begin{cases} 0, \text{ if the server is idle,} \\ 1, \text{ if the server processes the type-1 customer,} \\ 2, \text{ if the server processes the type-2 customer,} \end{cases}$$

ν_t is the state of the $MMAP$ underlying process ν, $\nu_t = \overline{0, W}$.

The Markov chain $\xi_t, t \geq 0$, has one denumerable component i_t and three finite components n_t, r_t, ν_t. Let us enumerate the states of this Markov chain in the lexicographic order and refer to the set of the states having the value i of the component i_t as to the *level i*. Let $Q_{i,l}$, $i,l \geq 0$, be the matrices of intensities of transition from the states included to the level i to the states included to the level l.

The block matrix Q with blocks $Q_{i,l}$, $i,l \geq 0$, is the generator of the Markov chain $\xi_t = \{i_t, n_t, r_t, \nu_t, \}$, $t \geq 0$.

Let us denote: $\bar{R} = (R+1)$, $\bar{W} = (W+1)$, $K = (R+1)(W+1)$, $\alpha_i = i\alpha$.

Lemma 1. Generator Q has the following block structure:

$$Q = \begin{pmatrix} Q_{0,0} & Q_{0,1} & O & O & \cdots \\ Q_{1,0} & Q_{1,1} & Q_{1,2} & O & \cdots \\ O & Q_{2,1} & Q_{2,2} & Q_{2,3} & \cdots \\ O & O & Q_{3,2} & Q_{3,3} & \cdots \\ \vdots & \vdots & \vdots & \vdots & \ddots \end{pmatrix} \tag{1}$$

where the non-zero blocks $Q_{i,i}$, $i \geq 0$, of size $3K \times 3K$ are defined by:

$$Q_{i,i-1} = \begin{pmatrix} O & \alpha_i I & O \\ O & O & O \\ O & O & O \end{pmatrix}, \ i \geq 1,$$

$$Q_{i,i+1} = \begin{pmatrix} O & O & O \\ O & I_{\bar{R}} \otimes D_1 & O \\ O & O & I_{\bar{R}} \otimes D_1 \end{pmatrix}, \ i \geq 0,$$

where $\otimes$ is the symbol of Kronecker product of matrices,

$$Q_{i,i} = \begin{pmatrix} Q_{i,i}^{(0,0)} & Q_{i,i}^{(0,1)} & Q_{i,i}^{(0,2)} \\ Q_{i,i}^{(1,0)} & Q_{i,i}^{(1,1)} & O \\ O & Q_{i,i}^{(2,1)} & Q_{i,i}^{(2,2)} \end{pmatrix}, \ i \geq 0,$$

where the matrices $Q_{i,i}^{(m,n)}$, $i \geq 0$, $m,n = \overline{0,2}$, of size $K \times K$ are given by:

$$Q_{i,i}^{(0,0)} = \begin{pmatrix} D_0 & O & \cdots & O \\ O & D_0 - \gamma I & \cdots & O \\ \vdots & \vdots & \ddots & \vdots \\ O & O & \cdots & D_0 - R\gamma I \end{pmatrix} - \alpha_i I,$$

$$Q_{i,i}^{(0,1)} = I_{\bar{R}} \otimes D_1, \quad Q_{i,i}^{(1,0)} = \mu_1 I, \quad Q_{i,i}^{(2,0)} = \mu_2 I,$$

$$Q_{i,i}^{(0,2)} = \begin{pmatrix} D_2 & O & \cdots & O \\ \gamma I & D_2 & \cdots & O \\ \vdots & \vdots & \ddots & \vdots \\ O & \cdots & R\gamma I & D_2 \end{pmatrix},$$

$$Q_{i,i}^{(1,1)} = \begin{pmatrix} D_0 & D_2 & \dots & O & O \\ \vdots & \vdots & \ddots & \vdots & \vdots \\ O & O & \dots & D_0 & D_2 \\ O & O & \dots & O & D_0 + D_2 \end{pmatrix} - \mu_1 I,$$

$$Q_{i,i}^{(2,2)} = \begin{pmatrix} D_0 & D_2 & \dots & \dots & O \\ \vdots & \vdots & \ddots & \vdots & \vdots \\ O & O & \dots & D_0 & D_2 \\ O & \dots & \dots & O & D_0 + D_2 \end{pmatrix} - \mu_2 I.$$

Here I and O are identity matrix and zero matrix of the appropriate dimension.

Theorem. Sufficient condition for ergodicity of the Markov chain $\xi_t,\ t \geq 0$, is the fulfillment of the inequality:

$$\rho = \lambda_1/\mu_1 < 1. \tag{2}$$

Proof. It can be verified that the Markov chain ξ_t belongs to the class of the asymptotically quasi-Toeplitz Markov chains $(AQTMC)$, see [6]. So we can use the results from [6] for derivation of condition (2).

Let R_i be the diagonal matrix having the positive diagonal entries coinciding, up to the sign, with the corresponding diagonal entries of the matrix $Q_{i,i}$. It is easy to see that the matrix $R_i,\ i \geq 0$, has the structure

$$R_i = \begin{pmatrix} R_0^{(i)} & O & O \\ O & R_1 & O \\ O & O & R_2 \end{pmatrix}$$

where the blocks R_1 and R_2 do not depend on i.

Let us introduce the matrices $Y_k, k = 0, 1, 2$, by

$$Y_0 = \lim_{i\to\infty} R_i^{-1} Q_{i,i-1}, \ Y_2 = \lim_{i\to\infty} R_i^{-1} Q_{i,i+1}, \ Y_1 = \lim_{i\to\infty} R_i^{-1} Q_{i,i} + I.$$

It is easy to see that the matrices $Y_k,\ k = \overline{0,2}$, have the form

$$Y_0 = \begin{pmatrix} O & I & O \\ O & O & O \\ O & O & O \end{pmatrix}, \ Y_2 = \begin{pmatrix} O & O & O \\ O & R_1^{-1}(I \otimes D_1) & O \\ O & O & R_2^{-1}(I \otimes D_1) \end{pmatrix},$$

$$Y_1 = \begin{pmatrix} O & O & O \\ Y_1^{(1,0)} & Y_1^{(1,1)} & O \\ Y_1^{(2,0)} & O & Y_1^{(2,2)} \end{pmatrix},$$

where

$$Y_1^{(1,0)} = R_1^{-1}\mu_1, \ Y_1^{(1,1)} = R_1^{-1}(I \otimes (D_0 - \mu_1 I) + \hat{I} \otimes D_2) + I,$$

$$Y_1^{(2,0)} = R_2^{-1}\mu_2, \ Y_1^{(2,2)} = R_2^{-1}(I \otimes (D_0 - \mu_2 I) + \hat{I} \otimes D_2) + I.$$

Here the square matrix $\hat{I}$ of order $\bar{R}$ has all zero entries except the entries $(\hat{I})_{r,r+1}$, $r = \overline{0, R-1}$, and $(\hat{I})_{R,R}$ which are equal to 1.

By [6], ergodicity condition for $AQTMC$ under consideration can be expressed in terms of the matrix generating function $Y(z) = Y_0 + Y_1 z + Y_2 z^2$, $|z| \leq 1$. This matrix has the following structure:

$$Y(z) = \begin{pmatrix} \tilde{Y}(z) & O_{2K \times K} \\ Y_{2,1}(z) & Y_{2,2}(z) \end{pmatrix}.$$

This means that the matrix $Y(1)$ is reducible and has a single stochastic diagonal block $\tilde{Y}(1)$. Then, according to Theorem 5 and Corollary 6 from [6], ergodicity condition is expressed in terms of the block $\tilde{Y}(z)$. Let us represent the matrix function $\tilde{Y}(z)$ in the form

$$\tilde{Y}(z) = \tilde{Y}_0 + \tilde{Y}_1 z + \tilde{Y}_2 z^2$$

where

$$\tilde{Y}_0 = \begin{pmatrix} O & I \\ O & O \end{pmatrix}, \quad \tilde{Y}_1 = \begin{pmatrix} O & O \\ Y_1^{(1,0)} & Y_1^{(1,1)} \end{pmatrix}, \quad \tilde{Y}_2 = \begin{pmatrix} O & O \\ O & R_1^{-1}(I \otimes D_1) \end{pmatrix}.$$

According to Corollary 6 from [6], sufficient condition for ergodicity of the chain $\xi_t = \{i_t, n_t, r_t, \nu_t\}$ is the fulfillment of the inequality

$$\mathbf{x}\tilde{Y}_0 \mathbf{e} > \mathbf{x}\tilde{Y}_2 \mathbf{e} \tag{3}$$

where the vector $\mathbf{x}$ is the unique solution to the system

$$\mathbf{x}(\tilde{Y}_0 + \tilde{Y}_1 + \tilde{Y}_2) = \mathbf{x}, \quad \mathbf{x}\mathbf{e} = 1. \tag{4}$$

By substituting the explicit form of the matrices $\tilde{Y}_k$ and partitioning the vector $\mathbf{x}$ as $\mathbf{x} = (\mathbf{x}_1, \mathbf{x}_2)$, where the vectors $\mathbf{x}_r$, $r = 1, 2$, have the dimension K, we reduce inequality (3) to the following inequality:

$$\mathbf{x}_2 R_1^{-1} \mathbf{e} > \mathbf{x}_2 R_1^{-1}(I \otimes D_1)\mathbf{e}, \tag{5}$$

and system (4) to the following system:

$$\mathbf{x}_2 R_1^{-1} \mu_1 = \mathbf{x}_1,$$
$$\mathbf{x}_1 + \mathbf{x}_2 R_1^{-1}(I \otimes (D_0 - \mu_1 I + D_1) + \hat{I} \otimes D_2) = \mathbf{0}.$$

By eliminating the vector $\mathbf{x}_1$ from this system, we get the following equation for the vector $\mathbf{x}_2$:

$$\mathbf{x}_2 R_1^{-1}(I \otimes (D_0 + D_1) + \hat{I} \otimes D_2) = \mathbf{0}. \tag{6}$$

Let us represent the vector $\mathbf{x}_2 R_1^{-1}$ in the form

$$\mathbf{x}_2 R_1^{-1} = \mathbf{y} \otimes \boldsymbol{\theta} \tag{7}$$

where the vector $\boldsymbol{\theta}$ is an invariant vector of the underlying process of the $MMAP$.

Using this presentation in (6), we get

$$\mathbf{y} \otimes \boldsymbol{\theta}(D_0 + D_1) + \mathbf{y}\hat{I} \otimes \boldsymbol{\theta}D_2 = \mathbf{0}.$$

Due to the fact that $\boldsymbol{\theta}(D_0 + D_1 + D_2) = \mathbf{0}$ this equation implies

$$\mathbf{y}(\hat{I} - I) \otimes \boldsymbol{\theta}D_2 = \mathbf{0}.$$

Because $\boldsymbol{\theta}D_2 \neq 0$, the vector $\mathbf{y}$ satisfies the equation $\mathbf{y}(\hat{I} - I) = \mathbf{0}$ which evidently has a solution $\mathbf{y} = (0, \ldots, 0, c)$ where c is some positive constant.

Using presentation (7) in (5) we get the following condition for ergodicity:

$$\mu_1(\mathbf{y} \otimes \boldsymbol{\theta})\mathbf{e} > (\mathbf{y} \otimes \boldsymbol{\theta})(I \otimes D_1)\mathbf{e}. \tag{8}$$

Taking into account that $\mathbf{y} = (0, \ldots, 0, c)$, and $\lambda_1 = \boldsymbol{\theta}D_1\mathbf{e}$, $\boldsymbol{\theta}\mathbf{e} = 1$, we rewrite inequality (8) as $\mu_1 c > c\lambda_1$ what implies formula (2). Theorem is proved.

Remark. Usually, condition of stability of the system is equivalent to the condition that the service rate is higher than the arrival rate when the system is overloaded. If the system under study is overloaded, the number of customers in the orbit-1 is huge, so these customers always succeed to catch the available server in competition with the customers from the finite orbit-2. Thus, stability condition assumes that service rate of type-1 customers is higher than their arrival rate, irrespectively to the arrival and service rates of type-2 customers. So, condition (2) is intuitively clear. However, as we see, its rigorous proof is not trivial.

Let us assume that condition (2) is fulfilled. Then, the stationary probabilities

$$\pi(i, r, n, \nu) = \lim_{t \to \infty} P\{i_t = i, r_t = r, n_t = n, \nu_t = \nu\},$$

$$i \geq 0, \ r = \overline{0, R}, \ n = 0, 1, 2, \ \nu = \overline{0, W},$$

exist. We combine the stationary probabilities $\pi(i, r, n, \nu)$ of the states, which belong to the level i, to the row vectors $\boldsymbol{\pi}_i$, $i \geq 0$. These vectors satisfy the system of Chapman-Kolmogorov equations

$$(\boldsymbol{\pi}_0, \boldsymbol{\pi}_1, \boldsymbol{\pi}_2, \ldots)Q = \mathbf{0}, \ (\boldsymbol{\pi}_0, \boldsymbol{\pi}_1, \boldsymbol{\pi}_2, \ldots)\mathbf{e} = 1.$$

Effective and numerically stable procedure for computing the infinite system of equations for vectors $\boldsymbol{\pi}_i$, $i \geq 0$, is presented in [6]. This algorithm elaborated in [6] for the generators of upper-Hessenberg form becomes more simple for the generators of the three-block diagonal form (1) were it is defined as follows (see, e.g., [3]).

Algorithm. Vectors $\boldsymbol{\pi}_i$, $i \geq 0$, are computed by:

$$\boldsymbol{\pi}_i = \boldsymbol{\pi}_0 F_i, \ i \geq 1,$$

where non-negative matrices F_i, $i \geq 1$, are defined by

$$F_0 = I, \ F_i = \prod_{l=1}^{i} Q_{l-1,l}[-(Q_{l,l} + Q_{l,l+1}G_l)]^{-1}, \ i \geq 1,$$

the vector $\boldsymbol{\pi}_0$ is the unique solution to the system

$$\boldsymbol{\pi}_0(Q_{0,0} + Q_{0,1}G_0) = \mathbf{0}, \quad \boldsymbol{\pi}_0\sum_{l=0}^{\infty} F_l\mathbf{e} = 1,$$

and the matrices G_i are computed from recursion

$$G_i = [-(Q_{i+1,i+1} + Q_{i+1,i+2}G_{i+1})]^{-1}Q_{i+1,i}, \; i \geq 0.$$

So, we can consider the vectors $\boldsymbol{\pi}_i, \; i \geq 0$, be computed.

4   Some Performance Measures of the System

Having the vectors $\boldsymbol{\pi}_i, \; i \geq 0$, been computed, we can easily compute the following performance measures of the system.

- The average number of customers in the orbit-1 is given by

$$L_1 = \sum_{i=1}^{\infty} i\boldsymbol{\pi}_i\mathbf{e}.$$

- The average number of customers in the orbit-2 is given by

$$L_2 = \sum_{i=0}^{\infty} \boldsymbol{\pi}_i(\mathbf{e}_3 \otimes \text{diag}\{0, 1, \ldots, R\} \otimes \mathbf{e}_{\bar{W}})\mathbf{e}_{\bar{R}}$$

  where $\text{diag}\{0, 1, \ldots, R\}$ denotes the diagonal matrix with the diagonal entries $0, 1, \ldots, R$.
- Probability that the server is idle is computed by

$$P^{(0)} = \sum_{i=0}^{\infty} \boldsymbol{\pi}_i(\text{diag}\{1, 0, 0\} \otimes I)\mathbf{e}.$$

- Probability that the server serves type-1 customer at an arbitrary time is computed by

$$P_{serv}^{(1)} = \sum_{i=0}^{\infty} \boldsymbol{\pi}_i(\text{diag}\{0, 1, 0\} \otimes I)\mathbf{e}.$$

- Probability that the server serves type-2 customer at an arbitrary time is computed by

$$P_{serv}^{(2)} = \sum_{i=0}^{\infty} \boldsymbol{\pi}_i(\text{diag}\{0, 0, 1\} \otimes I)\mathbf{e}.$$

- Probability that the arbitrary type-2 is lost (because the orbit is full at its arrival epoch) is computed by

$$P_{loss} = \frac{1}{\lambda_2}\sum_{i=0}^{\infty} \boldsymbol{\pi}_i(\text{diag}\{0, 1, 1\} \otimes \text{diag}\{0, 0, \ldots, 0, 1\} \otimes D_2)\mathbf{e}.$$

- Probability that the arbitrary type-k customer enters the service immediately upon the arrival is computed by

$$P_{imm}^{(k)} = \frac{1}{\lambda_k} \sum_{i=0}^{\infty} \boldsymbol{\pi}_i (\text{diag}\{1,0,0\} \otimes D_k)\mathbf{e}, \ k = 1, 2.$$

5   Numerical Example

Let the $MMAP$ be defined by the matrices $D_r, r = \overline{0,2}$, of the following form:

$$D_0 = \begin{pmatrix} -4.32575 & 0.32725 \\ 0.11675 & -0.477 \end{pmatrix}, \ D_1 = \begin{pmatrix} 2.161805 & 0.05958335 \\ 0.04277776 & 0.1573612 \end{pmatrix},$$

$$D_2 = \begin{pmatrix} 1.729446 & 0.0476667 \\ 0.0342222 & 0.125889 \end{pmatrix}.$$

Intensities λ_1 and λ_2 are equal to 0.823485 and 0.658788, total intensity of arrivals is equal to 1.48227298. Squared coefficient of variation of inter-arrival times is equal to 3.8606665. Squared coefficient of variation of times between type-1 customer arrivals is equal to 3.09133771. Squared coefficient of variation of times between type-2 customer arrivals is equal to 2.8165405. Coefficient of correlation of successive inter-arrival times is equal to 0.20012666. Coefficient of correlation of successive inter-arrival times of type-1 customers is equal to 0.133577911. Coefficient of correlation of successive inter-arrival times of type-1 customers is equal to 0.110613191.

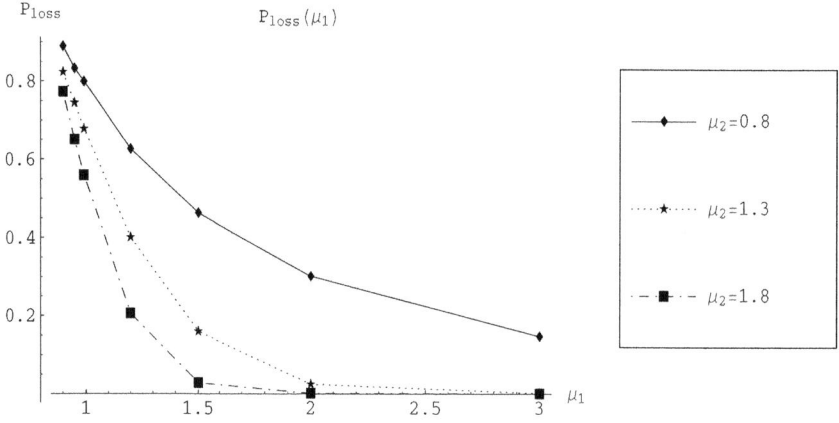

Fig. 2. Probability P_{loss} that an arbitrary type-2 customer is lost as a function of the intensity μ_1

Let us fix the orbit-2 capacity as $R = 50$, retrial intensities as $\alpha = 1.0$ and $\gamma = 0.9$. We fix three values of the intensity of the service of type-2 customers as $\mu_2 = 0.8$, $\mu_2 = 1.3$, $\mu_2 = 1.8$, and vary intensity μ_1 of the service of type-1 customers.

Figure 2 presents the dependence of the probability P_{loss} that an arbitrary type-2 is lost on the intensity μ_1 for three different values of the intensity μ_2.

Figure 2 shows that the probability P_{loss} quickly decreases with the growth of service intensities of both types of the customers.

Let now fix $\mu_1 = 1.1, \mu_2 = 1.0, R = 50$.

Figure 3 presents the dependence of probability P_{loss} that an arbitrary type-2 is lost on the retrial intensity α for different values of the retrial intensity γ.

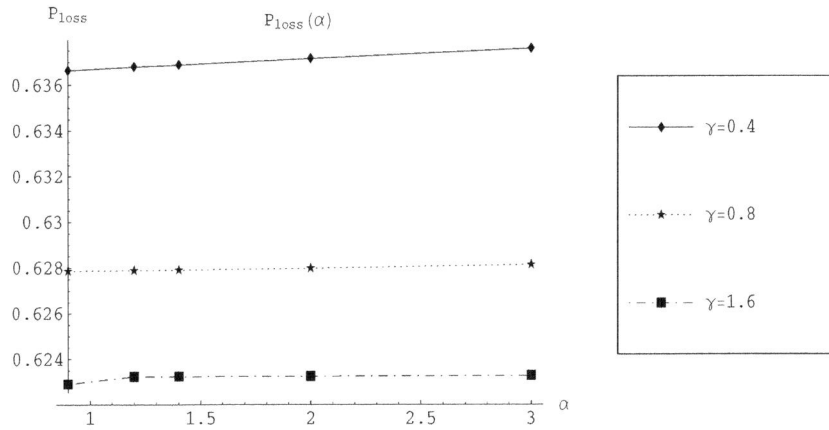

Fig. 3. Probability P_{loss} that an arbitrary type-2 customer is lost as a function of the retrial intensity α for different values of the retrial intensity γ

One can conclude that this probability essentially depends on the retrial intensity γ and only slightly increases when the retrial intensity α from the orbit 1 increases.

6   Conclusion

We have considered the single server retrial queue with two types of customers having different service rate and retrial rate. Behavior of the system is described by the four-dimensional asymptotically quasi-Toeplitz Markov chain. Ergodicity condition is derived, stationary distribution of the system states is computed based on [6], some performance measures are presented. Simple numerical results are given.

More detailed numerical study aiming to analyze effect of correlation between arrival of customers of both types as well as cross-correlation is in the progress. Problem of optimal adjusting retrial rates to get the fair sharing of the channel between two types of the customers and the balance of the number of customers in both orbits is also under consideration.

Results can be more or less easily extended to the cases when the service time distribution is of more general PH (Phase) type and when the system has several servers. Generalization to the case of more than two types of customers (with one infinite orbit and a set of the finite orbits) seems also be straightforward. Possibility of changing the type of a customer (and possible transition to another orbit) also can be incorporated into the model. It is worth to note that we intentionally used denotation α_i instead of $i\alpha$ for the total retrial rate from the orbit-1 when i customers stay in this orbit. All the presented results remain true for any dependence of the intensity α_i on the number i of customers in the orbit such as $\lim_{i\to\infty} \alpha_i = \infty$. In the case of the constant retrial rate ($\alpha_i = \alpha$ for all i, $i > 0$) only ergodicity condition is changed. All other results (the form of generator, performance measures) are the same as above or are even simpler (algorithm for computing the stationary distribution). Another generalization, which can be easily done, is the model with non-persistent customers where, with a fixed probability depending on the orbit, the customer returns to the orbit or leaves the system without the service.

References

1. Artalejo, J.R., Comez-Corral, A.: Retrial queueing systems: a computational approach. Springer, Heidelberg (2008)
2. Domenech-Benlloch, M.J., Gimenez-Guzman, J.M., Pla, V., Casarea-Giner, V., Martinez-Bauset, J.: Solving multiserver systems with two retrial orbits using value extrapolation: a comparative perspective. In: Al-Begain, K., Fiems, D., Horváth, G. (eds.) ASMTA 2009. LNCS, vol. 5513, pp. 56–70. Springer, Heidelberg (2009)
3. Dudin, S.A.: The $MAP/N/N$ retrial queueing system with time-phased batch arrivals. Problems of Information Transmissions 45, 270–281 (2009)
4. Gomez-Corral, A.: A bibliographical guide to the analysis of retrial queues through matrix analytic techniques. Annals of Operations Research 141, 163–191 (2006)
5. He, Q.M.: Queues with marked customers. Advances in Applied Probability 28, 567–587 (1996)
6. Klimenok, V.I., Dudin, A.N.: Multi-dimensional asymptotically quasi-Toeplitz Markov chains and their application in queueing theory. Queueing Systems 54, 245–259 (2006)

Laws of Conservation in the Queueing Theory

Yuri Ryzhikov

Institute of computer science and automation
of the Russian Academy of Sciences
Saint-Petersburg, Russia
ryzhbox@yandex.ru

Abstract. The history of laws of preservation of the queuing theory is considered. Additional formulations, their consequences and use at the calculation of complex queueing systems and networks are discussed. It is offered to use more actively conservation laws at training at the universities and at research work.

Keywords: queuing theory, conservation laws.

In physics, chemistry, engineering *the conservation laws* play an important role. Principles of conservation of the mass, energy, charge, moment of movement quantity etc. allow frequently to receive directly the end results or to reduce considerably necessary calculations. The statement of conservation laws is considered as a maturity of corresponding science.

The existence of stationary modes in queuing systems at a stationary entering flow induce some conservation laws between the quantities describing a state of the system. A merit of the author of [1] M.Krakovski is the collecting of some well-known results — equation (1) and its generalization on a heterogeneous case — as quantitative consequences from conservation laws formulated verbally. These laws and consequences from them were added and advanced by efforts of some other authors — see [2, 3, 4, 5]. They have clear physical interpretation, and their application simplifies queueing systems analysis. Last circumstance is especially important at the training in the queueing theory which — in one year of century of first Erlang's works — frequently does not dare to make for a long time the ripened step forward.

At seeming evidence of conservation laws it is possible to take from them not trivial consequences. Let's consider most important of these laws, accompanying them by verbal theses underlining their sense.

1 Conservation of Demands

The law of demands conservation is formulated in the following kind:

frequency of demands entering is on the average equal to frequency of exits.

A. Vinel et al. (Eds.): MACOM 2010, LNCS 6235, pp. 119–126, 2010.

Let's apply this principle to calculation of probability of a free state of one-channel system $GI/G/1$. Average frequency of demands arrival is $\lambda = 1/a$, average duration of service — b, where a and b — means of corresponding distributions. Frequency of exits is equal to probability of employment $1 - p_0$, divided by b. In a stationary mode should be $1/a = (1 - p_0)/b$, whence follows

$$p_0 = 1 - b/a. \tag{1}$$

The conclusion p_0 for a heterogeneous case is instructive also. The condition of balance of demands here looks like

$$\lambda_i = \frac{\lambda_i}{\Lambda} \frac{1 - p_0}{b_i},$$

where λ_i/Λ — probability of service by the occupied system of the i-th type demand. Multiplying both parts of this equality on b_i and summing results on all i, we receive

$$p_0 = 1 - \sum_i \lambda_i b_i.$$

These results also have been resulted in M.Krakovski's article. We shall receive some more additional consequences. For n-channel system the mean of number of the busied channels will make

$$\sum_{j=0}^{n-1} jp_j + n(1 - \sum_{j=0}^{n-1} p_j) = n - \sum_{j=0}^{n-1}(n - j)p_j,$$

and the condition of balance of demands is reduced to applied in [5]

$$\sum_{j=0}^{n-1}(n - j)p_j = n - \lambda b = n(1 - \lambda b/n). \tag{2}$$

Strict inequality of factor of loading

$$\rho = \lambda b/n < 1 \tag{3}$$

for opened systems with unlimited queue is a condition of a stationary mode existence.

The principle of demands conservation remains in force and at the analysis of systems with refusals (the limited length of queue), and also with "impatient" demands leaving from queue. In any case in the equation of balance all reasons of demand leaving from system should be taken into account. At the heterogeneous case this principle is fair for demands of each type separately. In queueing networks in the case of transforming demands its generalized variant is applied.

The principle of conservation operates and at processing a flow of intensity λ of demand *batches*. In this case factor of loading

$$\rho = \lambda b \bar{f}/n, \tag{4}$$

where $\bar{f}$ — average volume of a batch.

2   Conservation of a Queue

Let's record number of demands in the queue before arrival of the next demand and at the moment of its reception on service. It is obvious, that at the queueing discipline FCFS

> *distribution of the number of demands which have arrived during expectation of the beginning of service coincides with distribution of queue length before demand arrival.*

M.Krakovski has called this law as "conservation of system seniority". As obvious consequence from it he considered Little's formula connecting average waiting time w_1 with average queue length q_1:

$$w_1 = q_{[1]}/\lambda = q_1/\lambda, \tag{5}$$

and its strengthened analogue — for average sojourn time and the average number of demands in the system. Little was the first he proved them strictly in 1961. The variant of a proof is presented in [8], three variants — in [9]. Actually Little's formula is well-known from the first years of queueing theory existence, why sometimes is called as "the folklore theorem". It is easy to write down it for the closed systems — after averaging intensity of an entering flow.

As imitating experiment demonstrated, the formula (5) is applicable applied to *batch* arrivals. Thus in queue batches are taken into account only "untouched", and (5) will give an average waiting time for the head demand of a batch. Other demands will receive the additional delays determined by sequence of a choice of the demand from a batch.

The strengthened version of Little's formula is fair and for a queueing network (this was checked up by the author on imitating model).

The principle formulated above is true for system $GI/G/n$ and all its special variants. The same principle in some cases can be applied and to system as a whole. Then

> *distribution of the number of demands which have arrived during waiting of service ending coincides with distribution of the number of demands in the system before demand arrival.*

As a condition "First Come — First Served" should be kept, the second formulation is applicable to narrower systems class: $GI/G/1$ and $GI/D/n$, in which the coincidence of the order of service ending with the order of sampling from queue is guaranteed.

Above formulated laws allow connect distributions of the demand's waiting time and of the number of demands in the queue before arrival. Let

$\Pi(z)$ — generating function of the distribution of queue length (it can be interpreted as probability of that all demands in the queue are "red");

$w(t)$ — density of a waiting time of distribution before service beginning;

$\bar{A}_z(t)$ — complementary distribution function (CDF) of intervals between demands of the initial flow rarefying with probability of demand conservation $\bar{z} = 1 - z$.

Because $\Pi(z)$ can be interpreted as the probability of absence of "blue" demands in the system at moment of new arrival, from the principle formulated above an equality follows

$$\Pi(z) = \int_0^\infty \bar{A}_z t) w(t)\, dt. \tag{6}$$

At the Poissonian entering flow of intensity λ the rarefied flow also will be Poissonian, and (6) passes in

$$\Pi(z) = \int_0^\infty e^{-\lambda(1-z)t} w(t)\, dt. \tag{7}$$

Let's put $\lambda(1 - z) = s$. Then (7) is reduced to

$$\Pi(1 - s/\lambda) = \int_0^\infty e^{-st} w(t)\, dt = \sum_{k=0}^\infty \frac{(-s)^k}{k!} w_k, \tag{8}$$

where $\{w_k\}$ — the initial moments of distribution $w(t)$. It is possible to show that they are expressed through the factorial moments $\{q_{[k]}\}$ lengths of queue by Brumelle's formula [10]

$$w_k = q_{[k]}/\lambda^k, \qquad k = 1, 2, \ldots \tag{9}$$

In particular, at $k = 1$ it gives the Little's formula.

The formula (6) for a general recurrent flow was applied after replacement of a nucleus $\bar{A}_z(t)$ of integral equation by CDF of H_2-distribution constructed on the moments of distribution of intervals between demands of rarefied flow. Thus the required density $w(t)$ was replaced with Erlang- or gamma- density (if the account requires more than two moments — with a correcting polynomial). Through parameters of density an average waiting time and values of integral for several (depending on the number of free parameters of density) values $z \in (0, 1)$ were expressed. For the same values it was calculated $\Pi(z)$ — for calculation of distribution of the number of demands in system and in queue there are well fulfilled methods [3]. Thus the correction was entered into values of generating function on limitation of the amount of calculated probabilities, drawn through the relation of two last ones. One of equations was based on the Little's formula.

3   Conservation of State Probabilities

To this class of laws we shall relate conditions of balance in a stationary mode of opposite transitions through the cuts isolating one state or dividing groups of states. Such approach in essence was used still by Erlang, writing down the difference equations for state probabilities by one state. For the Markovian diagram with transitions only between the adjacent states ("birth-and-death" process) the requirement of transitions balance through dividing cut results directly in equality

$$p_{k-1}\lambda_{k-1} = p_k\mu_k. \tag{10}$$

So necessary probabilities can be determined recurrently:

$$p_k = \frac{\lambda_{k-1}}{\mu_k}p_{k-1}, \qquad k = 1, 2, \ldots, \tag{11}$$

being sent from initial probability p_0.

Let's specify some non-trivial applications of probabilities conservation law from an author's operational experience:

1. At the realization of Takahashi—Takami iterative method [5] the calculation of multi-phase queueing systems the balance of transitions through a horizontal cut between layers of the transitions diagram was used.

2. At the calculation of the same systems by a method of a matrix-geometrical progression after calculation of a matrix denominator of a progression R it is required to find probabilities of micro-states on initial layers from zero up to n-th inclusive (n is the number of channels). The corresponding system of the linear algebraic equations for $M/H_2/n$ system includes $(n+1)(n+2)/2$ unknown variables, and its decision for big n is rather labor-consuming. However the vectors $\{p_k\}$ of the probabilities of micro-states for $k = \overline{0, n-1}$ can be expressed with the help of initial transition matrices through p_n. Having written down conditions of balance for *vertical* cuts on the diagram and having added to them accordingly transformed equation of balance (2), we reduce a task to system of $(n+1)$ equations.

4   Conservation of Work Volume

Let's allocate a class so-called "conservative" disciplines service, in which:

1) all demands remain in system until full service completion;
2) interruption of service does not increase system loading.

Last condition means that expenses of time of the serving device for maintenance of interruption are negligible small, and renewal of the interrupted service is made without loss before spent resources.

> In the specified class of disciplines the distribution of the volume of outstanding work contained in queueing system, is constant and does not depend on a choice of discipline.

Let's show application of this principle to calculate an average demand's waiting time in one-channel system. First of all, in this case the average volume of work coincides with average waiting time w of new arrived demand. This time will consist of the time $\bar{f}_1$ of ending the current service and of the service time of before waiting demands. The average rest of the started service is equal $b_2/(2b_1)$ and should be taken into account with probability $\rho = \lambda b_1$ of the system employment:

$$\bar{f}_1 = \rho b_2/(2b_1) = \lambda b_2/2 = w^*.$$

Average of the number of demands in a queue on the basis of Little's formula will make λw, and each of them is on the average served b_1 time units. So, the law of volume of work conservation results in equality

$$w = \lambda b_2/2 + \lambda w b_1, \tag{12}$$

from which the Pollachek—Khintchine formula follows

$$w = \lambda b_2/[2(1 - \rho)] \tag{13}$$

for an average waiting time in $M/G/1$. With its help it is easy to demonstrate influence of factor of loading on an average waiting time and to show, that at real queueing systems calculations the accounting average values only of intervals between demands and service duration is insufficient.

Let us consider balance equation (12) of volume work for priority queueing system. Relatively to marked demand of j-th type the residual holding time should be taken average on all types of demands, and to components of a delay we must add an average service time of demands of types i, $i = \overline{1, j - 1}$, arriving during expectation of marked one. Received final formulas are well-known, however the accent on the stated principle of their derivation is represented rather important.

Multiplying both parts of the formula (13) on ρ, we shall receive

$$w\rho = \frac{\rho}{1 - \rho}\frac{\lambda b_2}{2} = \frac{\rho}{1 - \rho}w^*.$$

This equality in a heterogeneous case becomes

$$\sum_i \rho_i w_i = \frac{\rho}{1 - \rho}w^* = C. \tag{14}$$

It operates all possible redistributions of average waiting times between demands of different types in the case of priorities introduction. Its performance proves to be true by imitating and numerical calculations (including for multichannel models). However at non-exponential service with interruptions and the subsequent continuation the invariant (14) is broken (the error about the tenth shares of percent was observed).

Consider the generalization of Pollachek—Khintchine formula for n-channel system with service time distribution $B(t)$. The CDF of intervals between demands leaving such a system can be written as

$$\bar{B}_n(t) = \bar{B}(t)[\bar{B}^*(t)]^{n-1}.$$

Here $\bar{B}^*(t)$ is the CDF of residual service time. We shall approximate this disrtribution by Weibull's one:

$$\bar{B}^*(t) = e^{-t^k/T}.$$

Then

$$[\bar{B}^*(t)]^{n-1}] = e^{-(n-1)t^k/T}, \tag{15}$$

and relating moments can be computed as

$$b_m^* = \left(\frac{T}{n-1}\right)^{m/k} \Gamma(1+m/k), \quad m = 1, 2, \ldots$$

Now we can approximate our process as the problem of summation two flows with CDF's of interarrival time as initial $\bar{B}(t)$ and (15) respectively. The technique of flows summation based on hyperexponential interarrival approximation is described in [12]. Let b be mean service time, f — mean time between service completions, $\tilde{f}$ — its random modification. The probability of system full employment can be presented as $\pi = (\lambda b/n)^n$. Then the analogue of balance equation (12) will be $w = \pi \tilde{f} + \lambda w f$. Finally

$$w = \pi \tilde{f}/(1 - \lambda f).$$

5   Calculation of System $M/M/1$ under Conservation Laws

Let's apply queueing conservation laws to calculate basic characteristics of $M/M/1$ system. First of all we shall note, that in this case $\lambda(k) = \lambda$ and $\mu(k) = mu$ irrespective of k. Hence, for all k takes place $\mu p_{k+1} = \lambda p_k$, or $p_{k+1} = (\lambda/\mu)p_k = \rho p_k$, and stationary probabilities

$$p_k = (1-\rho)\rho^k, \qquad k = 0, 1, \ldots \tag{16}$$

form a geometrical progression with a denominator ρ. Condition of the existence of a stationary mode is the inequality $\rho < 1$. Probability $p_0 = 1 - \rho$, found here from a normalizing condition, coincides with expression (1) at the corresponding replacement of designations, that confirms the law of demands conservation.

Generating function of stationary probabilities

$$P(z) = \sum_{k=0}^{\infty}(1-\rho)\rho^k z^k = \frac{1-\rho}{1-\rho z}.$$

On the basis of the formula (8) the LST of the demand sojourn time distribution

$$\nu(s) = P(1 - s/\lambda) = \frac{1-\rho}{1-\rho(1-s/\lambda)} = \frac{\mu-\lambda}{\mu-\lambda+s}. \tag{17}$$

Hence, demand sojourn time in the system is described by exponential law with parameter $\mu - \lambda$. Average sojourn time

$$v = 1/(\mu - \lambda). \tag{18}$$

Average number of demands in the system

$$L = (1 - \rho)\sum_{k=1}^{\infty} k\rho^k \rho(1 - \rho)\sum_{k=1}^{\infty} k\rho^{k-1}$$
$$= \rho(1 - \rho)\frac{d}{d\rho}\frac{1}{1 - \rho}\frac{\rho}{1 - \rho} = \frac{\lambda}{\mu - \lambda}.$$

Relation of the type (5) between L and v appeared true, that will be coordinated to the law of queue conservation.

Conclusion

Taking into account naturalness and almost evidence of queueing conservation laws, it is surprising facts of proceeding application century prescription of Erlang's technology in textbooks on queueing theory — even such a high level, as [11].

References

[1] Krakowski, M.: Conservation Methods in Queuing Theory. Revue Française d'Automatique, Informatique et R écherhe Op érationnelle 1, 63–84 (1973)

[2] Kleinrock, L.: Queuing Systems. In: Computer Applications, vol. 2, Wiley, New York (1976)

[3] Ryzhikov, Y.I.: Machine Methods of Queuing Systems Calculation, p. 177. Military Cosmic academy under A.F.Mozhajsky, Leningrad (1979) (in Russian)

[4] Stepanov, S.N.: Numerical Methods of the Calculation of Systems with Repeated Calls. Science, 229 (1983) (in Russian)

[5] Takahashi, Y., Takami, Y.: A Numerical Method for the Steady-State Probabilities of a GI/G/c Queuing System in a General Class. J. of the Operat. Res. Soc. of Japan 19(2), 147–157 (1976)

[6] Ryzhikov, Y.I.: Algorithm of Calculation of Multi-Channel System with Erlangian Service. Automatics and Telemecvhanics 5, 30–37 (1980) (in Russian)

[7] Ryzhikov, Y.I., Khomonenko, A.D.: Iterative Method of Calculation of Multi-Channel systems with General Service Time Distribution. Problems of Control and Information Theory 3, 32–38 (1980)

[8] Matveev, V.F., Ushakov, V.G.: Queuing Systems: Manual on a special course "Applied mathematics", p. 239. Moscow State University (1984) (in Russian)

[9] Robertazzi, T.G.: Computer Networks and Systems: Queueing Theory and Performance Evaluation, p. 306. Springer, New York (1990)

[10] Brumelle, S.L.: A Generalization of $L = \lambda W$ to Moments of Queue Length and Waiting Times. Operat. Res. 20(6), 1127–1136 (1972)

[11] Bocharov, P.P., Pechinkin, A.V.: Queueing Theory: The textbook, p. 529. University under P.Lumumba (1995) (in Russian)

[12] Ryzhikov, Y.I., Khomonenko, A.D.: Computation of the Opened Queueing Networks with Flows Transformation. Automatics and Computer Technics 3, 15–24 (1989) (in Russian)

Intra-flow Interference Study in IEEE 802.11s Mesh Networks

Andrey Lyakhov and Ivan Pustogarov

Institute for Information Transmission Problems, RAS,
Bolshoy Karetny per., 127994 Moscow, Russia
{lyakhov,ivan.pustogarov}@iitp.ru
http://www.iitp.ru

Abstract. In wireless mesh networks, when a packet is relayed by a station, it can get into collision with another packet of the same flow. This effect is called intra-flow interference. In this paper we propose an analytical model to study the problem of intra-flow interference. Taking into account collisions caused by both neighbor stations and hidden stations, we estimate the saturated throughput of a given flow in the mesh network.

Keywords: Mesh, performance evaluation, IEEE 802.11s, intra-flow interference, throughput, modeling.

1 Introduction

A wireless mesh network is characterized by decentralized and autonomous operation, multi-hop communication capabilities and mobility. These features are suitable for fast, scalable, reliable and cost-effective network deployment under a wide range of applications such as monitoring, military, disaster relief, etc. This makes wireless mesh networks a key technology for next generation wireless networking and attracts more and more attention both from academia and industry. Nowadays, the most popular technology for mesh networks is described by IEEE 802.11s amendment [1]. The text of the amendment is almost ready for publication and is very unlikely to change in future. Due to such a popularity and wide range of applicability of mesh networks, an extensive research is being conducted now for almost every network aspect and performance evaluation seems to be the most critical.

Normally, each mesh network is deployed for some finite set of applications. These applications have particular requirements on network performance and it is vital that the network satisfies these requirements. This is the reason why the network performance should be estimated prior to the deployment. Otherwise, the network can become congested and should be redeployed. This leads to extra expenses.

Traditionally, there are two approaches to the problem of performance evaluation: simulation and analytical modeling. Simulation models are characterized by that almost all features of the analyzed protocol are taken into account. The

A. Vinel et al. (Eds.): MACOM 2010, LNCS 6235, pp. 127–138, 2010.

advantage of this is that such models can be used to obtain more performance indices with high accuracy. Nevertheless, one run of the simulation model can take a lot of time. To obtain statistically sound results, one should carry out many experiments and the time consumed becomes significant. Analytical models, on the other hand, are based on some assumptions and do not take into account some features of the protocol. Normally these models are designed to obtain a limited set of performance metrics, but these metrics can be obtained really fast (seconds compared to minutes or hours of simulations). This is the main advantage of analytical models.

Nowadays, many papers are written on analytical performance evaluation of both one-hop [2]-[6] and multi-hop [7]-[15] IEEE 802.11 networks. Though there are several models for one-hop IEEE 802.11 networks [2], [3] which are recognized by academy, the situation with IEEE 802.11s multi-hop networks is different. There is no commonly recognized model. Such a situation can be explained by that existing models are normally based on some assumptions that do not hold in real networks.

The model described in [7] being one of the first papers on performance evaluation of mesh networks, takes into account packet collisions due to hidden nodes but do not take into account packets retransmissions. In [8], the backoff procedure does not correspond to IEEE 802.11. In [9], collisions due to hidden terminal phenomenon are not taken into account. In [10], the model which predicts very accurately the performance of one-hop network under non-saturated traffic is described. Further, the model is enhanced for the multihop scenario, but has several drawbacks: the modeling of hidden nodes is not accurate, specifically the vulnerable period is equal to the duration of the successful transmission which is not true; the difference between the model and simulation is more than 10% for some configurations.

Several other papers describe analytical models for mesh networks with regular topology [11], [12] or based on time scheduling [13]-[15] which makes them inapplicable for real mesh networks.

The main disadvantage of models described in [7]-[15] is that they consider multi-hop network with one-hop flows. Nevertheless, real networks are characterized by that the source and destination are at least several hops away from each other. Due to one-hop flows assumption, these models:

- Do not take into account intra-flow interference effect: packets of the same flow get into collision when relayed by successive nodes.
- Network throughput is estimated as the sum of throughputs of all links rather than throughputs of flows.

These disadvantages lead to restricted applicability of models [7]-[15] for real networks. A fair approach is to consider mesh network with multi-hop flows. These flows take some routes which are constant for some time. Hence, we can consider the throughput of the whole network as sum of the throughputs of each flow.

In this paper we propose an analytical model for throughput estimation of one flow. We represent this flow as a chain of stations where the first station in the chain is the only source and the last station is the only destination. Other station just relay traffic. In our model we take into account collisions of packets of the same flow caused by both neighbor nodes and hidden nodes.

The rest of the paper is organized as follows. In Section 2, we provide the framework of our model. In Section 3, we present the model for the chain scenario. Numerical results can be found in Section 4. Section 5 concludes the paper.

2   Model Framework

The main reason of nonapplicability of one-hop networks models [2] is that in such networks all stations sense the channel identically. In multihop networks, stations sense medium differently. A station's behavior is determined by its neighborhood, which differs from station to station.

For the purpose of better understanding of MAC layer mechanisms impact on the network performance, we intentionally do not take into account some feature of PHY Layer. We use simplified protocol model [16] but modify it so that it takes capture effect into account. We assume that:

1. Transmission range R_T is fixed and the same for all stations. Station A can only receive a packet from station B if and only if station A is in the transmission range of station B.
2. Carrier sense range R_D is fixed and the same for all stations. Station C can only detect busy medium (and defer its own transmission) if it is in the carrier sense of the transmitter.
3. Capture effect is taken into account in a specific way. Let us consider transmitter S and receiver D. Whether a transmission of station C disrupts packet reception at station D depends on the distances between these three stations, so we define the notion of interference range $R_I(S \rightarrow D)$ for a link rather than for a station.
4. Ideal channel conditions are considered, so collision can only occur if two packets overlap.
5. EDCA scheme is used on each station. Nevertheless, in this paper we consider traffic of only one priority.

To formally describe how a station senses medium, we use the notion of virtual slot of a station [2] which is the time interval between two consecutive backoff counter changes of the station. Virtual slot duration of a station is variable and depends on how the station senses the medium. Virtual slot duration of station A can be (see Fig.1):

- Empty slot duration σ if the channel is sensed free and station A does not not transmit.
- Station A successful transmission duration T_s if station A transmits successfully.
- Duration T_c of collision if station A transmits and there is a collision on the receiver.
- Duration T_b of the transmission of a neighbor of station A. We refer to this slot as *busy slot*.

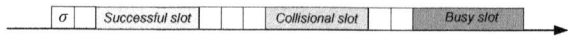

Fig. 1. Virtual slots classification

If we fix the packet length the durations of empty slot, successful slot and collisional slot are constant:

$$T_s = T_{DATA} + SIFS + T_{ACK} + AIFS,$$
$$T_c = T_{DATA} + T_{ACK\_timeout} + AIFS,$$

where T_{DATA} is the packet transmission duration at a given rate. Values $SIFS$, $AIFS$, T_{ACK} and $T_{ACK\_timeout}$ are defined by IEEE 802.11 standard. In the rest of the paper we assume that $T_{ACK\_timeout} = SIFS + T_{ACK}$, and hence $T_c = T_s$. Duration of busy slot is variable. We should estimate its average value T_b along with slots probabilities to find the network performance indices.

Following [2], we assume that the probability that an arbitrary chosen slot is empty, successful, collisional or busy is independent from the previous slot. Empty slot, successful slot, collisional slot and busy slot probabilities for station x are denoted by $P_e(x)$, $P_s(x)$, $P_c(x)$ and $P_b(x)$ correspondingly.

Let τ_x be the probability that station x starts its transmission at the beginning of its virtual slot and $p(x)$ be the probability that a transmission of station x gets into collision.

Let $b(x)$ be conditional probability that at the beginning of virtual slot of station x the medium becomes busy given that station x doesn't transmit at this slot.

Now we express slots probabilities using newly defined values:

$$P_e(x) = (1 - \tau_x)(1 - b(x)), \tag{1}$$
$$P_s(x) = \tau_x(1 - p(x)),$$
$$P_c(x) = \tau_x p(x), \tag{2}$$
$$P_b(x) = (1 - \tau_x)b(x).$$

Using the probabilities, we obtain the average virtual slot duration of station x:

$$T_{slot}(x) = P_e(x)\sigma + P_s(x)T_s + P_c(x)T_c + P_b(x)T_b = \tag{3}$$
$$= (1 - \tau_x)(1 - b(x))\sigma + \tau_x(1 - p(x))T_s + \tau_x p(x)T_c + (1 - \tau_x)b(x)T_b(x).$$

We define throughput $S(x)$ of station x as the average amount of bits transmitted successfully by the station per second.

$$S(x) = \frac{P_s(x)L}{T_{slot}(x)} = \frac{\tau_x(1 - p(x))L}{T_{slot}(x)}, \tag{4}$$

where L is the packet length. Evidently, the throughputs of stations that relay packets of the same flow are mutually dependent.

Also we define throughput of flow f with the source at station j and the sink at station k as the average amount of bits initially generated at the source and successfully delivered to the sink per second.

According to (1)-(4) to obtain the throughputs of each station we should estimate: 1) probabilities τ_x, $p(x)$, $b(x)$ and 2) the average duration $T_b(x)$ of busy virtual slot of station x.

3   Chain Scenario Model

In this section we consider a chain of n stations where $n \geq 4$ (see fig. 2a)). We assume that the distance D between two consecutive stations in the chain is the same and lies within $(R_T/2; R_T)$ interval. This assumption normally holds in real networks if all stations work at a fixed bit rate.

We consider only one flow f in the chain with the source at station 1, sink at station n and offered load λ. The flow packets are relayed through the intermediate stations. As long as $D \in (R_T/2; R_T)$ each packet is relayed consequently by each station.

Our objective is to find the saturated throughput of the chain. To achieve this goal we use the following approach. We continuously increase the value of offered load λ and the throughput increases accordingly. Obviously, there exists such a value λ^* that if $\lambda > \lambda^*$ the throughput of the chain does not change anymore due to packet losses beacuse of buffer overflow. We assume that the packets are lost exeptionally at station 1. Our assumption is based first, on simulation results and second, the following intuitive reasoning. The main cause of collisions is hidden stations. When we increase the load on the first station, the number of packets at station h which is hidden from station 1 also increases. This leads to more collisions on link $1 \rightarrow 2$. This in turn decrease the number of successfully delivered packets to station h. Hence station 1 always has more packets in its buffer than any other stations in the chain.

If $\lambda = \lambda^*$, station 1 starts to work in saturation. Following [2] we can express the probability that 1 starts to transmit:

$$\tau_1 = \frac{2(1 - 2p(1))}{(1 - 2p(1))(W + 1) + p(1)W(1 - (2p(1))^m)}, \tag{5}$$

where W is the minimum contention window and m is the maximum backoff stage. As in [2] we assume no retry limit.

As long as packets are lost exeptionally at station 1, the throughputs of all stations are the same, i.e.

$$\lambda = S(i) = S(j) \text{ for } i, j = \overline{1, n}. \tag{6}$$

The mutual dependency of the probabilities of successful slot of different stations follows from (3) and (5):

$$\frac{P_s(i)}{P_s(j)} = \frac{T_{slot}(i)}{T_{slot}(j)}. \tag{7}$$

If we express $P_s(i)$ and $T_{slot}(i)$, $i = \overline{1,n}$ as functions of τ_i, $i = \overline{1,n}$, (5) and (7) make up a system of equations, which depends only on τ_i $i = \overline{1,n}$. Thus, we can solve the system numerically and obtain the value of $S(i)$.

Let us consider a station x in the chain. As long as it transmits only to its neighbor y the throughput of station x equals to the throughput of link $x \to y$. For a given station we introduce several sets of stations.

Carrier sensing set $D(x)$ of station x. $s \in D(x) \Leftrightarrow distance(x,s) \leq R_D$

Interference set $I(x \to y)$ of link $x \to y$. $s \in I(x \to y) \Leftrightarrow distance(y,s) \leq R_I(x \to y)$. As long as the distance between two consecutive stations in the chain is the same, $R_I(x \to y)$ is also the same for all stations in the chain and depends on receiver y only. Hence, in the rest of the paper we use $I(x \to y)$ and $I(y)$ interchangeably.

Hidden nodes set $H(x \to y)$. This set contains such stations that: (1) do not sense a transmission of station x and (2) corrupt the frame received at y if start to transmit. This means that

$$H(x \to y) = I(y) \setminus D(x).$$

Collision set $C(x \to y)$. This set contains stations whose transmissions will be blocked if station x starts to transmit. But these station can start to transmit simultaneously with station x and corrupt the frame received at y.

$$C(x \to y) = I(y) \cap D(x).$$

Fig. 2 shows an example of the introduced sets. Fig. 2 a) shows an example of carrier sensing set os station x (stations within brackets). For the sake of simplicity of the example, we assume that the carrier sensing set of station y and the interference set of link $x \to y$ are coincide. Fig. 2 b) shows an example of: (1) interference set of link $x \to y$ and carrier sensing set of station y (stations within brackets), (2) hidden nodes set $H(x \to y)$ which contains h_1, and (3) collision set $C(x \to y)$ which contains stations c_1, c_2, c_3 and c_4.

Several conditions should be satisfied for transmission $x \to y$ to be successful:

1. No station from $H(x \to y)$ is transmitting when x starts to transmit to y (see left side on fig. 3).
2. No station from $H(x \to y)$ starts to transmit during transmission $x \to y$ which occupies $\frac{T_{DATA}}{\sigma}$ slots (see right side on fig. 3).
3. No station c in $C(x \to y)$ starts to transmit simultaneously with station x.

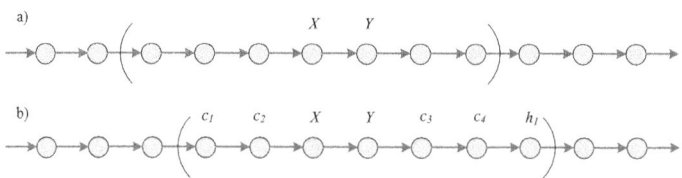

Fig. 2. Interference set, Physical carrier sensing set, Hidden nodes set and Collision set

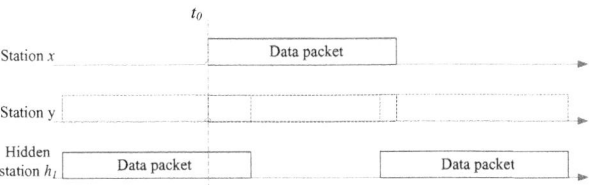

Fig. 3. Hidden node problem

First we find the probability of condition 3. We consider the discrete time axis of station $c \in C(x \rightarrow y)$ which is divided into virtual slots (see Fig.1). Let station x start to transmit at moment t_0. Station c can start to transmit simultaneously with x with probability τ_c only if its backoff counter is not frozen just before moment t_0. Note that in our case the condition that x transmits to y at t_0 is equal to the condition that stations in $D(x)$ were silent just before t_0. So, we find the probability that some station c in $C(x \rightarrow y)$ starts to transmit simultaneously with station x as follows:

$$P_{tr}^{coll}(c|x) = \tau_c P[\text{Backoff counter of station } c \text{ is not frozen}|x]. \qquad (8)$$

Thus, the probability of condition 3 can be found as $\prod_{c \in C(x \rightarrow y)}(1 - P_{tr}^{coll}(c|x))$. We find the probability $P[\text{Backoff counter of station } c \text{ is not frozen}|x]$ in (8) as the fraction of time during which station c counts empty slots. Its backoff counter can be frozen due to transmissions of its neighbors. As long as x starts to transmit to y at t_0, all stations in $D(x) \cup \{x\}$ were silent just before t_0. Hence, we exclude their transmissions from the time axis of station c in Fig.1 and do not consider these stations as candidates to freeze the backoff counter of station c. So the probability $P[\text{Backoff counter of station } c \text{ is not frozen}|x]$ in (8) can be estimated as follows:

$$P[\text{Backoff counter of station } c \text{ is not frozen}|x] = \frac{P_e(c|x)\sigma}{T_{slot}(c|x)}, \qquad (9)$$

where $T_{slot}(c|x)$ is the average duration of virtual slot of station c and $P_e(c|x)$ is the probability that its virtual slot is empty, given stations in $D(x) \cup \{x\}$ are silent.

We find $T_{slot}(c|x)$ as follows (remember that we exclude transmissions of stations from $D(x) \cup \{x\}$ and that $c \in D(x)$):

$$T_{slot}(c|x) = P_e(c|x)\sigma + P_b(c|x)T_b(c|x), \qquad (10)$$

where $P_b(c|x)$ and $T_b(c|x)$ are the probabilities of busy virtual slot of station c and its average duration correspondingly, given that stations from $D(x)$ do not transmit.

Now we find the probability of condition 1. For each station $h \in H(x \rightarrow y)$ we find the fraction of time during which station h transmits given that stations in $D(x) \cup \{x\}$ are silent. Since stations from $D(x)$ do not transmit we exclude their transmissions from the time axis of station h and do not take them into account when getting this fraction. Hence, we get the probability that station h is transmitting to station $h+1$ when x starts to transmit to y:

$$P_{tr}^{hid\_before}(h|x) = \frac{\tau_h(T_{DATA} + (SIFS + T_{ACK}) \cdot ind((h+1) \in D(y)))}{\tau_h T_s + P_e(h|x)\sigma + P_b(h|x)T_b(h|x)}, \quad (11)$$

where $ind(condition)$ is a function which equals 1 if $condition = true$ and equals 0 if $condition = false$. We use this function since ACK frame transmitted by $h+1$ might also result into collision with transmission $x \to y$. Thus, the probability of condition 1 can be found as $\prod_{\tilde{h} \in H(x \to y)}(1 - P_{tr}^{hid\_before}(\tilde{h}|x))$

Now we find the probability of condition 2. To get this probability we first find the probability $P_{tr}^{hid\_after}(h|x)$ that station $h \in H(x \to y)$ starts to transmit in a chosen slot during transmission $x \to y$. This event can only happen with probability τ_h if the backoff counter of station h is not frozen. We find this probability as the fraction of time during which its backoff counter is not frozen.

$$P_{tr}^{hid\_after}(h|x) = \tau_h P[\text{Backoff counter of station h is not frozen}|x] \quad (12)$$

To find probability $P[\text{Backoff counter of station h is not frozen}|x]$ we use the same approach as we used for stations in $C(x \to y)$ but also exclude station h transmission from the consideration:

$$P[\text{Backoff counter of station h is not frozen}|x] = \frac{P_e(h|x)\sigma}{T_{slot}(h|x)}, \quad (13)$$

We find $T_{slot}(h|x)$ exactly the same way we did for $c \in C(x \to y)$(remember that we exclude transmissions of stations from $D(x)$ and the transmission of station h itself):

$$T_{slot}(h|x) = P_e(h|x)\sigma + P_b(h|x)T_b(h|x), \quad (14)$$

Taking into account that transmission of station x lasts $\frac{T_{DATA}}{\sigma}$ slots, we get the probability that station h does not start to transmit during transmission $x \to y$ as $(1 - P_{tr}^{hid\_after}(h|x))^{T_{DATA}/\sigma}$. Thus the probability of condition 2 can be found as $\prod_{\hat{h} \in H(x)}(1 - P_{tr}^{hid\_after}(\hat{h}|x))^{\frac{T_{DATA}}{\sigma}}$.

Now we find the probability of collision of transmission $x \to y$ as follows:

$$p(x \to y) = 1 - \prod_{c \in C(x \to y)} (1 - P_{tr}^{coll}(c|x)) \times$$

$$\times \prod_{\tilde{h} \in H(x \to y)} (1 - P_{tr}^{hid\_before}(\tilde{h}|x)) \prod_{\hat{h} \in H(x)} (1 - P_{tr}^{hid\_after}(\hat{h}|x))^{\frac{T_{DATA}}{\sigma}}$$

We find the probability of busy slot of station x conditioned that it does not transmit as follows (since station x counts its backoff, all stations in $D(x)$ also do not transmit):

$$b(x) = 1 - \prod_{d \in D(x)} (1 - P_{tr}^{coll}(d|x)).$$

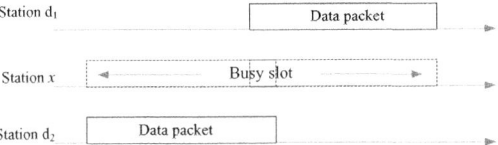

Fig. 4. Busy slot is occupied by more than one transmission

Conditioned probabilities $P_e(c|x)$ and $P_b(c|x)$ in (10) and (14) are found similarly to $P_e(x)$ and $P_b(x)$ in (1):

$$P_b(c|x) = (1 - \tau_c)b(c|x),$$
$$P_e(c|x) = (1 - \tau_c)(1 - b(c|x)).$$

where $b(c|x)$ is the conditional probability that some station in $D(c)$ starts to transmit in a virtual slot of station c given that station in $D(x) \cup \{x\}$ are silent:

$$b(c|x) = 1 - \prod_{\alpha \in D(c) \setminus (D(x)+x)} (1 - P_{tr}^{coll}(\alpha|c)), \tag{15}$$

Our final step is to find the average duration of busy slot of station x. Assume that some station $d_1 \in D(x)$ starts to transmit. It may seem that the duration of busy virtual slot is equal to the duration of successful or collisional slot (which are equal in our case). But it is not true as long as some station $d_2 \in D(x) \setminus D(d_1)$ may start to transmit during the transmission of d_1. Fig. 4 shows such a situation when busy slot $T_b(x)$ is increased by some number of slots.

$$T_b(x) = T_s + additional\_slots(x)$$

Taking into account that any station in $D(x)$ can start to transmit we get

$$additional\_slots(x) = \sum_{d_i \in D(x)} P_{tr}^{coll}(d_i|x) \times$$

$$\times \sum_{k=0}^{T_{DATA}/\sigma} \left[\prod_{\beta \in D(x) \setminus D(d_i)} (1 - P_{tr}^{coll}(\beta|d_i))^k \left(1 - \prod_{\beta \in D(x) \setminus D(d_i)} P_{tr}^{coll}(\beta|d_i) \right) k\sigma \right]$$

The values of $b(x)$ and $p(x \to y)$ depend on $b(r|x)$ and $T_b(r|x)$, where $r \in H(x \to y) \cup C(x \to y)$. The values of $b(r|x)$ and $T_b(r|x)$ depend on similar probabilities of neighbors and hidden stations of r and so on. We can find these probabilities consequently and hence get the values of $b(x)$ and $p(x \to y)$. As the result of iterative process we express all parts in system (5) and (7) in terms of τ_i, $i = \overline{1, n}$ and can solve it numerically. With known values of τ_i we can find the flow throughput.

4   Numerical Results

In this section we compare simulation results with those obtained by our analytical model. To evaluate the methods described above, we use discrete event simulator NS-3. Our simulations consist of chain of stations with the source at the first station and destination at the last station. Intermediate stations just relay packets. We use static routing, i.e. there is no signaling frames in the network like route requests and route responses.

With the numerical anlysis we assume 6 mbps bit rate. The distance between successive stations is 800 meters, the power of each transmitter is 16 dBm, carrier sense threshold for each station is −90 dBm.

We carry out experiments for chains of 4, 5, 6, 10 and 15 stations. For each of these chains we consider packets of 20 and 80 bytes which correspond to video traffic.

Fig. 5 shows the results for 20 byte packets. Fig. 6 shows the results for 80 bytes packet. The results obtained by our analytical model and simulations are almost coincide. This verifies the correctness of our analysis. Hence, the model is able to predict accurately the performance of IEEE 802.11s networks.

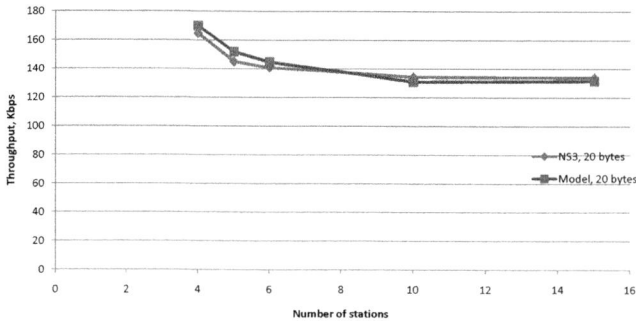

Fig. 5. Comparison of analytical model results with simulation, 20 bytes packet

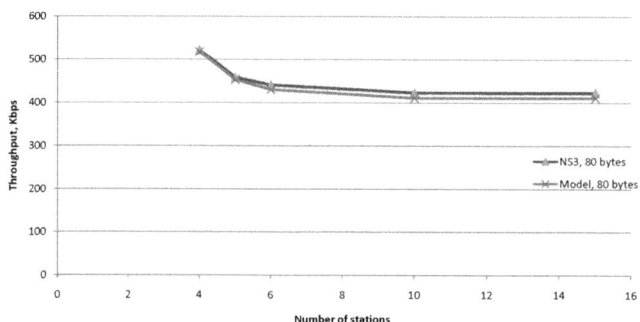

Fig. 6. Comparison of analytical model results with simulation, 80 bytes packet

The plots show that the throughput of chain decreases as the chain length increases up to ten station. Chains with more stations have the same throughput.

5   Conclusion

Nowadays, there are no commonly recognized analytical models for throughput estimation of mesh networks. This can be explained by that existing models are based on assumptions that do not normally hold in real networks. The main disadvantage of these models is that they consider one-hop flows. Obviously, it does not hold in real networks where source and destination are at least several hop away from each other. Due to one-hop flows assumption, these models do not take into account intra-flow interference and estimate the throughput of the network as the sum of throughputs of all links rather than throughputs of flows. It is more fair to consider mesh networks with multi-hop flows.

In this paper, we have presented an analytical model for saturated throughput estimation of one multi-hop flow that takes into account intra-flow interference. The model is suited for any sizes of carrier sense and interference ranges, and for any number of relay stations in the route. Using the model one can decide whether the network meets the requirements imposed by some application. Comparison with simulation results shows that the model is accurate in predicting the system throughput. Also, as far as our model more accurately describes the behavior of real networks, we hope that it will encourage future research to address the problem of performance evaluation of wireless mesh networks.

References

1. IEEE P802.11s/D3.02, Draft Amendment to Standard, Mesh Networking (March 2009)
2. Bianchi, G.: Performance analysis of the IEEE 802.11 distributed coordination function. IEEE Journal on Selected Areas in Communications 18, 535–547 (2000)
3. Cali, F., Conti, M., Gregori, E.: Dynamic tuning of the IEEE 802.11 protocol to achieve a theoretical throughput limit. IEEE/ACM Transactions and Networking 8(6), 785–799 (2000)
4. Manshaei, M.H., Cantieni, G.R., Barakat, C., Turletti, T.: Perfornamce Analysis of the IEEE of the IEEE 802.11 MAC and Physical Layer Protocol. In: WOWMOM 2005, pp. 88–97 (2005)
5. Foh, C.H., Zukerman, M.: Performance Analysis of the IEEE 802.11 MAC Protocol. In: EW 2002 Proceedings (2002)
6. Xiao, Y., Rosdahl, J.: Performance Analysis and Enhancement for the Current and Future IEEE 802.11 MAC Protocol. ACM SIGMOBILE, special issue on Wireless Home Networks 7(2), 6–19 (2003)
7. Chhaya, H., Gupta, S.: Performance modeling of asynchronous data transfer methods of IEEE 802.11 MAC protocol. Wireless Networks 3, 234–271 (1997)
8. Wang, Y., Aceves, J.: Performance of collision avoidance protocols in single channel ad-hoc networks. In: Proc. of ICNP, pp. 184–190 (November 2002)
9. Carvalho, M., Aceves, J.: Scalable model for channel access Protocol in multihop ad hoc networks. In: ACM Mobicom 2004 (September 2004)

10. Alizadeh-Shabdiz, F., Subramaniam, S.: Analytical models for single-hop and multi-hop ad hoc networks. Mobile Networks and Applications 11(1), 75–90 (2006)
11. Gupta, P., Kumar, P.R.: The Capacity of Wireless Networks. IEEE Trans. Inform. Theory 46(2), 388–404 (2000)
12. Li, J., Blake, C., et al.: Capacity of Ad Hoc Wireless Networks. In: ACM MobiCm 2001, Rome, Italy (July 2001)
13. Jangeun, J., Sichitiu, M.L.: The nominal capacity of wirelss mesh networks. IEEE Wireless Communications, 8–14 (October 2003)
14. Jain, K., et al.: Impact of Interference on Multi-hop Wireless Networks Performance. In: ACM MobiCom 2003, San Diego, USA (September 2003)
15. Kodialam, M., Nandagopal, T.: Characterizing the Achievable Rates in Multi-hop Wireless Networks: The Joint Routing and Scheduling Problem. In: ACM MobiCom 2003, San Diego, USA (September 2003)
16. Shi, Y., Thomas Hou, Y., Liu, J., Kompella, S.: How to correctly use the protocol interference model for multi-hop wireless networks. In: MobiHoc 2009: Proceedings of the Tenth ACM International Symposium on Mobile Ad Hoc Networking and Computing, pp. 239–248. ACM, New York (2009)

Simulation Study of VoIP Performance in IEEE 802.11 Wireless Mesh Networks

Kirill Andreev and Pavel Boyko

Institute for Information Transmission Problems
Bolshoy Karetny per. 19, Moscow, 127994, Russia
{andreev,boyko}@iitp.ru

Abstract. Two existing approaches to wireless mesh networking (WMN), namely IEEE 802.11s and FLAME, are compared from the point of view of their ability to support VoIP service. Simulation results on WMN voice capacity in different scenarios using different mesh protocols are presented. This data allows as to compare alternative meshing approaches and can serve as a baseline for future improvements. Mesh protocols behavior is analysed and their features limiting VoIP performance are identified.

1 Introduction

For many Internet users voice over IP (VoIP) service is already become a popular low cost alternative to the public switched telephone network. Problem of delivering voice traffic with low delay, jitter and packet loss is solved in the Internet backbone with recent developments in networking (e.g. differentiated service QoS schemes) and voice processing (e.g. modern voice codecs) technologies. Usually users expect the VoIP quality to be close to what is provided by conventional telephony.

IEEE 802.11 based wireless mesh networks (WMN)[1] may provide a good platform for the fast deployment of VoIP service in many application scenarios. Moreover, VoIP is commonly regarded as one of the killer applications for WMN [1]. At the same time it is clear that the delay-critical applications such as realtime audio and video transmission pose unique challenges to the performance characteristics of the wireless mesh network solutions. An inherent variability of the wireless channel and the network topology as well as the overheads of CSMA/CA channel access and mesh protocols can lead to significant delay, jitter and packet loss and, finally, to the unacceptable VoIP quality.

There are two basic approaches to define the VoIP quality metrics. The network level performance is defined in terms of average end-to-end packet delay, jitter and packet loss probability. The user level performance is measured in terms of human subjective voice perception quality which is usually defined as an empirical function of the network level metrics. The most known and widely

---

[1] We do not distinguish the wireless mesh networks and mobile ad-hoc networks (MANET) for the purposes of this work.

A. Vinel et al. (Eds.): MACOM 2010, LNCS 6235, pp. 139–150, 2010.

used function of this sort is a transmission rating factor (or simply R factor) defined by ITU's E-model [2]. We consider users' perception as more adequate measure of service performance and will use R factor as the basic quantity in this paper. As suggested by companion ITU recommendation [3] voice stream quality is considered to be acceptable if $R > 50$ (by definition $0 \leq R < 100$). Derivative measure is a VoIP *service availability*, which is defined as an average fraction of time when $R > 50$ for a given network and traffic configuration. In general, availability is a decreasing function of a number of simultaneously active VoIP connections when all other factors are fixed. We define network *VoIP capacity* as the maximal number of simultaneous VoIP streams that network can support with service availability not less than 95%.

Considerable amount of work has been done in the area of VoIP capacity of the 802.11 network in the infrastructure mode. Comprehensive simulation study of the VoIP capacity of an IEEE 802.11b network was presented in [4]. Simple but very useful analytical model predicting VoIP capacity in the single hop topology (base-station and clients with all VoIP connections between wireless clients and wired PCs) was proposed in [5]. It was shown, how inherent 802.11 channel access overhead essentially limits resulting network capacity. We will compare our simulation results with the estimates of [5] in the section 4.1.

For multihop networks the main factor influencing VoIP performance is the network topology. Both experimental and simulation results on VoIP capacity in chain and lattice topologies of various size were presented in [6]. It was shown that capacity drastically degrades with the number of nodes in the chain. OLSR routing protocol was used. We will compare our simulation results with the results of [6] in the section 4.2. VoIP performance in realistic emergency communication scenario network topologies was considered in [7], see also the references within. The performance evaluation was mainly based on the obstacle coverage, i.e. the percentage of network area covered by obstacles, while network size (50 nodes) and number of simultaneous VoIP connections (one) remain fixed. AODV routing protocol was used.

Another critical factor for VoIP performance is a choice and configuration of protocol(s) enabling multihop connectivity. Paper [8] provides some simulation results on VoIP performance using AODV, DSR and 802.11s HWMP on the same network of 50 nodes distributed randomly. Experimental comparison of AODV and OLSR on small wireless testbed is presented in [9]. A lot of work is dedicated to VoIP-specific protocols optimizations, see *e.g.* [10] (QOLSR), [11] (link quality metrics, frame aggregation) and the references within.

In this work we compare two existing and very different approaches to wireless mesh networking, namely IEEE 802.11s and FLAME, from the point of view of their ability to support VoIP service. The contribution of this paper is twofold. First, we present simulation results on WMN voice capacity in different scenarios and using different mesh protocols. This data allows us to compare alternative meshing approaches and can serve as a baseline for future improvements. Second, we analyse mesh protocols behaviour and identify their features limiting VoIP performance.

The rest of the paper is organized as follows: next section provides a brief overview of IEEE 802.11s and FLAME protocols. Unlike the previous studies, we also include an "ideal" meshing solution as the reference point to the study, see section 2.3 for details. Simulation setup and methodology are presented in Section 3 and simulation results are discussed in Section 4. Section 5 presents conclusions and a discussion of future work.

2 Protocols

2.1 IEEE 802.11s

IEEE 802.11s Draft Standard [12] is an extension of 802.11-2007 Standard enabling transparent frame forwarding in arbitrary multi-hop topologies. Each mesh station participating in the 802.11s wireless mesh network operates as link layer router and is responsible for cooperation with all the other mesh stations in packet delivery. This is done using several dedicated protocols, two of them being most important. Peering management protocol is responsible for neighbor mesh station (peer) discovery and management. Routing protocol is responsible for multi-hop path discovery and maintenance. Short overview of these two protocols with emphasis on important implementation details is given below, see [12] for remaining details.

The Peering management protocol is used to open, maintain and close links (or peerings) with neighbour mesh stations, choosing whether to open a new link and closing links when detecting their failures. To be visible to its neighbors, every mesh station periodically sends small one-hop management frames known as beacons, we use the default beacon interval of 0.5 second. A peer link open handshake starts when mesh station receives a beacon from a previously unknown mesh station. A Peering Open management frame used as an association request in the 802.11 infrastructure mode and a Peering Confirm management frame serves as an association response. A peer link is established only when both stations have sent Peering Open requests and received Peering Confirm replies; this requirement guarantees that all established links are bidirectional. We have forced to change peering open and confirm timeouts from recommended 40 ms to 2 seconds to avoid Peering Open congestion in dense networks.

When a peer link has been detected as broken, a Peering Close frame is sent to the peer station, which is analogous to disassociation in infrastructure network. It is important to note that link failure detection rules are not defined by the 802.11s Draft. We use the following rule: a link is detected to be broken if 5 successive beacons from the corresponding peer station were lost *or* 5 successive unicast data frames to that neighbor were dropped because retry counter has reached its limit.

Routing protocol is used to discover and maintain multi-hop paths in the mesh network. As usual, routing protocol uses path selection metric to choose the optimal one of different possible paths. The Hybrid Wireless Mesh Protocol (HWMP) is a mesh routing protocol inspired by AODV and tree routing. In this work we use HWMP in its reactive mode only. Path discovery is done on demand using broadcast Path Request – unicast Path Response lookup. The best route among multiple replies is chosen at source by path metric. Default airtime link metric estimates time needed to transmit the reference size frame over a particular link in accordance with overhead, data transmission bit rate and frame error rate. An established path is used for 5 seconds and then path request – reply procedure is repeated to try to find a better path. Note that the path discovery procedure is used even for direct neighbors known by Peering Management protocol. This is done since a single-hop path is not necessary the best available one. We use a dedicated queue with the best effort access category for the broadcast Path Request management frames.

When peering management protocol detects that a link is broken, a Path Error management frame is sent to all stations which are known to use the broken link in their routes.

2.2 FLAME

Forwarding Layer for Meshing (FLAME [13]) is a very different way to enable multihop connectivity in wireless networks. First, it works on "2.5" layer of ISO/OSI model on top of any MAC with Ethernet compatible addressing but still transparently for network layer. Second, FLAME does not use any route lookup mechanism at all. Instead every station continuously analyses incoming traffic and updates its forwarding table on the ground of its findings.

Main axiom of FLAME is that all links are symmetric, i.e. if A has a path to B through C that means that C may deliver data to A through B. Using this assumption, FLAME constructs a forwarding table that is used to find a next hop for a packet that has to be forwarded to a certain destination. Another role of forwarding table is to drop broadcast packets, that have already been received before. FLAME forwarding table is updated with every received and accepted packet. As the consequence of continuous forwarding table updating FLAME effectively uses random paths from source to destination trying to maximize packet delivery probability by using most up to date information.

The next hop address for locally originated packets is determined in accordance to the following rules:

1. For broadcast and multicast packets the next hop address is set to broadcast.
2. If there is a forwarding table entry for a given destination, the next hop address is set in accordance with it.
3. If there is no forwarding entry for a given destination, the next hop address is set to broadcast.
4. If the station did not send broadcasts more than 5 seconds, the next hop address is set to broadcast.

Forwarding rules for packets to be forwarded are:

1. Check that sequence number for a packet is the latest known for a given station (sequence number is stored in forwarding table).

2. If destination MAC address is unicast and the next hop is unknown, the packet shall be dropped, otherwise the packet is forwarded.
3. If destination MAC address is broadcast, the packet shall be forwarded with next hop address set to broadcast.
4. If receiving station is a destination, and it has not send broadcast more, than 5 seconds, it sends a Path Update – empty broadcast packet through all network.

2.3   Global Routing

Global routing is a simulation specific "ideal" mesh routing solution. It operates above a simple ad-hoc 802.11 network and does not send any management information (including beacons). Instead, network topology is recovered from nodes coordinates and estimated RX range. Using network topology global routing solves all pairs shortest path problem and populates global forwarding table which is used by each mesh station for multihop unicast forwarding. Broadcast frames are not supported by global routing in its current implementation.

We use global routing as a reference point for VoIP performance measurements as shown in the section 4.

3   Simulation Setup and Methodology

We use Network simulator 3 [14] version 3.7 for simulations. Ease of use and contributing, available high fidelity IEEE 802.11 MAC and PHY models together with real world design philosophy and concepts made NS-3 our platform for this work. Basic 802.11 model parameters used are listed in Tab. 1. Default log-distance propagation loss model is used, which results in RX range of 130 m and CS range of 175 m.

802.11s, FLAME and Global Routing mesh stacks are implemented from scratch, see Section 2 for details.

Table 1. 802.11 model parameters

Parameter	Value
Standard	802.11a
TX power, dBm	16
TX gain + RX gain, dBi	1 + 1
RX noise figure, dBm	7
Thermal noise, dBm	-101
PHY energy detection threshold, dBm	-94
PHY CCA mode 1 threshold, dBm	-96
RX range, m	130
CS range, m	175
Rate control algorithm	ARF

VoIP session is modeled as steady one-directional CBR stream. Stream intensity is 50 packets per second and each packet has 32 bytes of UDP payload (20 bytes for two 10 ms G.729 frames + 12 bytes for RTP header). In the "crowd" and "random" scenarios (see below) source and destination of voice streams are chosen randomly and uniformly among all stations. In the "chain" scenario source and destination are fixed at the opposite ends of the chain, and half of the voice streams are directed from the beginning of the chain to the end of the chain, and another half of streams are directed in the opposite direction.

For every stream, per-second statistics for delay and packet loss are available, which are converted to the time-dependent R factor as defined in [2]. For codec-dependent E-model parameters, we use ITU G.729 codec as the reference. From the time-dependent R factor the stream- and time- average VoIP availability is deduced as defined in Sect. 1. To model steady state network behavior, every simulation run lasts for 500 seconds of simulation time, first 50 of them are not used in analysis. We use 40 statistically independent runs for every simulation experiment and present run-average VoIP availability with 95% confidence interval below.

Only static mesh networks are considered in this work. We considered three following extreme cases of static mesh topology (or "scenarios").

In the *crowd* topology all stations are direct neighbors of each other. We considered crowds of 8, 16 and 32 stations. This scenario implements single-hop mesh network with minimal diameter and maximal node density. Distance between nodes is small enough to use the maximal bit rate of 54 Mbps. Analysing this scenario is deductive in several ways. It is possible to compare mesh network VoIP capacity with infrastructure mode results discussed in Section 1. Mesh protocols overhead is clear in this case. In spite of the source and destination may transmit data directly to each other, routing protocol may construct multihop routes, or, if we use FLAME, broadcasting a frame over the whole network in this scenario affect negatively to the network capacity. Simulation results for this scenario are presented in Section 4.1.

In the *chain* scenario stations are placed in a long 1-dimensional chain, and every station (except for the first and the last ones) has exactly two neighbors. We considered chains of 8, 16, and 32 stations and place stations at the distance of 100 m (remember that RX range is 130 m and CS range is 175 m). This scenario implements the fully connected mesh network with a minimal possible stations density and the maximal hidden stations ratio. Also, this is one of the most popular scenarios used in evaluation of VoIP behavior in mesh networks, because it can be easily validated and all negative effects of channel access procedure are revealed here. As soon as in this topology each station has one or two neighbors, the maximum channel bit rate can not be used (otherwise, more distant stations may be available using slower transmission speed). The maximal channel bit rate that can be achieved in this scenario is equal to 18 Mbps. Simulation results for this scenario are presented in Section 4.2.

In the *random* topology 100 stations are uniformly distributed in the square field of size 6.7 × 6.7 RX range (870 × 870 m). These parameters are taken from [15] to obtain topologies with average CBR stream length more that 4 hops and probability that two nodes are connected of 96% (see [15] for details). In contrast with previous scenarios, this one has a lot of alternative routes, a maximum variety of transmission speeds may be used and maximum possibilities for intelligent path quality metric are revealed. Also the number of stations is high enough to see potential scalability problems of different solutions. Simulation results for this scenario are presented in Section 4.3.

4   Results

4.1   Crowd

VoIP service availability in the crowd scenario as function of the number of simultaneous streams is presented at Fig. 1. For the ideal case of global routing crowd VoIP capacity is approx. 60 streams regardless of number of stations. This result is in full agreement with the estimate capacity of 64 streams from [5] (see Tab. III column "G.729" row "20 ms" within).

The overhead of each real mesh protocol can be the most distinctly seen at this scenario. VoIP availability using FLAME is 100% for subcritical number of streams and capacity is 50 streams (80% of global routing) for crowd of 8 nodes and degrades to 40 streams (66% of global routing) for 32 nodes. Capacity degradation in comparison with Global Routing is caused by regular broadcast transmissions used by all FLAME stations. The same reason causes capacity degradation with the number of nodes.

IEEE 802.11s Peering Management and HWMP protocols produce more overhead than FLAME, so the overall capacity is worse (from 66% of global routing for 8 stations to 33% of global routing for 32 stations). The main difference is that VoIP availability is less than 100% even for small number of streams far from capacity. This is explained by the fact that 802.11s needs successful path request/reply handshake for traffic delivery. Since broadcast PREQ messages can collide with high priority voice traffic and the number of PREQ retries is limited to 3, route lookup can fail even in crowd topology with moderate voice traffic. Another factor of capacity degradation is that HWMP is observed to use 2-hop paths at times, which is probably caused by imperfections of rate control algorithm and spontaneously closed peer links (due to beacon or voice packets lost in the dense traffic). Another feature of 802.11s we would like to mention here is a mandatory data packets buffering while PREQ is sent and PREP is not received. Since VoIP application produces new packet every 20 ms and route lookup can be as long as 200 ms (the default timeout), buffering VoIP traffic results in periodic (remember that by default route must be updated every 5 seconds) packet bursts useless because delay is too large but dangerous for other voice streams. To summarize, some advanced features of 802.11s protocols are found to be harmful for VoIP capacity in the crowd scenario.

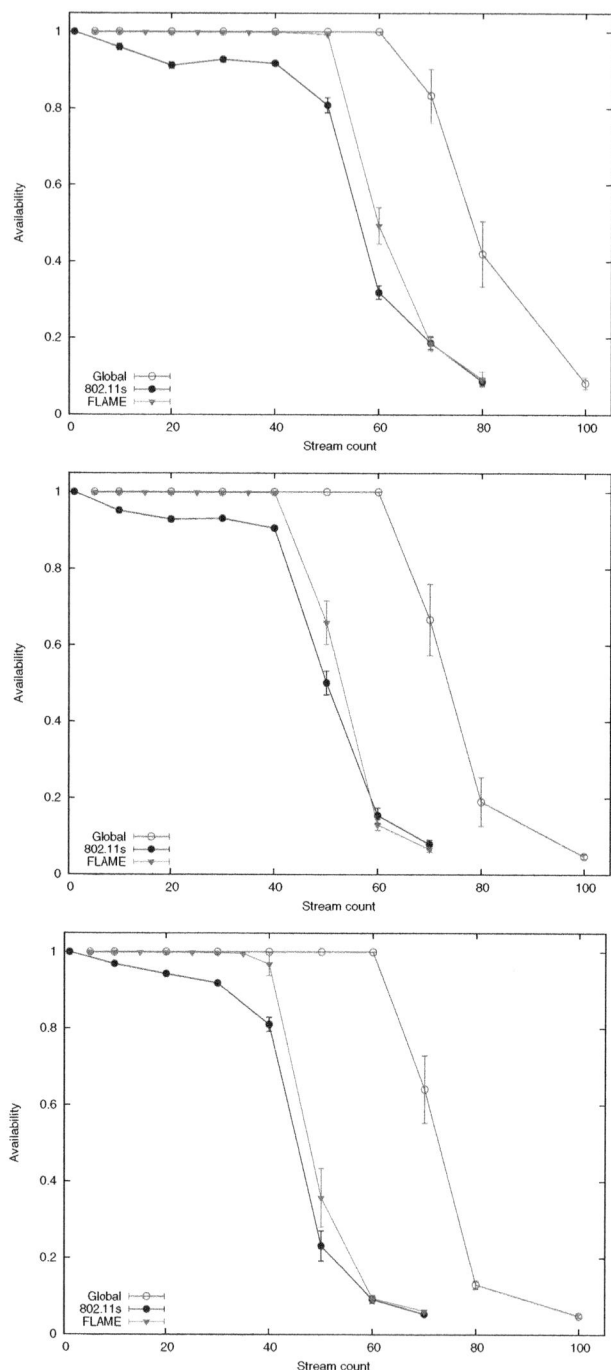

Fig. 1. Average voice availability as a function of number of streams for crowd topology. Top to bottom: 8 nodes, 16 nodes, 32 nodes.

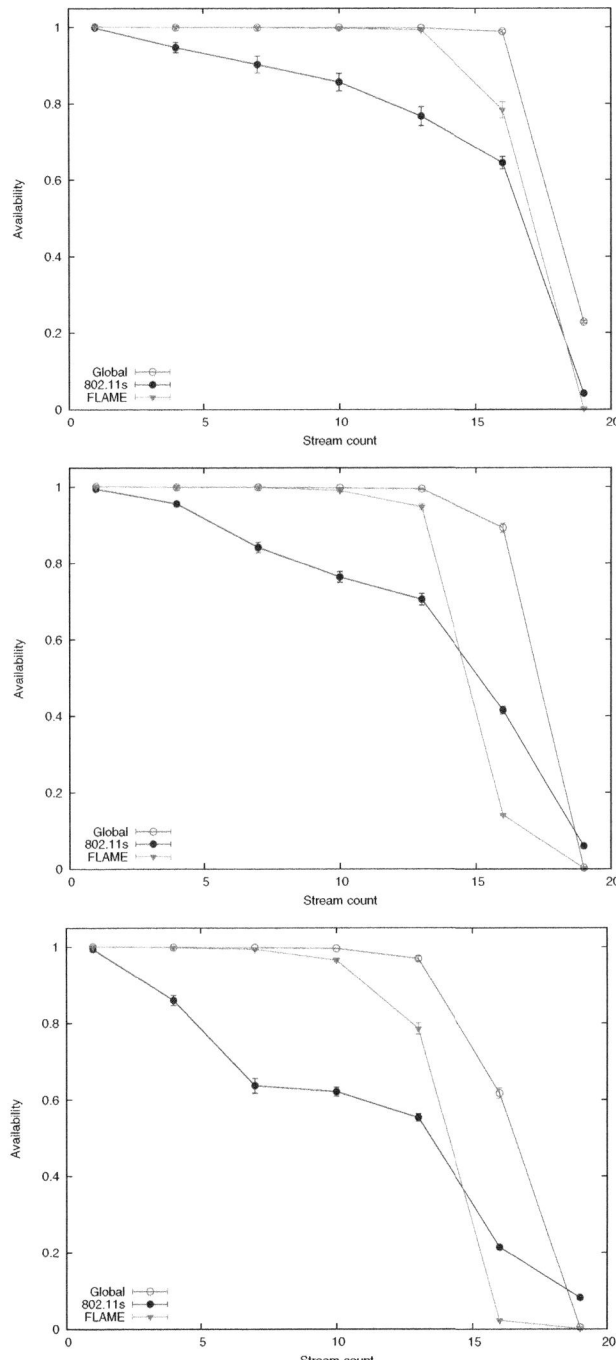

Fig. 2. Average voice availability as a function of number of streams for chain topology. Top to bottom: 8 nodes, 16 nodes, 32 nodes.

4.2   Chain

VoIP service availability in the chain scenario as function of the number of simultaneous streams is presented at Fig. 2. It is simple to explain why maximal chain VoIP capacity is much less in comparison with crowd scenario. First reason is that every frame after it has been received must be immediately forwarded, this cause at least twofold capacity degradation. Second, highest transmission rates are not available since stations are too far from each other, Third, a number of hidden stations is high which increases packet collision probability. Results are presented at Fig. 2. VoIP capacity shown by global routing is approximately four times smaller than the crowd capacity and slowly degrades with the chain size from 17 streams for 8 stations to 13 streams for 32 stations. This is in good agreement with results of [6] of approximately 10 streams in the chain of 8 stations using OLSR (see Fig. 1 and 2 in [6]).

As in the crowd scenario, FLAME utilizes 60–80% of maximal VoIP capacity and this number slightly decreases with network size. At the same time 802.11s shows much worse performance and VoIP capacity is effectively 1 stream for the chain of 32 nodes. The main reasons are high PREQ collision probability in the presence of dense high priority traffic and hidden stations and large route lookup delays causing packet bursts as explained in the previous section. It is interesting to note that for very large number of streams 802.11s shows better performance than FLAME or even global routing. This is because 802.11s effectively limits an average number of simultaneously active streams due to route lookup failures.

4.3   Random

Average voice availability as a function of of number of streams for random topology is presented at Fig. 3.

The VoIP capacity shown by global routing is 20 streams and lies between the chain (15) and crowd (60) values as expected. Since topology is random, all protocols show gradual degradation of average VoIP availability with the number of streams in contrast with sharp degradation at some critical point seen in crowd and chain scenarios. In this scenario the 802.11s shows better performance than FLAME, though the difference is not dramatic and capacities at the 90% average availability level are almost identical (approximately 10 streams which is 50% of global routing). This is a consequence of intelligent path selection metric and better scalability due to limited use of broadcast messages comparing to FLAME.

It is interesting that in this scenario both 802.11s and FLAME outperform global routing in the dense traffic, though the reasons are different. HWMP utilizes an airtime path selection metric to select better paths than hop-count based global routing. We observe that an average packet delay in case of single voice stream is almost 2 times less using 802.11s comparing to using global routing. Also it is important to note that when load is high, Peering Management protocol closes over-congested links (remember that 5 successive lost packets are enough) even in static network and this forces HWMP to find new paths avoiding

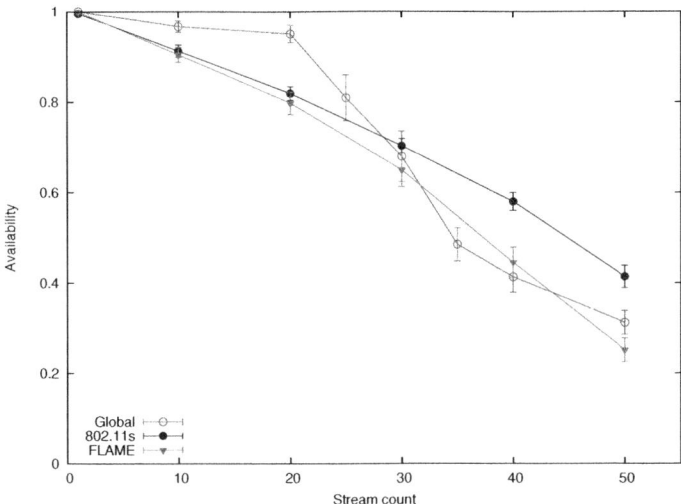

Fig. 3. Average voice availability as a function of number of streams for random topology

bottlenecks. On the other hand, FLAME uses almost random paths to forward traffic, but using most recent packet delivery information it also effectively avoids bottlenecks and over-congested paths.

5   Conclusion

In this paper we have presented the simulation results on VoIP performance in IEEE 802.11 based WMN. Two existing meshing solutions: IEEE 802.11s and FLAME were compared together with an "ideal" overhead-less routing in three static topologies: crowd, chain and random and in a wide range of traffic intensity.

In all considered scenarios all considered meshing solutions are able to support at least one VoIP stream with acceptable quality. FLAME is shown to outperform IEEE 802.11s in the crowd and chain scenarios. In both cases FLAME network shows 60%–80% of maximal voice capacity as implemented by global routing, while 802.11s network capacity is approximately 30%–60% of maximum in the crowd and can be as small as single VoIP stream in the chain of 32 stations. The main reason of poor voice capacity of the 802.11s network is large PREQ collision probability in the presence of dense high priority VoIP traffic. Another negative factors are queued packets bursts after long but successful route lookup and spontaneous peer link closures due to large total number of collisions. The benefits of route lookup and maintenance mechanisms of 802.11s are become apparent in the random topology scenario, where 802.11s slightly outperforms FLAME (and in the dense traffic even global routing). Both FLAME and 802.11s shows approximately 50% of maximal voice capacity in this case.

Apart of mesh protocols and network topologies, many important factors were fixed in the present work. Among them are potential wireless links variability and asymmetry, choice of voice codec, application level frame aggregation, mobility, mesh protocols parameters. Future research work will be carried to explore their influence on VoIP performance in WMN.

References

1. Wang, X., Patil, A., Wang, W.: VoIP over wireless mesh networks: challenges and approaches. In: WICON 2006: Proceedings of the 2nd Annual International Workshop on Wireless Internet, p. 6. ACM, New York (2006)
2. Bergstra, J.A., Middelburg, C.A.: ITU-T Recommendation G.107: The E-Model, a computational model for use in transmission planning. Technical report, ITU (2003)
3. ITU-T Study Group: ITU-T Recommendation G.109: Definition of categories of speech transmission quality. Technical report (1999)
4. Hole, D., Tobagi, F.: Capacity of an IEEE 802.11b wireless LAN supporting VoIP. In: 2004 IEEE International Conference on Communications, vol. 1, pp. 196–201 (June 2004)
5. Garg, S., Kappes, M.: Can I add a VoIP call? In: IEEE International Conference on Communications, ICC 2003, vol. 2, pp. 779–783 (May 2003)
6. Lee, B.H., Cai, G.Y., Ge, Y., Seah, W.: VoIP Capacity over Wireless Mesh Networks. In: Proceedings 2006 31st IEEE Conference on Local Computer Networks, pp. 551–552 (November 2006)
7. Birkos, K., Papageorgiou, C., Dagiuklas, T., Kotsopoulos, S.: User and Network Level Evaluation of VoIP over Emergency Ad-hoc Networks. In: MobiMedia 2009: 5th International Mobile Multimedia Communications Conference (2009)
8. Ksentini, A., Abassi, O.: A comparison of VoIP performance over three routing protocols for IEEE 802.11s-based wireless mesh networks (wlan mesh). In: MobiWac 2008: Proceedings of the 6th ACM International Symposium on Mobility Management and Wireless Access, pp. 147–150. ACM, New York (2008)
9. Armenia, S., Galluccio, L., Leonardi, A., Palazzo, S.: Transmission of VoIP Traffic in Multihop Ad Hoc IEEE 802.11b Networks: Experimental Results. In: International Conference on Wireless Internet, pp. 148–155 (2005)
10. Boi, F., Atzori, L.: Joint routing and playout buffering of IP telephony flows in MANETs. Mob. Netw. Appl. 13(3-4), 297–305 (2008)
11. Niculescu, D., Ganguly, S., Kim, K., Izmailov, R.: Performance of VoIP in a 802.11 Wireless Mesh Network. In: Proceedings of 25th IEEE International Conference on Computer Communications, INFOCOM 2006, pp. 1–11 (April 2006)
12. IEEE P802.11s/D3.0. Draft STANDARD for Information Technology Telecommunications and information exchange between systems Local and metropolitan area networks Specific requirements Part 11: Wireless LAN Medium Access Control (MAC) and Physical Layer (PHY) specifications Amendment: Mesh Networking (Electronic resource) (2009)
13. Elfrink, H.: Forwarding Layer for Meshing. Revision 2.0. Technical report, Twente Institute for Wireless and mobile Communications (2006)
14. The NS-3 network simulator, http://www.nsnam.org/
15. Kurkowski, S., Camp, T., Navi, W.: Minimal Standards for Rigorous MANET Routing Protocol Evaluation. Technical report, Colorado School of Mines (2006)

Modeling the Influence of the Real-Time Traffic on the Delay of the Non Real-Time Traffic in IEEE 802.16 Network

Zsolt Saffer[1], Sergey Andreev[2], and Yevgeni Koucheryavy[2]

[1] Budapest University of Technology and Economics (BUTE), Hungary
safferzs@hit.bme.hu
[2] Tampere University of Technology (TUT), Finland
sergey.andreev@tut.fi,
yk@cs.tut.fi

Abstract. In this paper we provide an analytical model for an efficient dynamic capacity allocation in the IEEE 802.16 wireless metropolitan area network, in which the non real-time traffic can utilize the bandwidth unused by the real-time traffic. We investigate the delay of the nrtPS service flow as a function of the capacity allocation of the rtPS (ertPS) and UGS service flows. Unicast polling is used for the bandwidth reservation for the nrtPS and rtPS (ertPS) service flows. The delay analysis counts for both the reservation and the scheduling delay components. The nrtPS packets arrive with Poisson process. The model enables asymmetric capacity allocation and asymmetric nrtPS traffic flows. The exact mean overall delay is obtained for the nrtPS service from the numerical solution of the appropriate queueing model with batch packet service by means of the embedded Markov chain technique. The analytical model is verified by means of simulation.

Keywords: IEEE 802.16, WMAN, performance evaluation, queuing model, batch service time.

1 Introduction

IEEE 802.16 standards family defines an air interface for Broadband Wireless Access (BWA) system. As a result of a recent revision the new base standard IEEE 802.16-2009 [1] consolidates the IEEE 802.16-2004 standard with several amendments. This wireless interface is recommended for Wireless Metropolitan Area Networks (WMANs), see [2]. The high-speed air interface specified by the IEEE 802.16 standards family supports multimedia services and provides several traffic types to ensure wide range of Quality-of-Service (QoS) requirements of the end users.

The standardization of metropolitan-scale wireless access is an ongoing activity performed by the IEEE 802.16 Working Group for BWA with the support of WiMAX Forum [3]. The applied scheduling, which is out of scope of the IEEE 802.16-2009 standard, has a major impact on ensuring QoS requirements of the

A. Vinel et al. (Eds.): MACOM 2010, LNCS 6235, pp. 151–162, 2010.

end users. As a consequence of it this problem is addressed by numerous research papers, like [4],[5] and [6], in which various frameworks are built and analyzed to guarantee a specified level of QoS.

Most of the analytical works in the literature do not count for both the reservation and the scheduling parts of the delay. An early general theoretic work on overall delay analysis of access-control systems is the fundamental paper of Rubin [7]. For more practical approach we refer to [8], in which the practical performance measures of IEEE 802.16 system are considered by various techniques. In [8] and [9] the overall system delay is approximated and verified. In our previous work [10], we established an analytical model for the exact overall delay of the nrtPS service flow with unicast polling in the IEEE 802.16 system.

In this paper, we continue the works in [9] and [10] by extending the analytical model to realize an efficient dynamic capacity allocation, in which the non realtime traffic of each Subscriber Station (SS) can utilize the bandwidth left free after the capacity allocation for the real-time traffic of the same SS. Thus the model incorporates the effect of the capacity allocation of the real-time polling service (rtPS), extended real-time polling service (ertPS) and unsolicited grant service (UGS) flows on the overall delay of the non real-time polling service (nrtPS) flow. However the study of this effect is out of scope of this paper and left open for future research. The analytical approach leads to a queueing model with batch packet service. The expression of the mean packet delay is given in terms of model probabilities, which are computed from the equilibrium distribution of a properly identified embedded Markov chain.

The rest of the paper is structured as follows. Section II gives a brief summary of the channel allocation schemes in IEEE 802.16. In Section III we provide the analytical model including the details of the capacity allocation and the uplink scheduling. The analysis of the queueing model follows in Section IV. We determine the mean overall delay of the nrtPS service flow in Section V. Finally the verification of the analytical results closes the paper in Section VI.

2   Channel Allocation Schemes in IEEE 802.16

In the centralized point-to-multipoint (PMP) IEEE 802.16 architecture there are one Base Station (BS) and one or more SSs. The packets are exchanged between BS and SSs via separate channels. The DownLink (DL) channel is used for the traffic from the BS to the SSs and the UpLink (UL) channel is used in the reverse direction. The standard defines two mechanisms of multiplexing DL and UL channels: Time Division Duplex (TDD) and Frequency Division Duplex (FDD). In FDD the the DL and the UL channels are assigned to different subband frequencies. In TDD mode the channels are differentiated by assigning different time intervals to them, i.e. MAC frame is divided between the DL part and the UL part. The border between these parts may change dynamically depending on the SSs bandwidth requirements. The SSs access the UL channel by means of Time-Division Multiple Access (TDMA). The structure of the MAC frame in TDD/TDMA mode is shown in Figure 1.

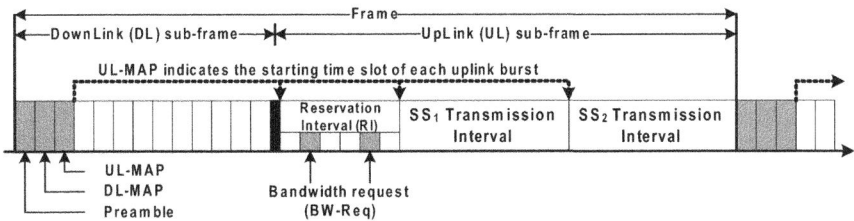

Fig. 1. IEEE 802.16 MAC frame structure in TDD/TDMA mode

3   Analytical Model and Notations

In the model all the five service flow types are allowed at each SS. For UGS, rtPS and ertPS services the QoS guarantees are ensured by means of the necessary capacity allocations. The nrtPS and Best Effort (BE) service flows use the remaining bandwidth, where the nrtPS service flow is prioritized over the BE traffic. Thus in the delay analysis model of the nrtPS service flow we count for the effects of the UGS, rtPS and ertPS service flows.

3.1   Restrictions of the Model

We impose several limitations on the IEEE 802.16 model. The operational mode is PMP and TDD/TDMA channel allocation scheme is used. Only the uplink traffic is considered as well as unicast polling is used for nrtPS, rtPS and ertPS services. Furthermore the uplink scheduler at the BS has an individual buffer for each SS to maintain the nrtPS packets. Piggybacking is not used.

3.2   General Model

There are 1 BS and N SSs in the system, which together comprise N+1 stations. Each SS maintains separate buffers with infinite capacity for the uplink packets of different service flows. The nrtPS packets arrive to SS i according to Poisson arrival process with arrival rate λ_i for $i = 1, \ldots, N$. Hence the overall nrtPS packet arrival rate is $\lambda = \sum_{i=1}^{N} \lambda_i$. We call the nrtPS packets arriving to SS i as i-packets. The arrival processes at the different SSs are mutually independent. The packet length is fixed and it equals η^{-1} bit, which includes data information and the header with packing/fragmentation overhead. The transmission rate of each channel is β bps. Therefore the transmission time of a data packet is $\tau = (\eta\beta)^{-1}$. All time durations are measured in seconds.

T_f denotes the duration of each frame. While all SSs are allowed to transmit on the uplink in one frame, they may be grouped by the reservation mechanism to reduce the polling overhead. In one frame only SSs assigned to one group are polled and are allowed to send BW-Req. Then the individual groups are polled in consecutive frames. P denotes the number of SSs in each group and hence the number of groups is $L = N/P$. The same SS group is polled in every L-th frame.

The period between two consecutive polling of the same SS group is called a cycle. Thus the length of a cycle is LT_f.

The duration of the DL and UL sub-frames are T_d and T_u, respectively. T_{ri} stands for the duration of the reservation interval and T_{ud} is the maximum available duration of the uplink data transmission in a frame. Therefore T_u is given by $T_u = T_{ri} + T_{ud}$.

The transmission time of a BW-Req is α. Hence $T_{ri} = P\alpha$ and T_{ud} can be expressed as $T_{ud} = T_u - P\alpha$.

3.3 Capacity Allocation

The total uplink capacity is distributed into disjunct parts among the SSs. As the packet transmission time has fixed length we measure the capacity in number of packets. C_i denotes the capacity assigned for SS i for total uplink transmission in a frame for $i = 1, \ldots, N$. Then $T_{ud} = \sum_{i=1}^{N} C_i \tau$ holds for the capacity allocation of the SSs.

Let C_i^u denote the fixed capacity assigned for SS i in a frame for the uplink UGS traffic for $i = 1, \ldots, N$. Similarly R_i denotes the variable capacity assigned for SS i in a frame for the uplink rtPS and ertPS transmissions together. The range of the discrete-time random variable R_i is given by

$$R_i^{min} \leq R_i \leq R_i^{max} < C_i - C_i^u, \quad i = 1, \ldots, N. \tag{1}$$

The available capacity for the nrtPS service flow is the variable capacity remaining after allocating the necessary capacity for the above three real-time traffic flows. Hence the variable capacity assigned for SS i in a frame for the uplink nrtPS, H_i, can be expressed as

$$H_i = C_i - C_i^u - R_i, \quad i = 1, \ldots, N. \tag{2}$$

Thus H_i is given in the dependency of the allocated capacity for the UGS, rtPS and ertPS services. Using (1) and (2) leads to the range of H_i as

$$H_i^{min} \leq H_i \leq H_i^{max}, \quad \text{where} \tag{3}$$
$$H_i^{min} = C_i - C_i^u - R_i^{max} \geq 1 \text{ and } H_i^{max} = C_i - C_i^u - R_i^{min}, \quad i = 1, \ldots, N.$$

Expression (3) shows that the capacity allocation for the nrtPS traffic is given by an upper limited discrete-time random variable, whose value is at least one. This ensures that the nrtPS traffic can not be blocked by the UGS, rtPS and ertPS traffic flows.

Finally the BE service flow utilizes the capacity, which is not used by the nrtPS traffic. This ensures an efficient capacity utilizing.

Thus this capacity allocation scheme enables asymmetric capacity allocation for the UGS, rtPS and ertPS services and asymmetric nrtPS traffic flows.

3.4 Uplink Scheduling

A BW-Req sent by an SS i represents the request for all nrtPS packets, which are accumulated in its outgoing buffer in the last cycle, i.e., since its last BW-Req sending. We leave the process of bandwidth requesting for rtPS and ertPS packets out of scope of this paper. We assume that the BS knows the number of rtPS and ertPS packets of SS i in each frame R_i and thus it can take them into account for calculating the actual available capacity for the nrtPS packets H_i. We note that the actual uplink transmission requirements represented by the rtPS and ertPS requests are always granted, since they are below the assigned capacity.

The disjunct partition of the total uplink capacity among the SSs enables mutually independent uplink scheduling for the nrtPS service flows of the individual SSs. Thus for the treatment of the BW-Req with request for the nrtPS packets, for each SS the BS maintains an individual BS grant buffer with infinite capacity. Let i-*polling slot* stands for the $(((i-1) \bmod P) + 1)$-th polling slot in the reservation interval of the frame, in which the group of SS i is polled. At the end of the i-polling slot the BS immediately handles the requests for the nrtPS packets from SS i, if any, and serves the individual BS grant buffer of SS i. We refer to the end of the i-polling slot as i-*reservation epoch*. The BS grant buffer of SS i is served also at the epochs following an i-*reservation epoch* by $T_f, 2T_f, \ldots, (L-1)T_f$ time. Hence all these epochs, including also the i-*reservation epochs*, are called i-*scheduling epochs*.

At receiving a request for the nrtPS packets from SS i at an i-*reservation epoch*, an individual BS grant is assigned to each nrtPS data packet of that request and then these BS grants are placed into the corresponding individual BS grant buffer of SS i according to their order in the request. Let the number of BS grants in the buffer of SS i be $S_i = 0, 1, \ldots$. During the service of the individual BS grant buffer of SS i at an i-*scheduling epoch* the BS takes the available BS grants from that buffer up to the available capacity for the nrtPS service flow of SS i (H_i) and schedules them for transmitting in the UL-MAP of the next frame. Let Y_i denote the number of these scheduled BS grants. Thus the number of i-packets transmitted in the next frame equals Y_i, which is given by

$$Y_i = min(S_i, H_i),$$

where $min(a, b)$ stands for the smallest value of set (a, b).

The BS uplink scheduling is illustrated in Figure 2.

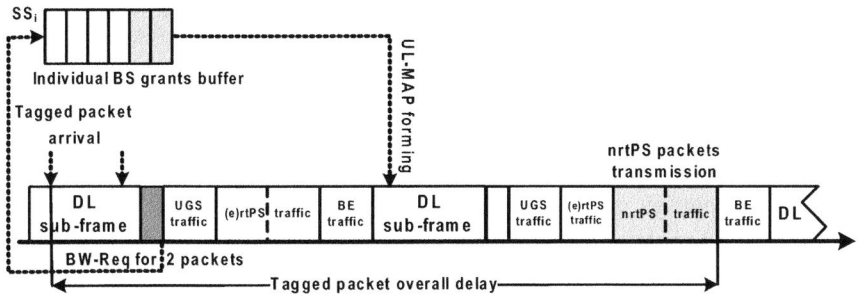

Fig. 2. BS uplink scheduling for a single SS

3.5 Model Assumptions

In statistical equilibrium the mean number of transmitted nrtPS packets equals the mean number of arriving nrtPS packets in a frame at each SS. This yields

$$E[Y_i] = \lambda_i T_f, \quad i = 1, \ldots, N. \tag{4}$$

The number of transmitted nrtPS packets is upper limited by the capacity allocated for them:

$$Y_i \leq H_i, \quad i = 1, \ldots, N. \tag{5}$$

Now we formulate the assumptions of our model:

A.1 Using (4), (5) and (2) implies that the following relation holds for the arrival rate of each SS i below the stability boundary:

$$\lambda_i T_f < C_i - C_i^u - E[R_i], \quad i = 1, \ldots, N. \tag{6}$$

This relation ensures the stability of the model.

A.2 The time of BS uplink scheduling is negligible.

A.3 The channel propagation time is negligible.

A.4 The transmission channels are error-free.

4 Queueing Analysis

The individual polling slot for each SS in a cycle and the independent uplink scheduling for the individual SSs together implies that the statistical behavior of the BS grant buffer of a particular SS is independent of the behavior of those of the other SSs. Therefore we model the stochastic behavior of the BS grant buffer of a particular SS by an individual queueing system.

In this queueing system the BS grants arrive to the BS grant buffer of SS i at i-*reservation epochs* and they are served at i-*scheduling epochs*.

4.1 The Content of the BS Grant Buffer at i-Reservation Epochs

Let $N_i(\ell)$ be the number of BS grants in the BS grant buffer of SS i at the ℓ-th i-*reservation epoch* for $\ell > 0$. The sequence $\{N_i(\ell), \ell > 0\}$ is an embedded Markov chain on the state space $\{0, 1, \ldots\}$. Let $[\boldsymbol{\Pi}_i]_{j,k}$ denote the probability of transition from state j to state k in the Markov chain and it is the (j, k)-th element of the $\infty \times \infty$ probability transition matrix $\boldsymbol{\Pi}_i$.

Let $H_i^{(m)}$ be the accumulated available capacity for the i-packets during m consecutive frames for $m = 0, \ldots, L$. The distribution of $H_i^{(m)}$ is given as the m times convolution of the distribution of H_i for $m = 1, \ldots, L$. The definition of $H_i^{(0)}$ implies that it takes the value 0 with probability 1. It follows that the minimum and maximum values of $H_i^{(m)}$ are mH_i^{min} and mH_i^{max}, respectively.

Let us consider the transition from state j to state k in the above defined Markov chain. The probability that the actual accumulated available capacity for the i-packets during a cycle is n equals $P(H_i^{(L)} = n)$. Assuming that $j \geq n$ the number of remaining BS grants in the BS grant buffer of SS i after its services during a cycle is $j - n$, which implies that $k \geq j - n$. Thus on one hand $n \geq j - k$ must hold and on the other hand $k - j + n$ i-packet arrivals occur during this transition. Hence this case contributes to $[\boldsymbol{\Pi}_i]_{j,k}$ with the probability

$$\sum_{n=j-k}^{j} P(H_i^{(L)} = n) \frac{(\lambda_i LT_f)^{k-j+n}}{(k-j+n)!} e^{-\lambda_i LT_f}.$$

Now assuming that $j + 1 \leq n$ implies that all the j BS grants are served during the cycle and thus k i-packet arrivals occur during this transition. Thus the contribution of this case to $[\boldsymbol{\Pi}_i]_{j,k}$ is the probability

$$\sum_{n=j+1}^{LH_i^{max}} P(H_i^{(L)} = n) \frac{(\lambda_i LT_f)^{k}}{(k)!} e^{-\lambda_i LT_f}.$$

Taking also into account the lower and upper limits of $H_i^{(L)}$ the transition probability $[\boldsymbol{\Pi}_i]_{j,k}$ can be expressed as

$$[\boldsymbol{\Pi}_i]_{j,k} = \sum_{n=max(LH_i^{min},j-k)}^{min(LH_i^{max},j)} P(H_i^{(L)} = n) \frac{(\lambda_i LT_f)^{k-j+n}}{(k-j+n)!} e^{-\lambda_i LT_f}$$

$$+ \sum_{n=max(LH_i^{min},j+1)}^{LH_i^{max}} P(H_i^{(L)} = n) \frac{(\lambda_i LT_f)^{k}}{(k)!} e^{-\lambda_i LT_f}, \quad j \geq 0. \quad (7)$$

where $max(a,b)$ stands for the largest value of set (a,b).

Let $[\boldsymbol{\pi}_i]_k$ denote the equilibrium probability of the state k in the Markov chain and it is the (k)-th element of the $1 \times \infty$ probability vector $\boldsymbol{\pi}_i$. Furthermore let $\mathbf{e}$ be the column vector having all elements equal to one.

Then the equilibrium probabilities of the Markov chain can be uniquely determined from the following system of linear equations:

$$\boldsymbol{\pi}_i \boldsymbol{\Pi}_i = \boldsymbol{\pi}_i, \quad \boldsymbol{\pi}_i \mathbf{e} = 1. \quad (8)$$

To keep the computation tractable an upper limit $K_i > H_i^{min}$ is set on the states, which results in finite number of unknowns and equations in the system of linear equations. An appropriate value of K_i depends on the required precision level, at which the probabilities $[\boldsymbol{\pi}_i]_k$ for $k > K_i$ can be neglected. These probabilities, $[\boldsymbol{\pi}_i]_k$ for $k > K_i$, are set 0.

4.2   The Content of the BS Grant Buffer at i-Scheduling Epochs

Let $[\boldsymbol{\pi}_i^+]_k$ denote the probability that the number of i-packets in the BS grant buffer of SS i at an arbitrary chosen i-scheduling epoch is exactly k and it is the (k)-th element of the $1 \times \infty$ probability vector π_i^+ for $k = 0, \ldots, K_i$. The probability that an arbitrary chosen i-*scheduling epoch* is the m-th after the last i-*reservation epoch* is $\frac{1}{L}$ for $m = 0, \ldots, L - 1$. Note that by definition the 0-th i-*scheduling epoch* after the last i-*reservation epoch* is that i-*reservation epoch*. At the m-th i-*scheduling epoch* after the last i-*reservation epoch* the i-packets in the BS grant buffer of SS i are those which remained after the last m services of the BS grant buffer. Hence the probability $[\pi_i^+]_k$ can be given as

$$[\pi_i^+]_k = \sum_{m=0}^{L-1} \frac{1}{L} \sum_{n=mH_i^{min}}^{mH_i^{max}} P(H_i^{(m)} = n)[\pi_i]_{n+k}, \quad 0 < k \le K_i, \quad \text{and}$$

$$[\pi_i^+]_0 = \sum_{m=0}^{L-1} \frac{1}{L} \sum_{n=mH_i^{min}}^{mH_i^{max}} P(H_i^{(m)} = n)\sum_{j=0}^{n}[\pi_i]_j. \tag{9}$$

4.3   The Content of the BS Grant Buffer at an Arbitrary Epoch

At an arbitrary epoch between two consecutive i-*scheduling epoch* the i-packets in the BS grant buffer of SS i are those which remained after the service of the BS grant buffer at the last i-*scheduling epoch*. Hence the probability of being exactly k packets in the BS grant buffer of SS i at an arbitrary epoch, p_k, is given by

$$p_k = \sum_{n=H_i^{min}}^{H_i^{max}} P(H_i = n)[\pi_i^+]_{n+k}, \quad 0 < k \le K_i - H_i^{min}, \quad \text{and}$$

$$p_0 = \sum_{n=H_i^{min}}^{H_i^{max}} P(H_i = n)\sum_{j=0}^{n}[\pi_i^+]_j. \tag{10}$$

4.4   The Size of the Transmitted i-Packet Batch

Let us consider the probability of transmitting exactly n i-packets in a frame for $0 \le n \le min(H_i^{max}, K_i)$. This can occur in two cases. In the first one the actual available capacity for the i-packets is exactly n and there are at least n BS grants in the BS grant buffer of SS i at i-scheduling epoch. The probability of this case is

$$\sum_{k=n}^{K_i} P(H_i = n)[\pi_i^+]_k.$$

In the other case the number of BS grants in the BS grant buffer of SS i at i-scheduling epoch is n, but the actual available capacity for the i-packets, k is greater than n. This has the following probability:

$$\sum_{k=n+1}^{H_i^{max}} P(H_i = k)[\pi_i^+]_n.$$

Taking also into account the lower limit of H_i the probability of transmitting exactly n i-packets in a frame can be expressed as

$$P(Y_i = n) = \sum_{k=n}^{K_i} P(H_i = n)[\pi_i^+]_k + \sum_{k=max(H_i^{min}, n+1)}^{H_i^{max}} P(H_i = k)[\pi_i^+]_n,$$

$$0 \leq n \leq min(H_i^{max}, K_i). \qquad (11)$$

5 Overall Delay Analysis

5.1 Overall Delay Definition

We define the *overall delay* (W_i) of the tagged i-packet as the time interval spent from its arrival into the outgoing buffer of SS i up to the end of its successful transmission in the UL. It is composed of several parts:

$$W_i = W_i^r + \alpha + W_i^s + W_i^t + \tau. \qquad (12)$$

Here W_i^r is the reservation delay, which is defined as the time interval from the i-packet arrival to SS i until the start of sending a corresponding BW-Req to the BS. α is the transmission time of a BW-Req. We define the *grant time of the tagged i-packet* as the *i-scheduling epoch* in the frame preceding the one, in which the tagged i-packet is transmitted. W_i^s is the scheduling delay, which is defined as the time interval from the end of sending a BW-Req of the tagged i-packet to its grant time. W_i^t is the transmission delay, which is defined as the time interval from the grant time of the tagged i-packet to the start of its successful transmission in the UL sub-frame. τ is the transmission time of an i-packet.

5.2 Reservation Delay

A bandwidth request can be sent for the nrtPS packets from SS i in the *i-polling slot* of every cycle. Thus an arriving i-packet waits for the reservation request until the end of the actual cycle and hence the mean reservation delay is given by

$$E[W_i^r] = \frac{LT_f}{2}. \qquad (13)$$

5.3 Scheduling Delay

The definition of the scheduling delay implies that the scheduling delay of the tagged i-packet is exactly the sojourn time of the BS grant assigned to the tagged i-packet in the BS grant buffer of SS i. Consequently the mean scheduling delay can be determined by applying Little's law on the mean number of i-packets in the BS grant buffer of SS i at an arbitrary epoch. Taking also into account the tractable computation of π_i the mean scheduling delay can be expressed as

$$E[W_i^s] = \frac{\sum_{k=1}^{\infty} k \, p_k}{\lambda_i} \approx \frac{\sum_{k=1}^{K_i - H_i^{min}} k \, p_k}{\lambda_i}. \qquad (14)$$

5.4 Transmission Delay

The transmission delay is a sum of the fixed time from the grant time of the tagged i-packet to the start of transmission of the i-packets in the next frame and the transmission times of the random number of i-packets preceding the tagged i-packet. Let y_i and $y_i^{(2)}$ be the first two factorial moments of the number of i-packets transmitted in a frame. The mean number of i-packets preceding the tagged i-packet is $\frac{y_i^{(2)}}{2y_i}$ (see [11]). Using it, the definitions of the first two factorial moments and taking into account the range of Y_i the mean transmission delay can be expressed as

$$
\begin{aligned}
E[W_i^t] = \; & T_f - \alpha\left(((i-1) \bmod P) + 1\right) + P\alpha + \sum_{k=1}^{i-1} C_k \tau + C_i^u \tau + E[R_i]\tau \\
+ \; & \frac{y_i^{(2)}}{2y_i}\tau = T_f + \alpha\left(P - ((i-1) \bmod P) - 1\right) + \sum_{k=1}^{i-1} C_k \tau + C_i^u \tau + E[R_i]\tau \\
+ \; & \frac{\sum_{k=2}^{min(H_i^{max}, K_i)} P(Y_i = k)k(k-1)}{2\sum_{k=1}^{min(H_i^{max}, K_i)} P(Y_i = k)k}\tau.
\end{aligned}
\tag{15}
$$

5.5 Mean Overall Delay

Taking the mean of (12) and substituting the expressions (13), (14) and (15) we obtain the expression of the mean overall delay as

$$
\begin{aligned}
E[W_i] = \; & \frac{L+2}{2}T_f + \frac{\sum_{k=1}^{K_i - H_i^{min}} k\, p_k}{\lambda_i} + \alpha\left(P - ((i-1) \bmod P)\right) \\
+ \; & \left(1 + \sum_{k=1}^{i-1} C_k\right)\tau + C_i^u \tau + E[R_i]\tau + \frac{\sum_{k=2}^{min(H_i^{max}, K_i)} P(Y_i = k)k(k-1)}{2\sum_{k=1}^{min(H_i^{max}, K_i)} P(Y_i = k)k}\tau.
\end{aligned}
\tag{16}
$$

6 Simulation Results

In order to validate the considered analytical model a simulation program for IEEE 802.16 MAC was developed. The program is an event-driven simulator that accounts for the discussed restrictions on the considered system model (see Section 3). For the sake of simplicity we restrict numerical results to the case of the symmetric arrival flows with the constant arrival rate λ for all the SSs. We also set constant capacity-related parameters C, C^u, R^{min} and R^{max} (see Subsection 3.3) thus avoiding explicit dependence on the SS number i. We illustrate the simplest case of the actual rtPS and ertPS capacity distribution, that is, uniform in the range $[R^{min}, R^{max}]$. The values for the other simulation parameters of IEEE 802.16 MAC and PHY are taken from [12]. The simulation parameters are summarized in Table 1.

Table 1. Basic IEEE 802.16 simulation parameters

Parameter	Value
PHY layer	OFDM
Frame duration (T_f)	$5\ ms$
DL/UL ratio	0:1
Channel bandwidth	$7\ MHz$
MCS	16 QAM $^3/_4$
Packet length	$256\ bytes$
BW-Req duration (α)	$0.17\ ms$
Total SS capacity per frame (C)	$5\ packets$
UGS capacity per frame (C^u)	$1\ packet$
Minimum (e)rtPS capacity per frame (R^{min})	$1\ packet$
Maximum (e)rtPS capacity per frame (R^{max})	$3\ packets$

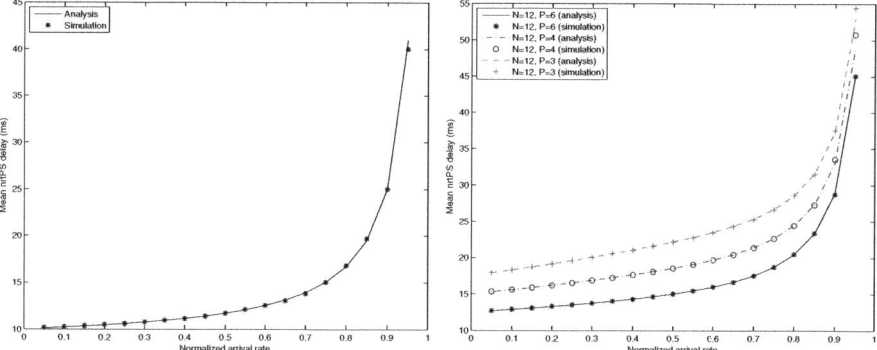

Fig. 3. Overall nrtPS delay depending on the arrival rate $N = 6$ (left side) and $N = 12$ with SS groups (right side)

The plots in Figure 3 compare the obtained simulation results with the analytical results given by the expression (16) for the above parameter settings. The plots show very good accordance of both analysis and simulation, for the cases both with and without SS groups, which verifies the constructed analytical model.

Besides of evaluating the mean overall delay the presented analytical model can be also used in network control for tuning the capacity parameters of the wireless system to the requirements according to given delay constraints.

References

1. IEEE 802.16-2009, Part 16: Air Interface for Broadband Wireless Access Systems, Standard for Local and Metropolitan Area Networks (May 2009)
2. IEEE 802.16.2-2004: IEEE Recommended Practice for Local and Metropolitan Area Networks – Coexistence of Fixed Broadband Wireless Access Systems (March 2004)

3. WiMAX Forum, home page, http://www.wimaxforum.org/
4. Paschos, G.S., Papapanagiotou, I., Argyropoulos, C.G., Kotsopoulos, S.A.: A heuristic strategy for IEEE 802.16 WiMAX scheduler for quality of service. In: 45th Congress FITCE (2006)
5. de Moraes, L.F.M., Maciel, P.D.: A variable priorities MAC protocol for broadband wireless access with improved channel utilization among stations. In: Int. Telecomm. Symp., vol. 1, pp. 398–403 (2006)
6. Chang, Y.J., Chien, F.T., Kuo, C.C.J.: Delay analysis and comparison of OFDM-TDMA and OFDMA under IEEE 802.16 QoS framework. In: IEEE Global Telecomm. Conf (GLOBECOM), vol. 1, pp. 1–6 (2006)
7. Rubin, I.: Access-control disciplines for multi-access communication channels: Reservation and TDMA schemes. IEEE Trans. Inf. Theory 25(5), 516–536 (1979)
8. Iyengar, R., Iyer, P., Sikdar, B.: Delay analysis of 802.16 based last mile wireless networks. In: IEEE Global Telecomm. Conf (GLOBECOM), vol. 5, pp. 3123–3127 (2005)
9. Saffer, Z., Andreev, S.: Delay analysis of IEEE 802.16 wireless metropolitan area network. In: Int. Workshop on Multiple Access Comm, MACOM (2008)
10. Andreev, S., Saffer, Z., Anisimov, A.: Overall delay analysis of IEEE 802.16 network. In: Int. Workshop on Multiple Access Comm, MACOM (2009)
11. Chaudhry, M.L., Templeton, J.G.C.: A First Course in Bulk Queues. John Wiley and Sons, New York (1983)
12. Sivchenko, D., Bayer, N., Xu, B., Rakocevic, V., Habermann, J.: Internet traffic performance in IEEE 802.16 networks. In: 12th European Wireless Conf. (2006)

Cross-Layer Channel-Aware Approaches for Modern Wireless Networks

Sergey Andreev[1], Olga Galinina[2], and Alexey Vinel[3]

[1] Tampere University of Technology (TUT), Finland
sergey.andreev@tut.fi
[2] Speech Technology Center, Russia
olga.galinina@gmail.com
[3] Saint-Petersburg Institute for Informatics and Automation
Russian Academy of Sciences (SPIIRAS), Russia
vinel@ieee.org

Abstract. In this paper we consider the most recent channel-aware performance optimization approaches primarily for cellular wireless networks. This survey is a summary of several core research works on the effective design and implementation principles of the advanced cross-layer resource allocation and link adaptation schemes. Following the original papers, we emphasize the three key performance metrics: spectral efficiency, energy efficiency and quality-of-service perception of wireless clients. Further, we present a systematic overview of relevant literature covering the main challenges and open problems of existing optimization techniques. We conclude that the state-of-the-art approaches should be extended further to take into account additional optimization parameters and aim at higher degree of control over trade-offs associated with the key metrics.

Keywords: spectral and energy efficiency, QoS provisioning, cross-layer channel-aware optimization, wireless cellular networks.

1 Introduction and Motivation

1.1 General Background

Wireless networks demonstrate worldwide proliferation, which has advanced recently with the introduction of the novel communication technologies [1], [2]. However, the future success of the wireless communication significantly depends on the solution to overcome the disproportion between the requested *quality of service* (QoS) and limited network resources.

Spectrum is a natural resource that cannot be replenished. As such, the need for its effective use introduces the problem of *spectral efficiency* (SE). On the other hand, *energy efficiency* (EE) is also becoming increasingly important primarily for small form factor mobile devices due to the growing gap between the available and the required battery capacity, which is demanded by the ubiquitous multimedia applications [32].

A. Vinel et al. (Eds.): MACOM 2010, LNCS 6235, pp. 163–179, 2010.

For the above reasons, resource allocation and management becomes critical for technologies, where multiple clients share the limited wireless spectral resources. Currently, the layered architecture dominates in networking design and each layer is operated independently to maintain transparency. Among these layers, the *physical* layer is responsible for the raw-bit transmission, whereas the *medium access control* layer arbitrates access of clients to the shared wireless resources.

However, wireless channels are commonly known to suffer from multipath fading. To make matters worse, the statistical channel characteristics of different clients are typically different. Therefore, traditional layer-wise architecture turns out to be inflexible and results in the inefficient wireless resource utilization. An integrated and adaptive design across different layers is thus required to overcome this limitation. As a consequence, *cross-layer* optimization across the physical and the medium access control layers is desired for wireless resource allocation and packet scheduling [66].

To enable cross-layer optimization, *channel-aware* approaches are introduced and developed to explicitly take into account wireless *channel state information* (CSI). Typically, a channel-aware technique adapts data transmission and dynamically allocates resources to ensure that a client with favorable channel conditions transmits a packet. Taking advantage of the independent channel variation across clients, channel-aware approaches are shown to substantially improve the network performance through multiuser diversity, whose gain increases with the number of clients [31].

This survey is aimed at indicating the key challenges and open problems of existing channel-aware techniques to improve spectral efficiency, energy efficiency, and quality of service perception of the wireless clients. It is basically a summary of [70], [45], and [27] providing background on cross-layer approaches for the most advanced wireless networking technologies, such as [1], [2], within one paper.

1.2   Physical Layer

The physical (PHY) layer is of primary importance in wireless communications due to the challenging nature of the communication medium. It focuses on raw-bit transmission over wireless channels and incorporates radio frequency (RF) circuits, modulation and coding schemes, power control algorithms, and other elements. Conventional wireless systems are typically built to transmit data on a fixed set of operating points [8], that is, without flexible power adaptation. This often results in excessive energy consumption or pessimistic data rate for peak channel conditions [47]. Hence, a set of PHY parameters should be flexibly adjusted to account for the actual client QoS requirements and for the state of the wireless medium to trade off energy and spectral efficiency.

As wireless medium is shared, the communication performance and energy consumption are affected not only by the layers comprising the point-to-point communication link, but also by the interaction between the individual links across the entire network. The above necessitates a more complex system approach. Furthermore, *orthogonal frequency division multiplexing* (OFDM) becomes a primary modulation scheme for next-generation wireless standards [1], [2]. From a resource management perspective, multiple channels in OFDM systems have the potential for more efficient medium access control design since sub-carriers may be dynamically assigned to different clients [16], [67]. To further improve the performance, adaptive power allocation on each sub-carrier may be applied [77], [68].

1.3   Medium Access Control Layer

The medium access control (MAC) layer ensures that wireless resources are efficiently allocated to maximize network-wide performance metrics, while maintaining individual client QoS requirements. Here, pessimistic medium access strategies that manage wireless resources to assure worst-case QoS may degrade network spectral and energy efficiency. Typically, medium access algorithms can be either centralized or distributed.

In distributed access schemes, MAC should be improved to reduce the number of wasted transmissions that are corrupted by the interference from other network clients, whereas in centralized access schemes efficient scheduling algorithms should exploit the variations across clients to maximize the overall network performance [46]. The MAC layer manages wireless resources for the PHY layer and they together directly impact overall network performance and energy consumption.

1.4   Cross-Layer Approaches

Clearly, spectral and energy efficiency are affected by all components of system design, ranging from RF circuits to applications. As mentioned earlier, the traditional layer-wise approach leads to independent design of different layers and may result in sub-optimal network performance. By contrast, cross-layer approaches exploit interactions between different layers and may significantly improve system performance as well as adaptability to service, traffic, and environment dynamics [70], [45], [27]. Cross-layer optimization for throughput improvement has been a popular research direction [71]. However, as wireless clients become increasingly mobile, the focus of recent efforts tends to shift toward energy consumption at all layers of communication systems [70], [45], [27], from architectures [10] to algorithms [64].

Since wireless channels are shared and highly dynamic, resource management is believed to be the most challenging element in the design of channel-aware systems. A scheduler to perform adaptive resource management should account for at least three primary performance metrics: overall system capacity (or spectral efficiency), energy consumption (or energy efficiency) of wireless clients, and their quality of service perception. It is also desirable to have a high degree of control over trade-offs associated with these metrics. Below we consider each of these important metrics and related trade-offs in more detail.

2   Spectral Efficiency

2.1   Background

Because of fading, the characteristics of a wireless channel vary with time, frequency, and client. In sub-section 1.2 we stressed that wireless channel is a shared medium and, as such, the communication performance depends both on individual link and, more importantly, on the interaction between the individual links across the entire network. Accounting for this fact, channel-aware medium access schemes have been proposed to adaptively transmit data and dynamically assign wireless resources based on CSI [47], [48]. With these schemes the wireless system spectral efficiency may be substantially improved [46].

The main principle of cross-layer channel-aware MAC is to schedule a client with favorable channel conditions to transmit with optimized link adaptation according to CSI [67], [68]. As mentioned above, medium access can be either centralized or distributed and we consider each option separately.

2.2   Distributed Medium Access

Random multiple access (RMA) algorithms provide the means to share network resources among clients subject to distributed control. Traditional contention-based RMA methods include pure, slotted, and reservation Aloha schemes, as well as carrier sense multiple access (CSMA) and CSMA with collision avoidance schemes, multiple access with collision avoidance for wireless [11] technique, and many others. Unfortunately, these MAC approaches do not use CSI explicitly. Hence, when MAC decides to transmit a frame, the wireless channel may be in a deep fade [45]. On the other hand, MAC may not transmit even though the client link is in a favorable state, which wastes channel resources.

Recently, so-called *opportunistic* RMA schemes have been studied in [21], [51], [9], [3], etc. to use CSI for performance improvement. With opportunistic random access, each client is aware of its own CSI and accounts for it during the contention behavior. Thus, clients with better channel conditions have higher contention probabilities and transmit more frequently. It is important to emphasize that all these opportunistic RMA approaches consider wireless networks where clients transmit to a common receiver, e.g., a base station. However, this popular scenario does not fit many practical wireless communication environments, such as sensor [4], ad-hoc [14], and mesh networks [5], which require separate consideration.

2.3   Centralized Medium Access

With a central controller, the best performance is known to be obtained by scheduling the client with the best channel conditions [67], [68]. However, CSI feedback to dynamically determine such a client sometimes incurs huge overhead, especially for densely-populated mobile networks, which results in poor network scalability. To reduce CSI feedback, distributed approaches are sometimes preferable. However, the use of a distributed MAC protocol is sometimes prohibited by the network topology.

Recently, the principles of the MAC design have evolved from the conventional point-to-point view to a multiuser network view. Special attention is paid to the fact that time-varying fading is a unique characteristic of a wireless channel. Previously, for a point-to-point link, using adaptive modulation and coding, the transmitter could send more data at a higher transmission data rate, when the channel quality is high [50]. However, the spectral efficiency remained unsatisfactory during the periods of deep fade. Currently, multiuser diversity has received much attention. Consequently, channel-aware dynamic packet scheduling was applied initially for *code division multiple access* (CDMA) systems [73].

The first results regarding multiuser diversity indicate that just the use of a simple channel-aware scheduler can significantly improve network spectral efficiency. Clearly, multiuser diversity follows from the independent channel variation across clients. With the increasing number of clients in a wireless network, the packets are likely to be transmitted at high data rates since different clients experience independent fading fluctuations. From a client perspective, packets are transmitted stochastically in the system using channel-aware scheduling, which is currently known as opportunistic communications [36].

2.4 Interference-Limited Scenario

Modern wireless networks, especially those with cellular topology [1], [2], are becoming increasingly interference-limited as more clients share the same spectrum to receive high-rate multimedia service. In modern cellular systems, *co-channel interference* (CCI) is expected to become one of the major performance-limiting factors, especially as these systems shift toward aggressive frequency reuse scenarios. While the overall spectral efficiency may indeed improve with aggressive frequency reuse, the performance of cell-edge clients degrades dramatically.

A popular CCI mitigation technique is to assign different sets of channels to neighboring cells [13] and a good summary of channel assignment can be found in [72]. In particular, a relatively novel approach to reducing interference for cell-edge clients is through *fractional frequency reuse* (FFR) [52]. With FFR, partial frequency reuse is specified for clients at cell edges, whereas full frequency reuse is applied for those at cell centers. Consequently, the throughput of cell-edge clients increases since they experience lower levels of interference.

To further improve the spectral efficiency with frequency reuse, CCI can be mitigated by advanced digital signal processing techniques [6]. However, these techniques are typically complex and therefore result in prohibitive implementation costs for mobile client devices. For downlink transmission, CCI can be reduced by joint encoding techniques among base stations [80], or completely avoided by using cooperative scheduling [15], both of which require exchange of additional instantaneous information. Recently, contention-based schemes have also been developed for CCI avoidance in addition to an intracellular centralized MAC protocol [54] to effectively combat CCI.

3   Energy Efficiency

3.1   Background

Energy efficiency is increasingly important for modern wireless networks due to the limited battery resources of mobile clients. Preventive measures are necessary to control the growing gap between the available and the required battery capacity, which is demanded by the ubiquitous multimedia applications [32]. For maximizing energy efficiency, so-called "bits-per-Joule" [63] or "throughput-per-Joule" [46] metrics are often considered. Several approaches exist to address energy efficiency, which include water-filling power allocation schemes that maximize throughput subject to a fixed overall transmit power constraint [67], [68], and adaptation of both overall transmit power and its allocation, according to the CSI [47], [48].

Again we emphasize the important fact mentioned in sub-section 1.2 that the client energy efficiency is affected not only by the layers of the point-to-point communication link, but also by the interaction between the individual links. As such, a cross-layer approach, including both transmission and multiuser resource management, is required for energy-efficient wireless communications. Additionally, energy-efficient techniques have the desirable benefit of reducing interference to other co-channel clients.

Information theorists have studied energy-efficient transmission for at least two decades [20]. Summarizing their efforts, for any transmission rate below the capacity per unit energy, error probability decreases exponentially with the total energy. So far, it was shown that the capacity per unit energy is achieved using an unlimited number of degrees of freedom per information bit, e.g., with infinite bandwidth [75] or long-duration regime communications [40]. As both are difficult to achieve in a realistic wireless network, more practical approaches to increase energy efficiency are addressed below.

3.2   Link Adaptation

As the quality of wireless channel varies with time, frequency, and client, link adaptation can be used to improve transmission performance. With link adaptation, modulation order, coding rate, and transmit power can be selected according to CSI. Earlier research on link adaptation focuses on power allocation to improve individual channel capacity [35], whereas state-of-the-art approaches emphasize the need for joint link adaptation and resource allocation [46].

More specifically, since channel frequency responses are different at different frequencies and for different clients, data rate adaptation over each sub-carrier, dynamic sub-carrier assignment, and adaptive power allocation can significantly improve the performance of OFDM networks. Through data rate adaptation, the transmitter can use higher transmission rates and reduced power consumption over the sub-carriers with better conditions so as to improve throughput and simultaneously to ensure an acceptable bit-error rate at each sub-carrier [76]. Despite the use of data rate adaptation [50], deep fading on some sub-carriers still leads to low channel capacity.

The information-theoretic results derived in [75] and [40] focus only on transmit power when considering energy consumption during transmission. Typically, a device will incur additional *circuit power*, which is relatively independent of the transmission rate [76], [18]. Thus, the circuit power consumption should be accounted for explicitly in maximizing energy efficiency. Consequently, the known method to transmit with the longest duration is not anymore the best since circuit energy consumption increases with transmission duration. Considering the impact of circuit power, the focus shifts toward using optimization theory for determining energy-optimal link settings [47].

Energy-efficient transmission problem is thus formulated as a trade-off among transmission energy, circuit energy, and transmission time [18]. Similarly, one may perform search of the optimal rate that minimizes the average power consumption subject to a constraint on average throughput [65]. Although power optimization plays a pivotal role in both interference management and energy utilization, little research has addressed joint interaction of link adaptation and resource allocation. An implicit discussion can be found in [42], which summarizes existing approaches that address either throughput or energy efficiency separately in the context of power control for CDMA networks. Other works address this joint limitation and investigate energy-efficient power optimization for OFDM communications [46], [47], [48].

3.3   Resource Allocation

Due to limited wireless resources, intricate performance trade-offs arise between an individual client and the entire network. The use of diversity across clients is likely to further reduce overall network energy consumption. Wireless resources can be managed in different domains to improve network energy efficiency. In this sub-section we consider *time* and *frequency* domains, whereas *spatial* domain is focused on in the following sub-section.

In the time domain, e.g., in a *time-division multiple access* (TDMA) network, the channel medium is shared through time division. Each client tends to extend its transmission time to save energy and thus contradicts the respective interests of other clients. Thus the allocation of transmission time across all clients is critical in determining network energy efficiency. To make the resource management scheme applicable, the scheduling is sometimes partitioned into a design-time phase and a run-time phase [39]. In the design-time phase, energy-performance representation can be derived for each client to capture the relevant trade-offs. In the run-time phase, a fast greedy algorithm may be used to tune the operating points to further improve energy efficiency.

While extensive efforts have been undertaken to improve energy-efficient resource management in time domain, little effort has been devoted to frequency domain. In the frequency domain, while increasing transmission bandwidth improves energy efficiency, the entire system bandwidth can not be allocated exclusively to one client in a multiuser system since this may hurt the energy efficiency of other clients, as well as that of the overall network [46]. Hence, frequency-domain resource management is critical in determining overall network energy efficiency. Frequency selectivity of broadband wireless channels further accentuates this necessity.

OFDM is known to divide an entire wireless channel into many orthogonal narrowband sub-channels (sub-carriers) to mitigate frequency-selective fading and to support higher data rate. Furthermore, in an OFDM-based wireless network, different sub-carriers can be allocated to different clients to provide a flexible MAC scheme and to exploit multiuser diversity. As channel characteristics for different client are almost mutually independent in multiuser environments [70], the sub-carriers experiencing deep fading for one client may be favorable for other clients. Therefore, each sub-carrier could be in a desirable condition for some clients in a multiuser OFDM wireless network. Thus, through dynamic sub-carriers assignment, the network can benefit from multiuser diversity.

Resource allocation issues and the achievable regions for multiple access and broadcast channels have been investigated in [34], which has proved that the largest data rate region is achieved when the same frequency range is shared with overlap by multiple clients in broadcast channels. However, when optimal power allocation is used, there is only a small range of frequency with overlapping power sharing [22]. Thus, optimal power allocation with dynamic sub-carrier assignment can achieve data transmission rate close to the channel capacity boundary.

3.4   Cooperative Networking

As more clients need to share the same spectrum for broadband multimedia communications and cellular networks move toward aggressive full-frequency reuse scenarios [1], [2], the performance of modern wireless networks is heavily impaired by interference. This motivates the use of multi-cell power control optimization for interference management [48].

Since wireless is broadcast, the transmission of one client interferes with that of neighboring clients and consequently reduces energy efficiency. However, clients can gain in energy efficiency, if cooperation among neighboring clients is allowed. Hence, spatial domain resource management is important to manage the behavior of clients at different spatial locations. On the other hand, cooperation requires additional signaling overhead and consumes extra energy. Cooperation can also cause transmission delay that may impact throughput adversely and thus hurt energy efficiency. However, delay can be exploited for energy-efficient link adaptation, as extending transmission duration may improve energy efficiency.

It has been demonstrated that significant energy savings can be achieved and they grow almost linearly with distance when either transmitter or receiver cooperation is allowed [17]. Furthermore, it is also shown that cooperation can even reduce delay within a certain transmission ranges since sometimes it enables higher order modulation and increases data rate. Similarly, when receiver cooperation is exploited, significant energy savings can be observed [24].

Besides transmitter and receiver cooperation, relay cooperation across neighboring clients is also effective in improving network energy efficiency. Since the energy for reliable data transmission grows exponentially with distance [72], it is more energy efficient to send data using several shorter intermediate hops than using a long hop, if the energy to compute the route is negligible [60]. However, client cooperation incurs delay and energy consumption of relay nodes. Therefore, in some scenarios, it is advantageous to use longer hops [23]. Hence, the optimal selection of relay nodes is a trade-off between source-node performance and relay cost to enhance overall network energy efficiency.

4   Quality-of-Service Perception

4.1   Background

Modern wireless networks are evolving to enable support for heterogeneous multimedia applications, e.g., at least best effort and QoS streaming traffic. Since the integration of different services within a single platform is expected to result in higher operator profits and at the same time to reduce network management costs, intensive research efforts have been invested into design principles of such networks [25], [43]. However, as wireless resources are limited and shared by clients experiencing time- and frequency-varying channels, service integration may be challenging [26], [44]. A key element in such systems is the packet scheduler, which typically needs to assure the satisfactory QoS level of clients. As discussed previously, such schedulers may be opportunistic, i.e., primarily serve clients with favorable channel condition. An attempt to combine opportunistic behavior, while meeting QoS requirements of wireless clients is a challenging research problem addressed by [12], [36], [37], [55], [56], and [57].

Most recent QoS-constrained opportunistic approaches for wireless cellular systems focus on *flow-level* performance and tend to consider stochastic traffic loads [30]. Particularly, new flows, either real-time sessions or file transfer requests, are initiated at random and leave the system after being served. As a consequence, the number of ongoing flows changes in time dynamically. This is usually referred to as the flow-level dynamics. Studying dynamic systems is helpful to better understand the performance of real-world systems, but it is also more difficult. As such, dynamic systems receive much less attention from the researchers than their static versions, i.e., with a fixed set of backlogged clients [27].

Each flow in a dynamic system is an abstraction of a stream of packets corresponding to a new file, web page, or real-time voice/video session. Poisson processes have initially been used to model flows generated by large population of independent clients, or new webpage download patterns from population of web browsing sessions [12]. Recently, flow-level models have been considered in studying statistical bandwidth sharing in wired networks [74]. In the context of wireless systems, it was observed that throughput seen by dynamic client population can be substantially different from that of a fixed number of clients [12]. Therefore, dynamic systems require more thorough treatment and some first promising steps in this direction are discussed below.

4.2   Opportunistic Resource management

Time- and frequency-varying channels are a distinguishing characteristic of wireless systems by contrast to wired systems. Throughout this paper we emphasized repeatedly that improper use of the channel variability may potentially result in poor system capacity. To prevent this from happening, channel-aware (opportunistic) scheduling was used, where system chooses to serve clients with favorable channel conditions. Opportunistic scheduling has been shown to substantially increase system capacity [36], [37] and thus constitutes a promising technique to achieve higher performance metrics.

However, unlike what is typically the case in wired systems, having more capacity does not necessarily implies better client-perceived QoS in an opportunistic system. This is due to the fact that the maximum achievable capacity in an opportunistic system is constrained by the individual QoS requirements of wireless clients [69]. Specifically, in order to sustain a minimum bandwidth requirement, it is sometimes necessary to sacrifice opportunism by scheduling clients with relatively bad channel state. The integration of heterogeneous multimedia applications may further limit the benefits of opportunism, which results in reduced stability region, degraded system throughput, and increased traffic delay. All these negative effects are referred to as a loss in opportunism due to integration.

Existing opportunistic scheduling schemes are mostly packet-level algorithms [36], [37], [55], and [56], which focus on static scenario with fixed client population and backlogged queues. However, this assumption is practical only in short time scales, where the client population does not change much. As such, existing static approaches may not capture the flow-level dynamics, in which new flows enter the system randomly and leave after being served.

In a realistic network, client population changes over time and it is sometimes important to evaluate time-averaged system performance metrics, such as mean throughput and mean traffic delay. The initial works, such as [12], did not consider the mixture of different traffic. Other attempts to integrate heterogeneous packet flows in wireless opportunistic systems have been recently done by [56]. However, these studies focused on packet-level performance only, whereas [30] addresses flow-level dynamics, but in a framework of the simplified system model.

4.3   Underutilization of Cellular Networks

Though future wireless systems promise to support higher capacity, this will typically be achieved for the cost of higher energy consumption resulting in shorter battery lifetimes for mobile clients. Therefore, improvement in energy efficiency has become, as we showed above, a critical and attractive research target. Unlike previous works on energy conservation in sensor and wireless local area networks [18], [19], [38], [58], and [61], new approaches address energy saving techniques for broadband cellular systems, e.g., WiMAX [1] or 3GPP-LTE [2]. Specifically, they focus on reduction in uplink RF transmission energy recognizing the fact that it is one of the main contributors to battery consumption [46].

It is important to highlight that wireless access networks are unlikely to be fully utilized all the time [29]. Indeed, as a result of time-varying, non-stationary and sometimes bursty loads, these networks are often 'over-designed' to be able to support a peak load condition, and, as such, are often underutilized. Thus, if a system has some spare capacity, which we will interpret as excess capacity relatively to a desired client-perceived performance [28], one may consider slowing down transmissions (so-called lazy scheduling [58]). This can be beneficial in terms of saving energy of mobile clients.

Due to the high variation in capacity that a wireless network can deliver to particular locations within its coverage area, we also expect high variability in the system load [7], [62]. Furthermore, in some cellular scenarios a substantial amount of bandwidth is set aside to ensure that client calls are not dropped during hand-offs [53]. This further contributes to the underutilization of the system, even when the loads are heavy and approach saturation. Therefore, underutilization of cellular networks should be accounted for explicitly by novel channel-aware techniques.

4.4   QoS-Aware Optimization

Earlier research on power adaptation mainly focused on controlling interference rather than reducing energy consumption [78]. Later, energy-efficient power control was first explored in the context of sensor networks [58]. The authors proposed lazy scheduling, where packets are transmitted as slowly as possible, while meeting packet delay constraints. This leads to significant energy savings.

The work in [61] and [41] further explores trade-offs between energy and delay under various scenarios. In fading environments, the use of opportunistic transmission to save energy was studied in [33] and [79]. However, the above work neglects circuit power, idling power and flow-level dynamics. Recent results show that if circuit power is taken into account, circuit energy consumption increases monotonically as the transmission time grows [18]. Thus, we cannot slow down the transmission rate arbitrarily, and indeed, there exists an energy-optimal transmission rate.

Solving this optimization problem, the work in [18] focuses on the physical modulation techniques for sensor networks. Cross-layer techniques are also proposed for small-scale sensor networks [19], wireless local area networks [11], [3], and further to the routing layer [38]. Energy-efficient transmission strategy for OFDM-based systems considering circuit power was proposed in [49]. However, previous work has addressed static saturated systems, and thus could not capture the coupling between power back-off and its impact on system dynamics. Advanced approaches address flow-level performance of the QoS-limited wireless systems [30]. However, they consider simplified TDMA system model and require further extension.

5   Open Problems and Conclusion

5.1   Future Work

Recent dynamic cross-layer channel-aware techniques are primarily based on TDMA system model [29] and thus somewhat simplify the challenges of a real-world wireless network, which is typically OFDM- or CDMA-based. However, these novel approaches may be suitable for a broader set of multiple access technologies, including CDMA and OFDM, and also could be extended to a *multi-cell scenario*. Another important observation is that advanced energy saving techniques effectively reduce the output power level of mobile clients, which is in turn beneficial to mitigate inter-cell interference. Consequently, even higher energy efficiency is expected.

The use of channel variation across clients so far has been extensively used in joint MAC-PHY design to improve spectral efficiency, energy efficiency, and QoS perception of wireless clients. Due to highly dynamical nature of modern wireless network, the issue of network *admission control* is becoming very challenging and important [29]. Existing work on efficiency, QoS, fairness, and stability of channel-aware techniques might benefit from accounting for this problem. The resulting complex research may result in both theoretical innovations and practical applications, as this topic may lead to rethinking the architecture of contemporary multimedia-over-wireless networks. Another important problem in multi-cell wireless networks is how to properly associate mobile clients with serving base stations. This problem is usually called *client association* [30].

Since opportunistic resource allocation schemes studied in recent works focus on centralized control, it would be interesting to extend existing research to channel- and QoS-aware multi-carrier *distributed MAC schemes* without a centralized controller. The related research should be based on deep understanding of specific properties of multi-carrier systems. Distributed channel- and QoS-aware MAC approaches are very promising for the following two major reasons [70]. Firstly, channel-aware-only MAC schemes are typically optimal for the total throughput rather than for diverse QoS requirements of different applications. Secondly, the queueing model of a single-carrier system is a single server with a queue. However, since multi-carrier systems can serve many clients simultaneously, there are multiple servers from a queueing-theoretical point of view. The consideration of multi-server systems would be advantageous to QoS provisioning.

5.2   Survey Summary

Throughout this survey we covered a wide range of problems associated with modern wireless networks: QoS awareness, energy savings, and efficient capacity use. We primarily focused on trade-offs among these three key metrics. The major target of respective complex research is to develop novel cross-layer channel-aware transmission and resource management algorithms to significantly improve client experience, system spectral efficiency, and energy efficiency. By exploiting channel state information of different clients, this approach leads to integrated algorithms, which utilize spectrum and energy resources both fairly and efficiently [45].

Summarizing, Shannon theory indicates that it is desirable to transmit a packet over a longer period of time to save transmit energy. However, when circuit energy, which is consumed by electronic components except amplifiers, is considered, this is no longer the case since the circuit energy increases with transmission duration [29]. Hence, optimal transmission duration, determined by the selected modulation order and power allocation, needs to be established to balance both transmit and circuit energy consumption. The frequency selectively in OFDM-based systems further complicates the problem, since different modulation orders and amounts of power can be applied on different sub-carriers.

Consequently, this approach requires deep and systematic study of cross-layer channel-aware techniques in modern wireless networks. It should not only propose a joint MAC-PHY wireless resource management architecture and corresponding scheduling algorithms that substantially improve the spectral efficiency, energy efficiency and effectively satisfy diverse performance objectives of heterogeneous traffic, but also provide deep understanding of fundamental mechanisms in advanced wireless resource management [27]. The joint study of spectral and energy efficiency, fairness, and stability would facilitate the design of future wireless multimedia networks that support diverse QoS requirements in a complicated environment, where multiple clients compete for shared channels with time-varying frequency-selective fading.

Acknowledgments

This work is supported by the Russian Foundation for Basic Research (projects #10-08-01071-a, #08-08-00403-a, #09-07-11004 and #10-08-90027) as well as by the Branch of Nano- and Information Technologies of Russian Academy of Sciences (project 2.3).

References

1. IEEE Std 802.16m, Amendment to IEEE Standard for Local and Metropolitan Area Networks – Part 16: Air Interface for Broadband Wireless Access Systems – Advanced Air Interface (IEEE Std 802.16-2009)
2. LTE Release 10 & beyond (LTE-Advanced)
3. Adireddy, S., Tong, L.: Optimal channel-aware Aloha protocol for random access in WLANs with multipacket reception and decentralized channel state information. IEEE Trans. Signal Proc. 56, 2575–2588 (2008)
4. Akyildiz, I., Weilian, S., Sankarasubramaniam, Y., Cayirci, E.: A survey on sensor networks. IEEE Commun. Magazine 40, 102–114 (2002)
5. Akyildiz, I., Wang, X.: A survey on wireless mesh networks. IEEE Commun. Magazine 43, S23–S30 (2005)
6. Andrews, J.: Interference cancellation for cellular systems: a contemporary overview. IEEE Wireless Commun. 12, 19–29 (2005)
7. Andrews, J., Ghosh, A., Muhamed, R.: Fundamentals of WiMAX. Prentice-Hall, Englewood Cliffs (2007)
8. Atheros Communications, White paper: Power consumption & energy efficiency (2003)
9. Bai, K., Zhang, J.: Opportunistic multichannel Aloha: distributed multi-access control scheme for OFDMA wireless networks. IEEE Trans. Veh. Tech. 55, 848–855 (2006)
10. Benini, L., Bogliolo, A., de Micheli, G.: A survey of design techniques for system-level dynamic power management. IEEE Trans. VLSI Syst. 8, 299–316 (2000)

11. Bhaghavan, V., Demers, A., Shenker, S., Zhang, L.: MACAW: A media access protocol for wireless LAN's. In: Proc. Sigcomm (October 1994)

12. Borst, S.: User-level performance of channel-aware scheduling algorithms in wireless data networks. In: Proc. IEEE INFOCOM, vol. 1, pp. 321–331 (2003)

13. Cao, G., Singhal, M.: An adaptive distributed channel allocation strategy for mobile cellular networks. J. Parallel and Dist. Comput. 60, 451–473 (2000)

14. Chlamtac, I., Conti, M., Liu, J.: Mobile ad hoc networking: imperatives and challenges. Ad Hoc Networks 1, 13–64 (2003)

15. Choi, W., Andrews, J.: Base station cooperatively scheduled transmission in a cellular MIMO TDMA system. In: Proc. Conf. Inf. Sci. Sys., pp. 105–110 (2006)

16. Chuang, J., Sollenberger, N.: Beyond 3G: Wideband wireless data access based on OFDM and dynamic packet assignment. IEEE Commun. Magazine, 78–87 (July 2000)

17. Cui, S., Goldsmith, A., Bahai, A.: Energy-efficiency of MIMO and cooperative MIMO techniques in sensor networks. IEEE J. Sel. Areas Commun. 22, 1089–1098 (2004)

18. Cui, S., Goldsmith, A., Bahai, A.: Energy-constrained modulation optimization. IEEE Trans. Wireless Commun. 4, 2349–2360 (2005)

19. Cui, S., Madan, R., Goldsmith, A., Lall, S.: Cross-layer energy and delay optimization in small-scale sensor networks. IEEE Trans. Wireless Commun. 6, 3688–3699 (2007)

20. Gallager, R.: Power limited channels: Coding, multiaccess, and spread spectrum. In: Proc. Conf. Inform. Sci. and Syst., vol. 1 (March 1988)

21. Ganesan, G., Song, G., Li, Y.: Asymptotic throughput analysis of distributed multichannel random access schemes. In: Proc. IEEE ICC, pp. 3637–3641 (2005)

22. Goldsmith, A., Effros, M.: The capacity region of broadcast channels with intersymbol interference and colored Gaussian noise. IEEE Trans. Inf. Theory 47, 219–240 (2001)

23. Haenggi, M., Puccinelli, D.: Routing in ad hoc networks: a case for long hops. IEEE Commun. Magazine 43, 112–119 (2005)

24. Jayaweera, S.: An energy-efficient virtual MIMO architecture based on V-BLAST processing for distributed wireless sensor networks. In: Proc. IEEE SECON, pp. 299–308 (October 2004)

25. Jiang, Z., Mason, H., Kim, B., Shankaranarayanan, N., Henry, P.: A subjective survey of user experience for data applications for future cellular wireless networks. In: Proc. of Symp. App. and the Internet, pp. 167–175 (2001)

26. Jiang, Z., Ge, Y., Li, Y.: Max-utility wireless resource management for best effort traffic. IEEE Trans. on Wireless Commun. 4(1), 100–111 (2005)

27. Kim, H.: Exploring Tradeoffs in Wireless Networks under Flow-Level Traffic: Energy, Capacity and QoS. PhD thesis, University of Texas at Austin (2009)

28. Kim, H., Chae, C.-B., de Veciana, G., Heath Jr., R.: A Cross-layer approach to energy efficiency for adaptive MIMO systems exploiting Spare Capacity. IEEE Trans. Wireless Commun. 8, 4264–4275 (2009)

29. Kim, H., de Veciana, G.: Leveraging Dynamic Spare Capacity in Wireless Systems to Conserve Mobile Terminals' Energy. IEEE Trans. on Networking (July 2010)

30. Kim, H., de Veciana, G., Yang, X., Venkatachalam, M.: Distributed alpha-Optimal User Association and Cell Load Balancing in Wireless Networks. IEEE Trans. on Networking (January 2010) (submitted)

31. Knopp, R., Humblet, P.: Information capacity and power controlling single-cell multiuser communications. In: Proc. IEEE Int. Conf. on Commun., pp. 331–335 (June 1995)

32. Lahiri, K., Raghunathan, A., Dey, S., Panigrahi, D.: Battery-driven system design: A new frontier in low power design. In: Proc. Intl. Conf. on VLSI Design, pp. 261–267 (January 2002)

33. Leung, K.-K., Sung, C.: An opportunistic power control algorithm for cellular network. IEEE/ACM Trans. Networking 14, 470–478 (2006)

34. Li, L., Goldsmith, A.: Optimal resource allocation for fading broadcast channels – part I: Ergodic capacity. IEEE Trans. Inf. Theory 47, 1083–1102 (2001)

35. Li, Y., Stuber, G.: OFDM for Wireless Communications. Springer, Heidelberg (2006)

36. Liu, X., Chong, E., Shroff, N.: Opportunistic transmission scheduling with resource-sharing constraints in wireless networks. IEEE J. Select. Areas Commun. 19, 2053–2064 (2001)

37. Liu, X., Chong, E., Shroff, N.: A framework for opportunistic scheduling in wireless networks. Computer Networks 41 (2003)

38. Madan, R., Cui, S., Lall, S., Goldsmith, A.: Modeling and optimization of transmission schemes in energy-constrained wireless sensor networks. IEEE/ACM Trans. Networking 15, 1359–1372 (2007)

39. Mangharam, R., Rajkumar, R., Pollin, S., Catthoor, F., Bougard, B., van der Perre, L., Moeman, I.: Optimal fixed and scalable energy management for wireless networks. In: Proc. IEEE INFOCOM 2005, vol. 1, pp. 114–125 (March 2005)

40. Meshkati, F., Poor, H., Schwartz, S., Mandayam, N.: An energy-efficient approach to power control and receiver design in wireless networks. IEEE Trans. Commun. 5, 3306–3315 (2006)

41. Meshkati, F., Poor, H., Schwartz, S., Balan, R.: Energy Efficiency and Delay Quality-of-Service in Wireless Networks. In: Proc. Inaugural Workshop of the Center for Information Theory and Its Applications (2006)

42. Meshkati, F., Poor, H., Schwartz, S.: Energy-efficient resource allocation in wireless networks. IEEE Commun. Magazine, 58–68 (May 2007)

43. Meshkati, F., Poor, H., Schwartz, S.: Energy-Efficient Resource Allocation in Wireless Networks: An Overview of Game-Theoretic Approaches. IEEE Sign. Proc. Magazine (May 2007)

44. Meshkati, F., Poor, H., Schwartz, S., Balan, R.: Energy-Efficient Resource Allocation in Wireless Networks with Quality-of-Service Constraints. IEEE Trans. Commun. 57(11), 3406–3414 (2009)

45. Miao, G.: Cross-Layer Optimization for Spectral and Energy Efficiency. PhD thesis, School of Electrical and Computer Engineering, Georgia Institute of Technology (2008)

46. Miao, G., Himayat, N., Li, Y., Swami, A.: Cross-layer optimization for energy-efficient wireless communications: A survey. Wiley J. Wireless Commun. and Mob. Comp. 9(4), 529–542 (2009)

47. Miao, G., Himayat, N., Li, Y.: Energy-efficient link adaptation in frequency-selective channels. IEEE Trans. Commun. (to appear 2009)

48. Miao, G., Himayat, N., Li, Y., Koc, A., Talwar, S.: Interference-aware energy-efficient power optimization. IEEE Trans. Commun. (February 2009) (submitted)

49. Miao, G., Himayat, N., Li, Y., Talwar, S.: Low-complexity energy-efficient scheduling for uplink OFDMA. IEEE Trans. Commun. (March 2009) (submitted)

50. Nanda, S., Balachandran, K., Kumar, S.: Adaptation techniques in wireless packet data services. IEEE Commun. Magazine, 54–64 (January 2000)

51. Naware, V., Mergen, G., Tong, L.: Stability and delay of finite-user slotted Aloha with multipacket reception. IEEE Trans. Inf. Theory 51, 2636–2656 (2005)

52. Necker, M.: Coordinated fractional frequency reuse. In: Proc. ACM Symp. on Mod., Anal., and Sim. of wireless and Mob. Syst., pp. 296–305 (2007)
53. Oliveira, C., Kim, J., Suda, T.: An adaptive bandwidth reservation scheme for high-speed multimedia wireless networks. IEEE Jour. Select. Areas in Comm. 16, 858–874 (1998)
54. Omiyi, P., Haas, H.: Improving time-slot allocation in 4th generation OFDM/TDMA TDD radio access networks with innovative channel-sensing. In: Proc. IEEE ICC, vol. 6, pp. 3133–3137 (2004)
55. Patil, S., de Veciana, G.: Feedback and opportunistic scheduling in wireless networks. IEEE Trans. Wireless Commun., 4227–4238 (December 2007)
56. Patil, S., de Veciana, G.: Managing resource and quality of service in heterogeneous wireless systems exploiting opportunism. IEEE/ACM Trans. Networking 15, 1046–1058 (2007)
57. Patil, S., de Veciana, G.: Measurement-based opportunistic scheduling for heterogeneous wireless systems. IEEE Trans. Commun. (2009)
58. Prabhakar, B., Biyikoglu, E., Gamal, A.: Energy-efficient transmission over a wireless link via lazy packet scheduling. In: Proc. IEEE INFOCOM, vol. 1, pp. 386–393 (2001)
59. Qin, X., Berry, R.: Exploiting multiuser diversity for medium access in wireless networks. In: Proc. IEEE INFOCOM, pp. 1084–1094 (April 2003)
60. Rabaey, J., Ammer, J., da Silva Jr., J., Patel, D.: PicoRadio: Ad-hoc wireless networking of ubiquitous low-energy sensor/monitor nodes. In: IEEE VLSI, pp. 9–12 (2000)
61. Rajan, D., Sabharwal, A., Aazhang, B.: Delay-bounded packet scheduling of bursty traffic over wireless channels. IEEE Trans. Inf. Theory 50, 125–144 (2004)
62. Rengarajan, B., de Veciana, G.: Architecture and abstraction for environment and traffic aware system-level coordination of wireless networks: the downlink case. In: Proc. IEEE INFOCOM, pp. 1175–1183 (2008)
63. Rodoplu, V., Meng, T.: Bits-per-Joule Capacity of Energy-limited Wireless Networks. IEEE Trans. Wireless Commun. 6(3), 857–865 (2007)
64. Schurgers, C.: Energy-Aware Wireless Communications. PhD thesis, University of California Los Angeles (2002)
65. Schurgers, C., Srivastava, M.: Energy optimal scheduling under average throughput constraint communications. In: Proc. IEEE ICC, vol. 3, pp. 1648–1652 (2003)
66. Shakkottai, S., Rappaport, T., Karlsson, P.: Cross-layer design for wireless networks. IEEE Commun. Magazine 41, 74–80 (2003)
67. Song, G., Li, Y.: Cross-layer optimization for OFDM wireless networks – part I: theoretical framework. IEEE Trans. Wireless Commun. 4(2), 614–624 (2005)
68. Song, G., Li, Y.: Cross-layer optimization for OFDM wireless networks – part II: algorithm development. IEEE Trans. Wireless Commun. 4(2), 625–634 (2005)
69. Song, G., Li, Y.: Utility-based resource allocation and scheduling in OFDM-based wireless networks. IEEE Commun. Magazine 43(12), 127–135 (2005)
70. Song, G.: Cross-Layer Optimization for Spectral and Energy Efficiency. PhD thesis, School of Electrical and Computer Engineering, Georgia Institute of Technology (2005)
71. Song, G., Li, Y.: Asymptotic throughput analysis for channel-aware scheduling. IEEE Trans. Commun. 54(10), 1827–1834 (2006)
72. Stuber, G.L.: Principles of Mobile Communication. Kluwer Academic Publishers, Norwell (2001)

73. Tenhonen, K., Hamalainen, J., Wichman, R., Horneman, K.: On the effect of channel-aware scheduling to CDMA uplink capacity. In: Proc. of IEEE PIMRC (September 2006)
74. de Veciana, G., Lee, T.-J., Konstantopoulos, T.: Stability and performance analysis of networks supporting elastic services. IEEE/ACM Trans. Networking 9, 2–14 (2001)
75. Verdu, S.: Spectral efficiency in the wideband regime. IEEE Trans. Inf. Theory 48, 1319–1343 (2002)
76. Wang, A., Cho, S., Sodini, C., Chandrakasan, A.: Energy efficient modulation and MAC for asymmetric RF microsensor system. In: Proc. Int. Symp. Low Power Electronics and Design, pp. 106–111 (2001)
77. Wong, C., Cheng, R., Letaief, K., Murch, R.: Multiuser OFDM with adaptive subcarrier, bit, and power allocation. IEEE J. Select. Areas Commun. 17, 1747–1758 (1999)
78. Yates, R.: A framework for uplink power control in cellular radio systems. IEEE Jour. Select. Areas in Comm. 13, 1341–1347 (1995)
79. Zafer, M., Modiano, E.: A calculus approach to energy-efficient data transmission with quality-of-service constraints. IEEE/ACM Trans. on Networking (2007)
80. Zhang, H., Dai, H.: Cochannel interference mitigation and cooperative processing in downlink multicell multiuser MIMO networks. EURASIP J. Wireless Commun. and Networking, 222–235 (February 2004)

On the Probabilistic Description of an Asynchronous DHA FH OFDMA System with Threshold Noncoherent Reception

Dmitry Osipov

Institute for Information Transmission Problems RAS
B. Karetny Lane 19/1, Moscow, Russia
d_osipov@iitp.ru

Abstract. In what follows an asynchronous DHA FH OFDMA (Dynamic Hopset Allocation Frequency Hopping) system with threshold noncoherent reception is considered. DHA FH OFDMA systems are a novel class of FH systems that combines frequency hopping with positional coding. DHA FH OFDMA with noncoherent reception is the most promising technique since it turns out to be much more jamming proof than other known multiple access systems including those utilizing conventional frequency hopping. An analytical expression of the probability density function of the decision statistics utilized in the considered system has been obtained using characteristic function apparatus.

1 Introduction

In recent years the value of information transmitted via wireless channels increased drastically due to the extension of the range of applications using such channels. Thus, the problem of protecting information transmitted via wireless channels from undesirable activity (i.e. eavesdropping or jamming) is a key one for the up-to-date communication systems.

FH CDMA has been considered to be a fundamental technique for solving this problem since the time of its introduction. Unfortunately FH systems are vulnerable to jamming technique implemented in the last-generation jammers. To solve this problem the Dynamic Hopset Allocation FH OFDMA (DHA FH OFDMA) concept has been introduced (see [1]). In [1] the probability density function of the decision statistics utilized in a coherent DHA FH OFDMA system was obtained. However, noncoherent DHA FH OFDMA with threshold reception considered in [2] is a much more promising technique due to its higher jamming proofness. Hereinafter the problem of obtaining the probability density function of the decision statistic utilized in a noncoherent DHA FH OFDMA system with threshold reception will be considered.

In Sect. 2.1 a short overview of a classical FH OFDMA technique complemented by a description of an asynchronous noncoherent DHA FH OFDMA system will be given. In Sect. 2.2 a probabilistic model of an asynchronous noncoherent DHA FH OFDMA system with threshold reception will be presented

A. Vinel et al. (Eds.): MACOM 2010, LNCS 6235, pp. 180–187, 2010.

and the probability density function of the decision statistics utilized in an asynchronous noncoherent DHA FH OFDMA system with threshold reception will be obtained.

2 Main Part

2.1 A DHA FH OFDMA System with Noncoherent Reception

In conventional frequency hopping the entire available frequency band is divided into Q subbands or subchannels (e.g. by means of OFDM). Following the terminology used in [3] we shall further on refer to the set of all subbands available to the user as a *hopset*. Each user's transmitter chooses one of the Q subbands (using a specialized subband numbers generator) and transmits a signal via the chosen subband using a conventional modulation technique (for the most part FSK is used). In the multiple access theory it is common to refer to each change of the subband in use as a hop. The process of switching between the subbands (which can also be interpreted in terms of assigning subbands to the users) is called frequency hopping. The sequence of subbands numbers chosen by the aforesaid generator of the user under consideration can be interpreted as a code sequence. Note that in conventional FH OFDMA frequency hopping technique is used to protect the transmitted information from undesirable activity and separate different users, whereas for data transmission proper conventional modulation techniques are used. In contrast in DHA FH OFDMA frequency hopping is information driven, i.e. the subbands (subchannels) used by a certain user depend on the transmitted information as well as on the code sequence in use.

Let us now consider the DHA FH OFDMA concept in more detail and particularly the basics of the scheme under consideration. Consider a multiple access system where K active users transmit data to the base station through a channel split into Q frequency subchannels; the transmission is asynchronous and uncoordinated (i.e. none of the users has information about the others). It is assumed that all the users transmit binary $c-$tuples. In the course of the transmission of each consecutive tuple the subchannel number generator assigned to the user under consideration chooses (in a random manner) $q = 2^c$ subchannels out of Q subchannels. Each tuple (or a part of the tuple) to be transmitted by the aforesaid user within the frame is mapped into the number of the subchannel, via which the signal is transmitted. Note that the vector of q subchannels is the instantaneous hopset, i.e. after each hop each user is allocated a new instantaneous hopset.

In what follows we shall assume that in the system under consideration optimal power control is used. The latter means that the powers of all the signals from distinct users are equal at the receiver side (hereinafter without loss of generality we shall assume that the powers of all the signals is equal to one i.e. $E_j = 1 j = 1 : K$). It is assumed that the base station is equipped with the subchannel numbers generator synchronized with that of the active user. The latter means that within the scope of the reception of the respective tuple the subchannel numbers generator of the base station produces the very same subchannel numbers vector that has been generated. Note, that this assumption is not restrictive since synchronized generators are an essential part of any conventional FH CDMA system. Thus, we simply replace a generator producing random numbers with a generator producing random vectors.

In noncoherent DHA FH OFDMA within the scope of the reception of a certain tuple sent by the user under consideration the receiver measures the values of power for all the signals received through the subchannels chosen by the subchannel numbers generator of the user under consideration obtaining q statistics in that manner. Thus, the receiver is to decide, which subchannel has actually been used by the active user under consideration. To do so the receiver compares each statistic value with a certain threshold (here we assume that the value of the threshold has been chosen in advance and is not to be changed.) If threshold crossing is detected in only one subchannel, the tuple corresponding to the subchannel where the threshold crossing was registered is accepted. Otherwise, an erasure decision is made. If a tuple other than the transmitted one is accepted, we say that an error has occurred. The block scheme of the system is shown in Fig.1

2.2   Decision Statistics Utilized in a Noncoherent DHA FH OFDMA System with Threshold Reception: A Probabilistic Description

Let us consider a transmission of a tuple by a certain user in more detail. Due to the above-made assumption of the availability of ideal power control in the system, the signal of each user at the receiver end can be represented by a vector of a unit amplitude with a random phase uniformly distributed on the circle $[0, 2\pi]$. Since all the users transmit data asynchronously, the duration of interaction of a fixed user with other active users (we call them "interfering" users) is a random variable uniformly distributed on $[0, T]$ (where T is the duration of the transmitted signal) . Therefore, components that are due to other users interference can be represented by vectors with amplitudes uniformly distributed on $[0,1]$ and random phases uniformly distributed on $[0, 2\pi]$.

In [1] it has been shown that for parameters that are of practical interest (i.e. for $K < (Q/2) - 1$) the probability of collision of multiplicity one is much greater than that of the collisions of higher multiplicity. Therefore hereinafter we shall consider only collisions of multiplicity two. Let us consider an active user that has transmitted a signal via the j^*-th subcnannel. The demodulated signal obtained via the subchannel j is then given by

$$\bar{r}_j = \bar{s}_j + \bar{y}_j + \bar{n} = \bar{x}_j + \bar{y}_j \qquad (1)$$

here $\bar{s}_j$ is the vector corresponding to the signal transmitted via the j-th subchannel.

$$|\bar{s}_j| = \begin{cases} 1 & j^* = j \\ 0 & j^* \neq j \end{cases}$$

$\bar{y}_j$ is a random vector with random phase uniformly distributed on $[0, 2\pi]$ and amplitude uniformly distributed on $[0,1]$ and $\bar{n}$ is the vector corresponding to

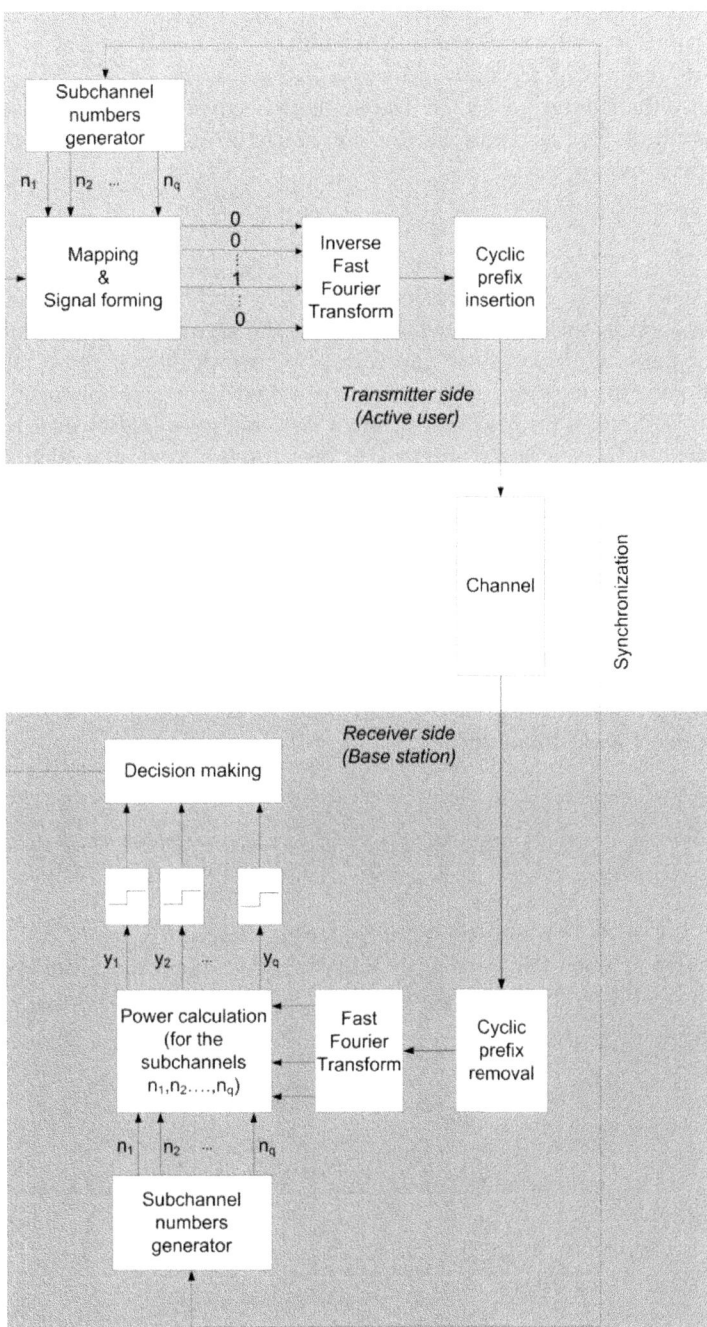

Fig. 1. Uplink transmission in an asynchronous DHA FH OFDMA system with non-coherent threshold reception

the additive white Gaussian noise. Thus the power of $\bar{x}_j = \bar{s}_j + \bar{n}$ has non-central χ^2 distribution with two degrees of freedom if $j = j^*$ and central χ^2 distribution with two degrees of freedom otherwise [3]. Since the phase of the vector $\bar{y}_j$ is uniformly distributed on $[0, 2\pi]$ the angle ϕ between $\bar{x}_j$ and $\bar{y}_j$ is also uniformly distributed on $[0, 2\pi]$. Thus the decision statistics in question (i.e. the power of $\bar{r}_j$) is given by

$$z = x^2 + 2xy\cos\phi + y^2 \tag{2}$$

where y has uniform distribution on $[0,1]$, ϕ is also uniformly distributed on $[0, 2\pi]$, and x has Rician distribution if $j = j^*$ (i.e. if the subchannel under consideration is the subchannel, via which the signal has been transmitted by the user under consideration) and Raleigh distribution otherwise [3]. Note that the problem in question is very close to a well known problem of finding the amplitude of the sum of several random vectors (see [5]). However this problem has been solved only for special cases (e.g. for the case of random phases and deterministic amplitudes or for the case of random components number) and to the best of authors knowledge none of the special cases for which a decision has been obtained up to the present moment corresponds to the problem under consideration.

To obtain the probability density function we shall apply the characteristic function apparatus. By definition the characteristic function of the random variable is simply the expectation of the function $\Phi(z) = e^{i\xi z}$. Since z in turn is the function of random variables x, y and ϕ the expectation of $\Phi(z)$ can be defined in a straightforward manner

$$g(\xi) = \int\limits_{0}^{+\infty} \int\limits_{0}^{1} \int\limits_{0}^{2\pi} e^{i\xi z(x,y,\phi)} f(x,y,\phi)\, d\phi\, dy\, dx \tag{3}$$

here $f(x, y, \phi)$ is the joint probability density function x, y and ϕ. Since x, y and ϕ are independent their joint probability density function is simply the product of the probability density functions of the respective variables.

Note that

$$f(z) = \int\limits_{-\infty}^{+\infty} e^{i\xi z} g(\xi)\, dz \tag{4}$$

where $f(z)$ is the probability density function that is of interest to us. Thus, $f(z)$ is given by

$$f(z) = \int\limits_{-\infty}^{+\infty} e^{-i\xi z} \int\limits_{0}^{+\infty} \int\limits_{0}^{1} \int\limits_{0}^{2\pi} \frac{1}{2\pi} e^{i\xi\left(x^2 + y^2 + 2xy\cos(\phi)\right)} f_\rho(x)\, d\phi\, dy\, dx\, dz \tag{5}$$

where $f_\rho(x)$ is the probability density function of the variable x (Rician if $j = j^*$ and Raleigh otherwise).

To find the analytical expression denoted by the equation (5) we shall consider a modified function

$$\Im(z, \beta) = \int_{-\infty}^{+\infty} e^{-i\xi z} \int_0^{+\infty} \int_0^1 \int_0^{2\pi} \frac{1}{2\pi a} e^{i\xi\beta x^2 + i\xi y^2 + 2i\xi xy \cos(\phi)} f_\rho(x) \, d\phi dy dx dz \quad . \quad (6)$$

Let us now consider the integral

$$\Im_0 = \int_0^{2\pi} \frac{1}{2\pi a} f_\rho(x) e^{i\xi\beta x^2 + i\xi y^2 + 2i\xi xy \cos(\phi)} d\phi = \frac{e^{i\xi\beta x^2 + i\xi y^2}}{a} f_\rho(x) \int_0^{2\pi} \frac{1}{2\pi} e^{2i\xi xy \cos(\phi)} d\phi.$$

$$(7)$$

Note that the integral

$$J_0(\mu) = \int_0^{2\pi} \frac{1}{2\pi} e^{2i\xi xy \cos(\phi)} d\phi = \frac{1}{(\Gamma(1))^2} \int_0^\pi e^{i\mu \cos(\phi)} d\phi \qquad (8)$$

is nothing else but a zero-order Bessel function of the first kind [?] of the variable $\mu = 2\xi xy$. Thus,

$$J_0(\mu) = \sum_{k=0}^\infty (i)^{2k} \frac{\mu^{2k}}{2^{2k}(k!)^2} = \sum_{k=0}^\infty (i)^{2k} \frac{(2\xi xy)^k}{2^{2k}(k!)^2} = \sum_{k=0}^\infty (i)^{2k} \frac{(\xi x)^{2k} y^{2k}}{(k!)^2} \qquad (9)$$

and $\Im_0$ is given by

$$\Im_0 = f_\rho(x) \sum_{k=0}^\infty (i)^{2k} \frac{e^{i\xi\beta x^2} e^{i\xi y^2} (\xi x)^{2k} y^{2k}}{(k!)^2}. \qquad (10)$$

Simplifying expression (10) by applying the Taylor series expansion to the multiplier $e^{i\xi y^2}$ we obtain

$$\Im_0 = f_\rho(x) \sum_{k=0}^\infty (i)^{2k} \frac{e^{i\xi\beta x^2} (\xi x)^{2k}}{(k!)^2} \sum_{m=0}^\infty \frac{i^m \xi^m y^{2m+2k}}{m!} \qquad (11)$$

Let us now consider the next integral

$$\Im_1(x, \xi) = \int_0^1 \Im_0(x, y) dy = f_\rho(x) e^{i\xi\beta x^2} \sum_{k=0}^\infty \frac{(i)^{2k} (\xi x)^{2k}}{(k!)^2} \sum_{m=0}^\infty \frac{i^m \xi^m}{(2m + 2k) m!}. \qquad (12)$$

$\Im_1(x, \xi)$ can be rewritten in the following form

$$\Im_1(x, \xi) = \sum_{k=0}^\infty \sum_{m=0}^\infty \frac{i^{m+k} \xi^{m+k}}{(2m + 2k)(k!)^2 m!} (i\xi x^2)^k f_\rho(x) e^{i\xi\beta x^2}. \qquad (13)$$

Note that

$$\Im_2\left(\xi,\beta\right)=\left(i\xi x^2\right)^k f_\rho\left(x\right)e^{i\xi\beta x^2}=\dfrac{\partial^k\int\limits_0^{+\infty}f_\rho\left(x\right)e^{i\xi\beta x^2}dx}{\partial\beta^k}. \tag{14}$$

Thus,

$$\Im\left(z,\beta\right)=\sum_{k=0}^{+\infty}\sum_{m=0}^{+\infty}\dfrac{1}{\left(2m+2k\right)\left(k!\right)^2 m!}\int\limits_{-\infty}^{+\infty}i^{m+k}\xi^{m+k}e^{i\zeta z}\Im_2\left(\xi,\beta\right)d\zeta. \tag{15}$$

Equation (15) can be rewritten in the following form

$$\Im\left(z,\beta\right)=\sum_{k=0}^{+\infty}\sum_{m=0}^{+\infty}\dfrac{1}{\left(2m+2k\right)\left(k!\right)^2 m!}\int\limits_{-\infty}^{+\infty}\dfrac{\partial^{m+k}\left(e^{i\zeta z}\Im_2\left(\xi,\beta\right)\right)}{\partial z^{m+k}}d\zeta \tag{16}$$

Therefore by substituting $\Im_2\left(\xi,\beta\right)$ with (14) we obtain

$$\Im\left(z,\beta\right)=\sum_{k=0}^{+\infty}\sum_{m=0}^{+\infty}\dfrac{1}{\left(2m+2k\right)\left(k!\right)^2 m!}\dfrac{\partial^{m+k}\partial^k}{\partial z^{m+k}\partial\beta^k}\left(\int\limits_{-\infty}^{+\infty}e^{i\zeta z}\int\limits_0^{+\infty}f_\rho\left(x\right)e^{i\xi\beta x^2}dxd\zeta\right). \tag{17}$$

Note that

$$\tilde{g}\left(\xi\right)=\int\limits_0^{+\infty}f_\rho\left(x\right)e^{i\xi\beta x^2}dx \tag{18}$$

is the characteristic function of the variable $\psi=\beta z^2$which is simply a power of vector $\bar{x}_j$ multiplied by a constant value β.

Therefore the function

$$f\left(z,\beta\right)=\int\limits_\infty^\infty e^{-i\xi z}\int\limits_0^{+\infty}f_\rho\left(x\right)e^{i\xi\beta x^2}dxd\xi \tag{19}$$

is the probability density function of the variable ψ and is given by

$$f\left(z,\beta\right)=\begin{cases}\dfrac{1}{\beta\sigma^2}e^{\left(-\frac{\left(z+\beta E_i\right)}{2\beta\sigma^2}\right)}I_0\left(\dfrac{\sqrt{E_i}z}{\beta\sigma^2}\right)&if j^*=j\\[2ex]\dfrac{1}{\beta\sigma^2}e^{\left(-\frac{z}{2\beta\sigma^2}\right)}&\text{otherwise}\end{cases} \tag{20}$$

(note that we have assumed that $E_i=1$ and thus the equation (20) can be simplified by omitting E_i). Therefore

$$\Im\left(z,\beta\right)=\sum_{k=0}^{+\infty}\sum_{m=0}^{+\infty}\dfrac{1}{\left(2m+2k\right)\left(k!\right)^2 m!}\dfrac{\partial^{m+k}\partial^k}{\partial z^{m+k}\partial\beta^k}f\left(z,\beta\right) \tag{21}$$

and the probability density that is of interest to us is given by

$$f\left(z\right)=\Im\left(z,\beta\right)|_{\beta=1} \tag{22}$$

3   Conclusion and Future Work

Hereinabove the problem of giving a probabilistic description of an asynchronous DHA FH OFDMA system with threshold noncoherent reception has been considered. An analytical expression for the probability density function of the decision statistics utilized in the considered system has been obtained using the characteristic function apparatus. The expression obtained is well suited only for the case of authorized users' activity. However, the problem of giving a probabilistic description of the users' activity for the case of intentional jamming in the system under consideration is also of great interest. This problem is a subject for future work.

References

1. Zyablov, V.V., Osipov, D.S.: On the optimum choice of a threshold in a frequency hopping OFDMA system. Problems of Information Transmission 44(2), 91–99 (2008)
2. Zyablov, V.V., Osipov, D.S.: Equidistant code-based signal-code construction in multiple-access system with concatenated coding. In: Proc. of the ITAS 2009 Bekasovo, December 15-18, pp. 145–151 (2009) (in Russian)
3. Zigangirov, K.S.: Theory of Code Division Multiple Access Communication. IEEE Press, Piscataway (2004)
4. Watson, G.N.: Theory of Bessel Functions. Cambridge Press, Cambridge (1945)
5. Abdi, A., Hashemi, H., Nader-Esfahani, S.: On the PDF of the Sum of Random Vectors. IEEE Transactions on Communications 48(1), 7–12 (January)

Adaptive Channel Estimation for STBC-OFDM Systems Based on Nature-Inspired Optimization Strategies

Leandro D'Orazio[1], Claudio Sacchi[2], and Massimo Donelli[2]

[1] SIEMENS S.p.A. Industry Sector, Viale Piero e Alberto Pirelli 10
I-20128 Milano (Italy)
leandro.dorazio.ext@siemens.com
[2] University of Trento
Department of Information Engineering and Computer Science (DISI)
Via Sommarive 14, I-38050, Povo (Trento), Italy
{sacchi,donelli}@disi.unitn.it

Abstract. In this paper, we propose an adaptive channel estimation methodology for Space-Time Block-Coded (STBC) OFDM systems, aided by nature-inspired evolutionary optimization strategies, namely: Genetic Algorithm (GA) and Particle Swarm Optimization (PSO). The use of GA and PSO allows at increasing the convergence of adaptive channel estimation to the optimal MMSE solution with respect to state-of-the-art optimization methodologies based on the concept of deterministic gradient. As a result, system performances are greatly improved, with a clear advantage taken by PSO, both in terms of channel estimation accuracy, implementation ease, and reduced computational effort.

Keywords: Multiple Input–Multiple Output (MIMO) systems, Orthogonal Frequency Division Multiplexing (OFDM), Channel Estimation, Genetic Algorithm, Particle Swarm Optimization.

1 Introduction

Orthogonal Frequency Division Multiplexing (OFDM) is one of the most emerging standards for wireless communications since early '90s. OFDM is based on frequency diversity, being able at splitting a frequency-selective channel into multiple flat-fading channels. ISI can be completely removed by inserting a short cyclic prefix [1]. However, the effects of frequency-selective fading can involve relevant symbol error rates in correspondence of those subcarriers that experience the largest channel attenuation. In order to increase the system diversity without increasing bandwidth, Multiple Input – Multiple Output (MIMO) techniques and Space-Time (ST) processing are often adopted in combination with OFDM [2]. A well-known example of conceptually simple, computationally efficient and mathematically elegant ST scheme based on orthogonal block coding (STBC) has been proposed by Alamouti in [3]. Optimal STBC decoding in MIMO-OFDM systems would require the ideal knowledge of the Channel State Information (CSI) [2] [4]. This would be a serious issue in

A. Vinel et al. (Eds.): MACOM 2010, LNCS 6235, pp. 188–198, 2010.

practical system implementation, as CSI is generally unknown. Therefore, the problem of channel estimation in STBC-OFDM systems is a key point to be addressed. State-of-the-art contributions concerning this topic can be found in [5-7]. The basic methodology for Minimum Mean Squared Error (MMSE) channel estimation is presented by Li *et.al.* in [5]. In [6], Barhumi, Leeus and Moonen analyzed the Least Square (LS) channel estimation together with its recursive implementation, namely: Recursive Least Square (RLS). In a recent work, Mohammadi, Ardabilipour *et.al.* thoroughly analyzed performance of gradient-descent based channel estimation methodologies (Least-Mean Square – LMS and already mentioned RLS) for STBC-OFDM [7].

In this paper, we are proposing novel methodologies for channel estimation for MIMO-OFDM systems working over non quasi-static channels. Such methodologies are based on nature-inspired optimization algorithms. In particular, we are considering adaptive MMSE channel estimation algorithms, based on Genetic Algorithm (GA) [8] and Particle Swarm Optimization (PSO) [9]. GA and PSO rely on the concept of evolutionary adaptation of biological species to environmental mutations. Some applications of GAs to multicarrier communications can be found in literature. In [10], Jiang, Akhtman and Hanzo proposed a GA-based joint channel estimator and multiuser detector for MIMO-OFDM systems based on Space Division Multiple Access (SDMA). Also some authors of this paper applied GA-based optimization to ML detection of nonlinearly distorted OFDM symbols [11] and to adaptive MC-CDMA linear MMSE multi-user detection [12]. In such applications, GAs provided near-optimal results, reached with affordable computational complexity. PSO can take some advantages with respect to GA, mainly in terms of reduced computational effort [13]. Some recent works evidenced a growing interest of researchers to the use of PSO to solve estimation and optimization problems related to OFDM. In [14] Carro *et.al.* proposed a digital adaptive OFDM pre-distorter, which is based on the optimization of pre-distortion coefficients obtained by means of a PSO. In [15], the comparative use of GA and PSO is discussed in order to solve the subcarrier allocation problem in multi-user OFDM systems. Finally, in [16], PSO is proposed as an effective and affordable solution to the critical problem of carrier frequency offset estimation in uplink OFDMA transmission.

For these reasons, we consider the utilization of GA and PSO in order to assist adaptive MMSE channel equalization in STBC-OFDM systems. The claimed aim is to improve the accuracy of channel estimation with respect to gradient-descent methodologies, while keeping affordable the computational burden. The paper is structured as follows: Sect.2 describes the STBC-OFDM system, Sect.3 details the proposed evolutionary channel estimation, experimental results are discussed in Sect.4 and, finally, conclusions are drawn in Sect.5.

2 System Description

In this paper, a 2x1 STBC-OFDM system with Alamouti's coding [3] using maximum-diversity combination at the receiver side has been considered in the non-quasi static channel case. The received signal after OFDM demultiplexing can expressed as follows [2]:

$$\begin{cases} \underline{Y}(t) = H_1(t)\underline{A}^1 + H_2(t)\underline{A}^2 + \underline{\Upsilon}(t) \\ \underline{Y}(t+1) = -H_1(t+1)\left(\underline{A}^2\right)^* + H_2(t+1)\left(\underline{A}^1\right)^* + \underline{\Upsilon}(t+1) \end{cases} \quad (1)$$

Let's denote with $\underline{A}^1$ and $\underline{A}^2$ the (Nx1) vectors of the BPSK symbols transmitted by the antenna element 1 and 2 respectively (N is the number of orthogonal subcarriers). As discussed in [2], the channel matrix $H_i(t) \triangleq diag\left(h_0^i(t),\ldots,h_{N-1}^i(t)\right)$ is constant during the transmission of $\underline{A}^i$ ($i \in \{1,2\}$), but changes to $H_i(t+1)$ at the next transmission interval. $\underline{\Upsilon}$ denotes the zero-mean random white Gaussian complex noise vector. Some literature on STBC (see e.g. [4]) ignored the difference between $H_i(t)$ and $H_i(t+1)$ when the received signal samples are recombined to estimate the transmitted symbols. In case of $H_i(t) \neq H_i(t+1)$, the estimates obtained with the conventional linear decoder for Alamouti's scheme are no longer orthogonal [2]. This fact results in an unpleasant error-floor, able to severely bound the system performances. In [2], an alternative combiner is proposed, namely maximum diversity combiner, suitable for non-quasi-static (yet not fast) fading channels. The combination rule is given as follows:

$$\begin{cases} \hat{\underline{A}}^1 = \left(H_1(t)\right)^* \underline{Y}(t) + H_2(t+1)\left(\underline{Y}(t+1)\right)^* \\ \hat{\underline{A}}^2 = \left(H_2(t)\right)^* \underline{Y}(t) - H_1(t+1)\left(\underline{Y}(t+1)\right)^* \end{cases} \quad (2)$$

The diversity combination at the received side would theoretically require the exact knowledge of the four channel matrices, i.e.: $H_1(t)$, $H_2(t)$, $H_1(t+1)$, and $H_2(t+1)$. Of course, such a hypothesis is not realistic in practical applications and computationally-efficient channel estimation is required.

3 Evolutionary-Based Adaptive MIMO-OFDM Channel Estimation

In this work, MMSE channel estimation presented in [5] is considered. Adapting the approach of [5] to the non-quasi-static case, we can derive the following expression for the MSE metric:

$$J = \left\| \underline{Y}(t) - \hat{H}_1(t)\underline{A}^1 - \hat{H}_2(t)\underline{A}^2 \right\|^2 + \left\| \underline{Y}(t+1) + \hat{H}_1(t+1)\left(\underline{A}^2\right)^* - \hat{H}_2(t+1)\left(\underline{A}^1\right)^* \right\|^2 \quad (3)$$

The exact computation of the channel matrix estimates $\left(\hat{H}_1(t)\right)^{opt}$ and $\left(\hat{H}_1(t+1)\right)^{opt}$ with respect to the MMSE criterion would require the inversion of KxK, matrices, where K is given by [5]:

$$K = 2 \left\lceil \frac{NT_G}{T_{OFDM}} \right\rceil \tag{4}$$

being T_G the duration of the cyclic prefix (generally equal to the delay spread of the channel) and T_{OFDM} the duration of the OFDM symbol. N is generally chosen quite large in order to increase the symbol duration. Therefore, the inversion of the aforesaid matrices is computationally intensive. This is the reason why alternative channel estimation techniques are investigated in the literature. In [5], the size of the matrices to be inverted is reduced by estimating the number of significant taps of the channel. This number is really a free parameter that should be chosen mainly on the basis of *SNR* values. In [6], Barhumi, Leeus and Moonen proposed a Least-Square (LS) channel estimation methodology. LS algorithm requires the insertion of pilot sequences. The right choice of these sequences is crucial in order to reduce the estimation error. Moreover, LS requires an inverse matrix computation for every training sequence. As shown in [6], the size of the inverse matrix can be relevant. Therefore, in order to avoid a direct matrix inversion, an iterative approach has been proposed in [6] (Recursive LS or RLS). In [7], Least-Mean Square (LMS) and already-mentioned RLS are discussed and their performance compared. It has been shown in [6] and [7] that performances and convergence rates of RLS and LMS are strongly influenced by the choice of the updating parameters. These weakpoints can hinder the employment of these techniques in time-varying fading channels, making them more suitable for static channels.

In order to overcome such limitations, we focus our attention on the employment of evolutionary strategies in order to derive near-optimal MMSE channel estimation, namely Genetic Algorithms and Particle Swarm Optimization. Both algorithms are inspired by Biology, both them start with a given population, both have fitness values to evaluate population. Both update the population and search for the optimum with random techniques. Nevertheless there are some substantial differences between them. GA is based on a Darwinian concept of "natural selection", where only the "best-suited" individual survives and the other ones are suppressed [8]. GAs operate on a group (namely: *population*) of trial solutions (namely: *individuals*) in parallel. A positive number (namely: *fitness*) is assigned to each individual representing a measure of goodness. At each iteration, called *generation*, the genetic operators of *selection*, *crossover* and *mutation* are applied to chromosomes with probability α and γ, respectively, in order to generate new solutions belonging to the search space [8]. The population generation process terminates when a satisfactory solution is reached or when a fixed number of iterations (namely: generation number) are completed. On the other hand, PSO relies on sociality and cooperation between individuals. The individuals of a society hold information that is part of a "belief space". Individuals share this information and may modify it by considering three aspects [9]:

- The knowledge of the environment (namely: fitness value);
- The individual previous history of states (namely: memory);
- The previous history of states of the individual neighborhood.

The definition of neighborhood configures the "social network" of the individuals. Several neighborhood topologies exist depending on whether an individual interacts with all the other ones, or only one of the rest of the population. Following some given rules of interaction, the individuals in the population adapt their scheme of belief to those ones that are more successful among their social network. Over the time, a culture arises, in which the individuals hold opinions that are closely related [9]. The PSO algorithm requires tuning of some parameters, namely: the individual and social weights (c_1 and c_2), the inertia factor (w), the epoch number (E), and the swarm dimension (S). A large inertia factor favors global search, while a small inertia weight favors local search [9]. Both theoretical and empirical studies are available to help in the selection of proper values of these parameters [9]. In literature, examples of comparisons of GA and PSO in the framework of specific applications are shown. In the work of Boeringer and Werner [13], a particle swarm optimizer is implemented and compared to a genetic algorithm for phased-array synthesis. Applied to such a specific problem, quantitative performances of two optimizers are, sometimes, rather similar and, sometimes one optimizer outperforms the other one (this depends on channel scenarios and array configurations). Nevertheless, as stated in [13], PSO takes advantage by a much simpler implementation with respect to GA. Similar considerations have been presented in [15]: GA and PSO can handle large allocation of subcarriers without significant degradation. However the performance of PSO is found to be better in terms of execution time, simplicity and convergence.

The main advantages of the PSO over the GA can be summarized as follows:

- *The algorithmic simplicity* – The GA considers three genetic operators and one has to choose the best configuration among several options of implementation. On the other hand, PSO consider one simple operator, which is the velocity updating;
- *The PSO parameters are very easy to manipulate* – As far as the GAs are concerned, some parameters have to be accurately calibrated, namely: population dimension, crossover and mutation rate. Whereas, PSO requires the selection of the swarm dimension, the inertial weight, and the two acceleration coefficients. If the number of control parameters is the same, it is then certainly easier to manipulate the PSOs parameters than evaluating values among various operators;
- *The capability of the PSO to prevent stagnation* – In GAs, the stagnation occurs when the individuals of the population assume a genetic code close to that one of the fittest chromosome of the overall population. In such a situation, the crossover operator has little effect and only a lucky mutation could move a new individual in another attraction basin. On the other hand, in the PSO, a suitable control of the inertial weight and of the acceleration coefficients allows to find new fittest locations in the solution space.

Taking into account these features, the PSO has been employed with success in several problems in the framework of applied and computational electromagnetic, where PSO obtained better results than GAs. For this reason, we regard GA and PSO in competition to provide a reliable solution to the challenging problem of channel estimation in STBC-OFDM systems.

Having assumed the MSE metric of (3) as fitness function for both algorithms, the adopted channel estimation procedure consists of two algorithmic steps:

1) *Training step:* during this step, an *L* bit-length binary training sequence is transmitted. The fitness function is computed by filling the data vectors $\underline{A}^1$ and $\underline{A}^2$ with the training bits. The training step is repeated with a period equal to the coherence time of the channel. During this step, the evolutionary optimizers are parameterized in order to ensure a fast convergence to the optimal solution of the problem;

2) *Decision-directed step:* the output of the training step consists of the channel matrices obtained by the evolutionary optimized parameterized in such a way to "learn" the channel in reliable way. During a coherence time period, the stochastic values of the channel coefficients acting over each subcarrier are strongly correlated. By this, a decision-directed updating step should be reasonably forecast (see e.g. [12]). In order to conveniently reduce the impact of possible decision errors on the channel estimate, it is reasonable to think to a parameterization of the evolutionary optimizer during this step in a "lighter" updating sense, as already pointed in literature [12]. The entire procedure is summarized in the flowchart of Fig. 1. The controller/enabler, which is driven by a proper temporization, sends a control signal (SWITCH) to allow switch between training sequence and symbol decision and between the different parameterization vectors used by the evolutionary optimizer during the training step and the decision-directed step. Moreover, it sends an enable signal (EN), at the end of the training step, in order to allow to fed the combiner with the estimated channel matrix, so to start the symbol decision procedure.

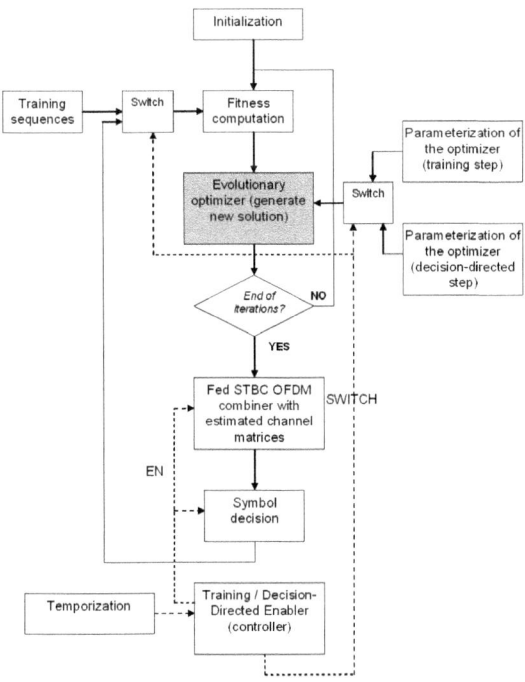

Fig. 1. Evolutionary-based channel estimation algorithm for MIMO-OFDM systems

4   Experimental Results

In order to assess the adaptive MMSE channel estimation for STBC-OFDM based on evolutionary optimization strategies, intensive simulation trials have been performed in MATLAB-SIMULINK environment. The bit-rate has been fixed to 1024Kb/s, the number of subcarriers N equals to 32. A Rayleigh multipath channel has been simulated by following the delay profile of COST 259 Hilly Terrain [17]. The delay spread of the channel is 18 μsec. We supposed mobility of the terminal, resulting in a Doppler spread of 100Hz. A cyclic prefix of suitable length has been inserted in order to remove ISI and ICI. The performance of the adaptive, receiver aided by the stochastic-gradient evolutionary optimizers, is compared with the best-known adaptive optimization strategies based on the deterministic gradient descent, i.e.: LMS and RLS.

In Tab. 1 the parameterization of GA and PSO optimizers is shown. Such a parameterization has been derived by keeping into account the best tradeoff between computational load and achieved results. One can note that GA and PSO work with the same number of iterations and the same population size. This choice is motivated by the necessity of fixing the common parameters of the two algorithms at the same values, in order to make a significant comparison among them, both in terms of performances and in terms of computational requirements. As far as the parameterization of LMS and RLS is concerned, we selected the best choice both of the step-size and of the forgetting factor for the training step and the decision-directed step by means of devoted simulation trials.

Results of such a comparison are shown in Fig. 2, showing the Estimation Error Variance (EEV), and Fig. 3, showing BER results at the output of the maximum diversity combiner of (2). One can note that both the GA-assisted adaptive channel estimation and the PSO-assisted one provide much better results than deterministic gradient-based channel estimation methodologies. On the other hand, LMS and RLS exhibit not convincing performances in the presence of relevant channel Doppler, confirming the claims of [6] and [7]. It is very interesting to note from Fig. 3 that PSO-based channel estimator approaches the theoretical lower bound on BER performance, which is related to the maximum diversity combiner provided with the actual channel matrices and not the estimated ones, slightly better than the GA-based channel estimator.

We can say that quantitative PSO and GA performances are not very far, but the advantage taken by PSO in such a specific application is actually not negligible. As far as computational complexity issues are concerned, GA and PSO are characterized by a computational burden proportional to the product between the population (swarm) dimension and generation (epoch) number [8][9] that is greatly reduced with respect to the ideal MMSE channel estimation proposed in [5] and based on the matrix inversion (see Tab.2). Considering the same approach of [8], where selection, cross-over and mutation are regarded as "elementary operations", we can say that the computational load required by PSO is approximately 1/3 of that one required by GA. In the same perspective, comparing GA and PSO with LMS and RLS (the number of elementary operations required by RLS is almost triplicate with respect to LMS [18]), we can say that evolutionary strategies are slightly disadvantaged with respect to deterministic gradient-based optimization (however, still affordable considering the computational capabilities of advanced digital signal processors). In our opinion, the achieved performance improvement yielded by GA and PSO optimizers can justify their adoption in real-world MIMO-OFDM receivers.

Table 1. Evolutionary-based optimizers parameterization

GA-based optimizer		PSO-based optimizer	
Training step	*Decision-directed step*	*Training step*	*Decision-directed step*
Generation number: 30	Generation number: 10	Epoch number: 30	Epoch number: 10
Population size: 30	Population size: 10	Swarm dimension: 30	Swarm dimension: 10
Cross-over prob. 0.99	Cross-over prob. 0.99	Inertia weight: linearly decreasing from 0.9 to 0.5	Inertia weight: linearly decreasing from 0.9 to 0.5
Mutation prob. 0.09	Mutation prob. 0.09	Maximum particle velocity: 0.01	Maximum particle velocity: 0.01
-	-	Individual and society weights $c_1=c_2=1.6$	Individual and society weights $c_1=c_2=1.6$

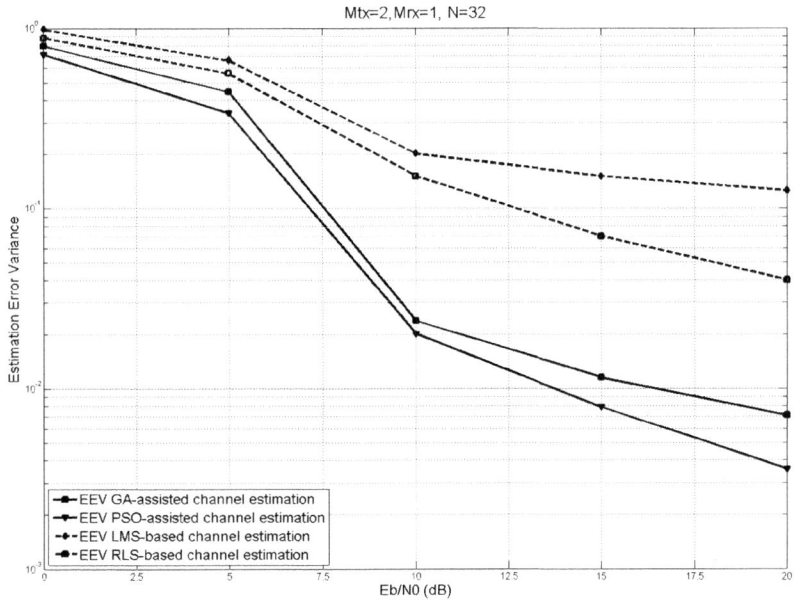

Fig. 2. Estimation Error Variance (EEV) achieved by the different MIMO-OFDM channel estimation algorithms

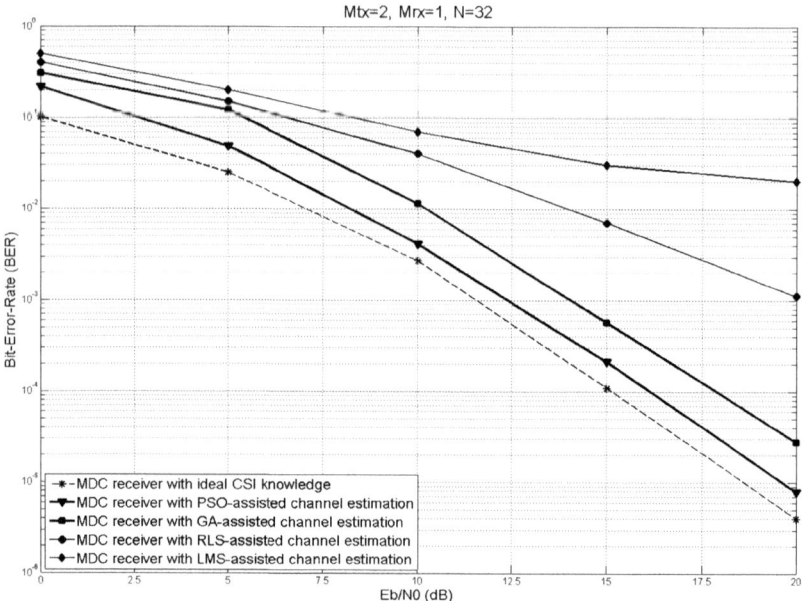

Fig. 3. Bit-Error-Rate (BER) achieved by the different channel MIMO-OFDM estimation algorithms assessed in the paper

Table 2. Comparison in terms of computational complexity of the different MIMO-OFDM estimation algorithms (algorithmic parameters: P=population size, G=generation number, E=epoch number, S=swarm dimension)

CHANNEL ESTIMATION ALGORITHM	ORDER OF COMPUTATIONAL COMPLEXITY	# OF ELEMENTARY OPERATIONS REQUIRED FOR ESTIMATING CHANNEL MATRICES
GA-assisted MMSE channel estimation	$(\alpha + \gamma)PG$	9.72×10^2
PSO-assisted MMSE channel estimation	$ES/3$	3×10^2
LMS-based channel estimation	$2N$	6.4×10^1
RLS-based channel estimation	$6N$	1.92×10^2
Ideal MMSE channel estimation	$(2K)^3$	46.6×10^3

5   Conclusion

In this paper, two Nature-inspired evolutionary optimization strategies (i.e.: Genetic algorithm and Particle Swarm Optimization) have been applied in order to solve the problem of channel estimation in Space-Time Block Coded MIMO-OFDM systems.

BER performances achieved by GA-based and PSO-based channel estimators are quite close to those achieved in the case of ideal CSI knowledge. Comparing results yielded by the two tested evolutionary strategies, we can say that PSO can take a not-dramatic, however noticeable advantage with respect to GA. As compared to deterministic-gradient based LMS and RLS optimization strategies, both GA and PSO yield a relevant performance improvement, confirmed also by the dramatic reduction of the EEV. As far as computational complexity issues are concerned, PSO exhibits a reduced computational burden with respect to GA, while both GA and PSO are more computationally-demanding with respect to LMS and RLS. In any case, the computational effort required by evolutionary optimization still appears reasonable. Considering these results, we can conclude that nature-inspired evolutionary optimization strategies may be regarded as valuable solutions also in applications related to multi-carrier MIMO-STBC reception, with potential benefits on the diffusion of MIMO technologies in wireless standards based on multicarrier modulations (e.g. IEEE 802.16x).

Acknowledgements

This work has been partially supported by the Italian Ministry of University and Scientific Research, under the framework of SALICE (COFIN 2007RFTYY7_002) research project.

References

1. Hanzo, L., Webb, W., Keller, T.: Single and Multi-carrier Quadrature Amplitude Modulation. Wiley, Chichester (2000)
2. De Abreu, G.F.T., Ochiai, H., Kohno, R.: Linear Maximum Likelihood Decoding of Space-Time Block-Coded OFDM Systems for Mobile Communications. In: IEE Proc. Commun., vol. 151(5), pp. 447–459 (October 2004)
3. Alamouti, S.: A Simple Transmit Diversity Technique for Wireless Communications. IEEE J. Selec. Areas in Comm. 16(8), 1451–1458 (1998)
4. Ling, Y.G., Chang, J.C., Sollenberg, N.R.: Transmitter diversity for OFDM systems and its impact on high-rate data wireless networks. IEEE Jour. Sel. Areas Comm. 17(7), 1233–1243 (1999)
5. Li, Y., Seshadri, N., Ariyavisitakul, S.: Channel estimation for OFDM systems with transmitter diversity in mobile wireless channels. IEEE Jour. Sel. Areas in Comm. 17(3), 461–471 (1999)
6. Bahrumi, I., Leus, G., Moonen, M.: Optimal Training Design for MIMO OFDM Systems in Mobile Wireless Channels. IEEE Trans. on Signal Process. 51(6), 1615–1624 (2003)
7. Mohammadi, M.A., Ardabilipour, M., Moussakhani, B., Mobini, Z.: Performance Comparison of RLS and LMS Channel Estimation Techniques with Optimum Training Sequences for MIMO-OFDM Systems. In: Proc. of 2008 IEEE Wireless and Opt. Comm. Networks (WOCN 2008), Surabaya (IY), May 5-7, pp. 1–4 (2008)
8. Goldberg, D.E.: Genetic Algorithms in Search, Optimization and Machine Learning. Addison-Wesley, Reading (1999)

9. Kennedy, J., Eberhart, R.: Particle Swarm Optimization. In: Proc. of IEEE Int. Conf. on Neural Networks, Perth (AUS), November 27- December 1, pp. 1942–1945 (1995)
10. Jiang, M., Akhtman, J., Hanzo, L.: Iterative Joint Channel Estimation and Multi-User Detection for Multiple-Antenna Aided OFDM Systems. IEEE Trans. on Wireless Comm. 6(8), 2904–2914 (2007)
11. Sacchi, C., Donelli, M., De Natale, F.G.B.: Genetic-Algorithm Assisted Maximum-Likelihood Detection of OFDM Symbols in the Presence of Nonlinear Distortions. IEEE Trans. on Comm. 55(5), 854–859 (2007)
12. Sacchi, C., Donelli, M., D'Orazio, L., Fedrizzi, R., De Natale, F.G.B.: A Genetic Algorithm-based MMSE Receiver for MC-CDMA Systems Transmitting over Time-Varying Mobile channels. Electronics Papers 43(3), 172–173 (2007)
13. Boeringer, D.W., Werner, D.H.: Particle Swarm Optimization Versus Genetic Algorithms for Phased Array Synthesis. IEEE Trans. on Ant. And Propagat. 52(3), 771–779 (2004)
14. Carro, P.L., Ducar, P.G., De Mingo, J., Valdovinos, A.: Nonlinear Distortion Cancellation Using Particle Swarm Optimization (PSO) based Predistortion in OFDM Systems. In: Proc. of the 16th IST Mobile and Wireless Communications Summit, Budapest (H), July 1-5 (2007) (CD-ROM available)
15. Ahmed, I., Majumder, S.P.: Adaptive Resource Allocation Based on Modified Genetic Algorithm and Particle Swarm Optimization for Multiuser OFDM Systems. In: Proc. of 5th Int. Conf. on Electrical and Computer Engineering (ICECE 2008), Dhaka (Bangladesh), December 20-22, pp. 211–216 (2008)
16. Tan, T.-H., Huang, Y.-F., Tsao, J.-Y.: Estimation of Carrier Frequency Offset for Generalized OFDMA Uplink Systems Using Particle Swarm Optimization Algorithms. In: Proc. of 10th Int. Symp. on Pervasive Systems, Algorithms and Networks (ISPAN 2009), Kaoshiung (Taiwan R.O.C.), December 14-16, pp. 442–447 (2009)
17. 3rd Generation Partnership Project: Technical Specification Group Radio Access Networks; Deployment aspects, 3GPP TR 25.943, v4.2.0 (2002)
18. Proakis, J.G.: Digital Communications, new edn. McGraw-Hill, New York (2000)

Fractional Frequency Reuse Scheme
in Cooperative Relaying
For Multi-cell OFDMA Systems

Abdelhalim Najjar[1], Noureddine Hamdi[2], and Ammar Bouallegue[1]

[1] Communication Systems Lab, ENIT
Manar University, Tunisia
[2] Departement of Computer Science and Mathematics, INSAT
Carthage University, Tunisia
abdelhalim.najjar@gmail.com,
noureddine.hamdi@ept.rnu.tn,
ammar.bouallegue@enit.rnu.tn

Abstract. This paper addresses the co-channel cell interference (CCI) mitigation in multi-cell OFDMA systems through cooperative relaying and frequency reuse partitioning. In the considered system, each cell is divided into two regions: the central region and the edge region. The frequency reuse factor (FRF) is set to 1 in the central region and fractional reuse factors of 7/3 or 7/4 are used in the edge region by dividing it respectively into three or four sectors. A fixed relay station (RS) by sector, which amplifies and forwards the received signal to the mobile, is placed at the border base station side of the central region. Simulation results are used to show the improvement of the proposed cooperative scheme compared to the performance of similar architecture without relays.

Keywords: Fractional Frequency Reuse, Cooperative Relaying, OFDMA.

1 Introduction

Orthogonal frequency division multiplexing access (OFDMA) is considered as a high spectral efficiency multi-access technique for present generation of wireless packet access networks. Besides of the high spectral efficiency, OFDMA features a scalable bandwidth and a powerful solution for frequency selective fading. But, in multi-cell systems, if the same frequency resource is reused in each cell of a given network, the users at the cell edge inevitably suffer from the co-channel interference (CCI). Efficient solutions of CCI are proposed in the literature which are based on several frequency reuse schemes. In [1], a cooperative scheme using a frequency reuse factor (FRF) equal to 1 is proposed and it can achieve an average (CCI) level in the cell edge almost similar to (CCI) of non-cooperative schemes with FRF of 3. In [2] a frequency reuse scheme for a fixed two-hop OFDMA relaying network is proposed, where CCI minimum is achieved by adjusting relay station and base station transmission power radio. In [3] and [4], the FRF of 3 can reduce the amount of the CCI in the boundary of the cell. While the

A. Vinel et al. (Eds.): MACOM 2010, LNCS 6235, pp. 199–210, 2010.

traditional FRFs are fixed at 1, 3 or 7, some of fractional number like 7/3 and 7/4 are used in [5] and [6]. [7] presents two cyclic delay optimization schemes based on linear approximation of the channel phase and the strongest path for multi-cell OFDMA system with cooperative relay.

In this paper, we propose a fractional frequency reuse scheme for the downlink of a multi-cell OFDMA system with cooperative relaying by using amplify and forward fixed relays. In the considered scheme, the same FRF design has been proposed in [5] and [6]. In the considered cellular network, each cell is divided into a central region and an edge region. The FRF is set to 1 in the central region. In the edge region, according to the difference set notion [8], the FRF of 7/3 and 7/4 are applied respectively with three and four sectors. A single fixed relay station (RS) by sector is placed at the limit of the central region and the edge region. Downlink cooperation is triggered with the Non orthogonal Amplify and Forward (NAF) cooperation protocol with one relay and two time slots [9].

The main contribution in this paper is to compensate the waste of power at both the base station (BS) and the mobile station, and the available bandwidth compared to one region sectored cell. By using fixed relays, transmit power and CCI can be controlled efficiently. The proposed scheme has been compared with schemes without cooperation given in [4], [5] and [6]. Simulation results show that the proposed cooperative scheme appreciably outperforms traditional frequency reuse scheme without relaying.

The remainder of this paper is organized as follows. While, the system model is introduced in section 2, the mathematical analysis for first and second time slot is described in section 3, the proportional fair scheduling algorithm is given in section 4, simulation results are shown in section 5 and we conclude by section 6.

2   System Model

We consider a downlink 19-cell OFDMA structure where each cell is divided into a central region and an edge region. In the central region, an omnidirectional antenna is adopted. A single RS per sector is used as shown in Fig.1. Depending on the number of sectors, each relay uses a directional antenna with 120° or 90° of coverage. The cell of our interest is the cell 0 (the central cell in the system layout), where a given user moves away from the BS to the edge region. We assume that the relays operate in two phases, where in the first phase (first time slot), the BS transmits whereas all RS remain silent and the signal is received both at RS and the mobile, and in the second phase (during the second time slot) all BS are silent and each relay introduces a pre-determined cyclic delay, amplifies the previously received signal, and then forwards it to the mobile. To achieve the best performance, full channel state information should be reported to RS by the mobile. However, this perfect knowledge comes at the expense of a huge feedback channel information. To alleviate this problem, we assume that the feedback from each mobile to the serving relay would be limited to the time delay of the strongest path only.

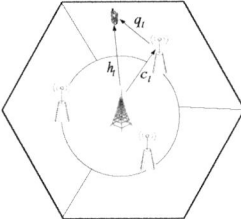

Fig. 1. Cooperative relaying communication within one 3-sectored cell

2.1 Difference Set

To provide a specific distribution of the available channels between neighboring cells, the difference set theory is used (see the Appendix). As shown in [8], the difference set can be used to allocate the same number of channels to each cell while fixing the number of shared channels between any two neighboring cells.

2.2 Scheme with FRFs of 1 and 7/3

In this scheme, the total bandwidth is divided into two parts corresponding to the two regions. In the central region the FRF is set to 1 and the mobile n is subjected to the interference of 18 cells. In the edge region, by dividing the matching bandwidth into seven breakdowns and using the $(7,3,1)$ difference set, the FRF of 7/3 can be achieved. This reuse provides a significant reduction of the number of interfering sectors (cells with channels $(6,5,1)$ and $(7,3,1)$).

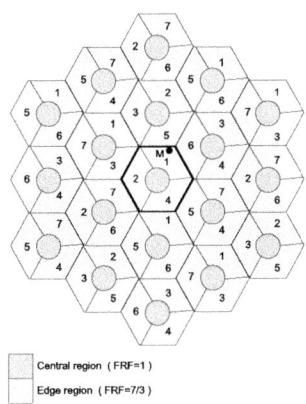

Fig. 2. system layout of cooperative relaying communication, $FRF = 7/3$ by $(7,3,1)$ difference set

2.3   Scheme with FRFs of 1 and 7/4

In this scheme, the FRF is equal to 1 in the central region. The edge region is divided into four sectors and using the $(7, 4, 2)$ difference set, the FRF of $7/4$ can be applied. Practically this reuse scheme can be used to mitigate the interference from all the sectors.

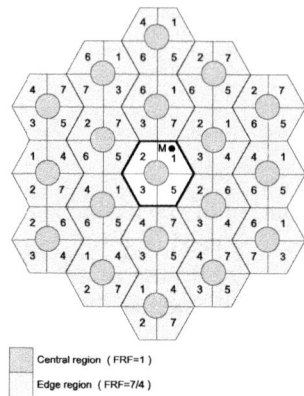

Central region (FRF=1)

Edge region (FRF=7/4)

Fig. 3. system layout of cooperative relaying communication, $FRF = 7/4$ by $(7, 4, 2)$ difference set

3   Mathematical Analysis

3.1   Phase 1 (First Time Slot)

Considering the cell 0, the received signal at the mobile n directly from all BS during the first time slot can be given by

$$r_{1,n}(t) = \sum_{i=0}^{I} \sum_{l=0}^{L-1} h_l^{(i)} s_{i,n}(t - v_l^{(i)}) + Z_1(t) \tag{1}$$

where I is the number of co-channel cells. $h_l^{(i)}$ is the channel impulse of the lth path within the cell i and v_l is the corresponding time delay. $Z_1(t)$ is the noise component. We assume that all BS are synchronized. The demodulator output of the kth OFDM symbol at subcarrier m for user n can be written as

$$R_{1,n}(k, m) = \sum_{i=0}^{I} H_{1,n}^{(i)}(k, m) s_{i,n,m} + z_{1,m} \tag{2}$$

where $z_{1,m}$ is the additive noise which is distributed as a zero mean complex Gaussian process with a power spectral density of N_0. $H_{1,n}^{(i)}(k, m)$ is the channel transfer function of the kth OFDM symbol for mobile n at subcarrier m which can be formulated as

$$H_{1,n}^{(i)}(k,m) = \sum_{l=0}^{L-1} h_l^{(i)}(kT_s)\exp(-2\pi jm\Delta f v_l^{(i)}) \tag{3}$$

where T_s and $\Delta f = T_s^{-1}$ are respectively the OFDM symbol duration and subcarrier spacing. In the following and without loss of generality, we denote $H_{1,n}^{(i)}(k,m)$ as $H_{1,n,m}^{(i)}$. The received signal at the relay within co-channel cell i is given by

$$y_{1,i}(t) = \sum_{l=0}^{L-1} c_{l,i}s_{i,n}(t - \tau_{l,i}) + z_i(t) \tag{4}$$

$$i = 1,2,...,I \tag{5}$$

where $c_{l,i}$ is the channel gain between BS i and the corresponding relay. The received signal at the relay within cell 0 from different BS can be written by

$$y_{1,0}(t) = \sum_{l=0}^{L-1} c_{l,0}s_{0,n}(t - \tau_{l,0}) + \sum_{i=1}^{I}\sum_{l=0}^{L-1} c_{l,i}'s_{i,n}(t - \tau_{l,i}') + z_0(t) \tag{6}$$

For $i = 1,2,...,I$, $c_{l,i}'$ is the channel gain of the lth path between BS i and the relay within cell 0. $\tau_{l,i}'$ is the corresponding time delay. The decibel path-loss and shadow attenuation of mobile n at the distance d_n from the serving BS are given by[10]

$$PL_{dB}(d_n) = 46.3 + 33.9\log_{10}(f_c) - 13.82\log_{10}(h_t) \tag{7}$$
$$-a(h_m) + (44.9 - 6.55\log_{10}(h_t)log_{10}(d_n)) + SH_\sigma(dB)$$

where f_c, h_t and h_m are respectively the carrier frequency, the base station antenna height, the mobile antenna height. $a(h_m)$ is the correction factor for the mobile antenna hight and it given by

$$a(h_m) = [1.1\log_{10}(f_c) - 0.7]h_m - 1.56\log_{10}(f_c) - 0.8 \tag{8}$$

The shadowing fading term $SH_\sigma(dB)$ denotes a log-normal distribution with a standard deviation σ. The channel gain between the serving BS and mobile n on subcarrier m can be written as

$$g_{1,n,m}^{(0)} = 10^{-PL_{dB}(d_n)/10}|H_{1,n,m}^{(0)}|^2 \tag{9}$$

The received (SINR) for mobile n on subcarrier m during the first time slot can be expressed as

$$\Gamma_{1,n,m} = \frac{g_{1,n,m}^{(0)}p_{n,m}^{(0)}}{N_0\Delta_f + \sum_{i=1}^{I} g_{1,n,m}^{(i)}p_{n,m}^{(i)}} \tag{10}$$

where $p_{n,m}^{(0)}$ and $p_{n,m}^{(i)}$ are the transmit power of useful signal on subcarrier m for mobile n allocated respectively by its serving cell and the ith co-channel cell.

3.2   Phase 2 (Second Time Slot)

Each relay amplifies and forwards its received signal while the BS are silent. Consequently, the received signal at the mobile n from the relays during the second time slot can be developed as

$$r_{2,n}(t) = \sum_{i=0}^{I} \sum_{l=0}^{L-1} \beta_i q_{l,i} y_{1,i}(t - \nu_{l,i}) + Z_2(t) \tag{11}$$

$$= \sum_{l=0}^{L-1} \beta_0 q_{l,0} y_{1,0}(t - \nu_{l,0})$$

$$+ \sum_{i=1}^{I} \sum_{l=0}^{L-1} \beta_i q_{l,i} y_{1,i}(t - \nu_{l,i}) + Z_2(t)$$

$$= \sum_{l=0}^{L-1} \sum_{p=0}^{L-1} \beta_0 q_{l,0} c_{p,0} s_{0,k}(t - \tau_{p,0} - \nu_{l,0})$$

$$+ \sum_{i=1}^{I} \sum_{l=0}^{L-1} \sum_{p=0}^{L-1} \beta_0 q_{l,0} c'_{p,i} s_{i,k}(t - \tau'_{p,i} - \nu_{l,0})$$

$$+ \sum_{i=1}^{I} \sum_{l=0}^{L-1} \sum_{p=0}^{L-1} \beta_i q_{l,i} c_{p,i} s_{i,k}(t - \tau_{p,i} - \nu_{l,i})$$

$$+ \sum_{i=0}^{I} \sum_{l=0}^{L-1} \beta_i q_{l,i} z_i(t - \nu_{l,i}) + Z_2(t)$$

where $q_{l,i}$ is the channel impulse of the lth path between BS i and the relay of cell 0. $\nu_{l,i}$ is the corresponding time delay. $z_i(t)$ and $Z_2(t)$ are assumed as a zeros-mean complex Gaussian noise process with power density of N_0. β_i is the amplification factor used at the relay within cell i and given by the following expression

$$\beta_i = \frac{1}{\sqrt{\sum_{l=0}^{L-1} |c_{l,i}|^2 + \frac{N_0 \Delta f}{E_s}}}, \tag{12}$$

E_s denotes the average energy per transmitted symbol. The demodulated signal sample of the nth OFDM symbol at subcarrier m for user n can written as

$$R_{2,n}(n, m) = H_{2,n,m}^{(0)} s_{0,n,m} + \sum_{i=1}^{I} H_{2,n,m}^{(i)} s_{i,n,m} + z_{2,m} \tag{13}$$

where

$$H_{2,n,m}^{(0)} = \beta_0 Q_{n,m}^{(0)} C_{n,m}^{(0)} \tag{14}$$

$$H_{2,n,m}^{(i)} = \beta_0 Q_{n,m}^{(0)} C_{n,m}^{(i)} + \beta_i Q_{n,m}^{(i)} C_{n,m}^{(i)} \tag{15}$$

For $i = 0, 1, 2, ..., I$, $C_{n,m}^{(i)}$ denotes the channel transfer function between BS i and its RS at the mobile n on subcarrier m. $Q_{n,m}^{(i)}$ is the channel transfer function between the relay of BS i and mobile n on subcarrier m. The channel gain between the relay of cell i and mobile n on subcarrier m can be expressed by

$$g_{2,n,m}^{(i)} = 10^{-PL_{dB}(d_{n,i})/10} |H_{2,n,m}^{(i)}|^2 \qquad (16)$$

$d_{n,i}$ denotes the distance between mobile n and the relay of cell i. During the second time slot, the SINR for mobile n on subcarrier m is given by the following formula

$$\Gamma_{2,n,m} = \frac{g_{2,n,m}^{(0)} p_{n,m}^{(0)}}{N_0 \Delta_f + \sum_{i=1}^{I} g_{2,n,m}^{(i)} p_{n,m}^{(i)}} \qquad (17)$$

The received SINR at the mobile n on subcarrier m in time slot 1 and 2 are maximum ratio combined (MRC) as follows

$$\Gamma_{n,m} = \Gamma_{1,n,m} + \Gamma_{2,n,m} \qquad (18)$$

To efficiently allocate the available resources and guaranteeing fairness among users, the proportional scheduling algorithm is considered which is described in the following section.

4   Proportional Fair Scheduling Algorithm

In this paper, we propose a proportional fair scheduling (PFS) algorithm to allocate the available subcarriers for users [11]. The PFS scheme allocates resources to user that experiences strong channel level weighted by his stochastic average rate. In mathematical terms, the index of the picked user satisfies

$$n_m = \underset{n}{\text{argmax}} \left(\frac{w_{n,m}(t)}{W_n(t-1)} \right) \qquad (19)$$

where $w_{n,m}(t)$ is the instantaneous transmittable rate of the subcarrier m when transmitted for user n at the current slot t. $W_n(t-1)$ is the stochastic average rate observed by the nth user, processed on a predefined window of t_c slots. $w_{n,m}$ can be expressed as

$$w_{n,m}(t) = log_2(1 + \frac{\Gamma_{n,m}(t)}{\gamma}) \qquad (20)$$

where γ is the SINR gap related to the target BER given by the following expression [12]

$$\gamma = \frac{-Log(5BER)}{1.5}. \qquad (21)$$

In each epoch t, the BS updates user's rates as follows

$$W_{n_m}(t+1) = (1 - \frac{1}{t_c})W_{n_m}(t) + \frac{w_{n_m,m}(t)}{t_c} \qquad (22)$$

$$W_n(t+1) = (1 - \frac{1}{t_c})W_n(t), n \neq n_m \qquad (23)$$

We assume that the information on the instantaneous transmittable rate of each downlink subcarrier is reported by the mobile.

5   Simulation Results

From Fig.4 to Fig.6, we consider a single frequency multi-cell OFDMA system and we assume that all available subcarriers are transmitted with equal power allocation. In our simulation, the distance of $600m$ is considered between the BS and the RS. The proposed cooperative scheme is compared with three other schemes: the first scheme refers to [5] uses the FRFs of 7/3 and 7/4 with the difference set but without cooperation, nor sectorization technique. The second scheme, given in [6], uses the same FRF designs adopted in this paper i.e $(1, 7/3)$ and $(1, 7/4)$ but without cooperation. The third one given in [4] exploits the reuses $(1, 3)$ without relays. A summary of simulation parameters is given in Table 1.

Table 1. Simulation parameters

Parameters	Values
channel bandwidth	10 MHz
Carrier frequency	2.5 GHz
FFT size	512
Number of subcarriers	350
Subcarrier spacing	15 KHz
White noise power density	-174 dBm/Hz
Fast fading model	Cost 231-Hata model
Lognormal shadowing	σ=8dB
BS transmit power	43dBm
Relay transmit power	33dBm
Minimum mobile to BS distance	100m
BER	10^{-6}
The cell radius	1.5Km
Inter-cell distance	2.8Km
BS height	32m
Mobile terminal height	1.5m

Fig.4 illustrates the received SINR versus the distance between the serving BS and the considered mobile n by using the proposed scheme, the schemes proposed in [4], [5] and [6]. The switching between the central region and the edge region is based on the distance threshold which is equal to $600m$ in our simulations. We can see that the received SINR values gradually decrease as user moves away from the BS due to the path-loss and CCI of adjacent cells. In the central region, all schemes provide similar performance by using the same reuse pattern.ie FRF=1. In this region, the received SINR is decreasing according to equation (10) with $I = 18$. The distance values are continually inspected by the serving BS. When the distance between the serving BS and the considered mobile exceeds the distance threshold, he would be considered in the edge region and is subjected to CCI of two and one cell respectively with the FRFs of 7/3

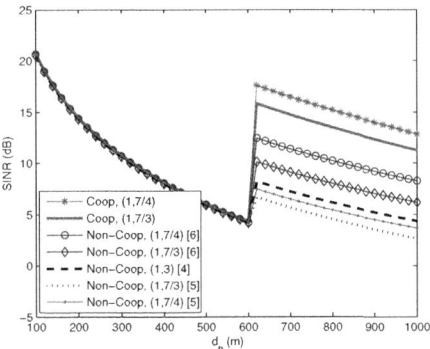

Fig. 4. The received SINR for the proposed scheme, schemes given in [4], [5] and [6]

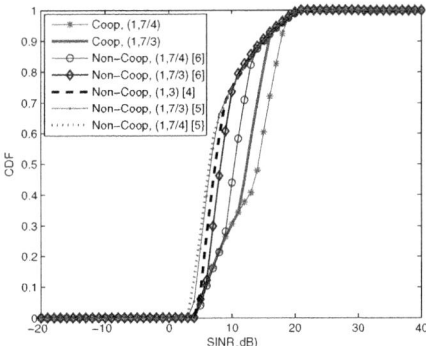

Fig. 5. Cumulative distribution function of the received SINR

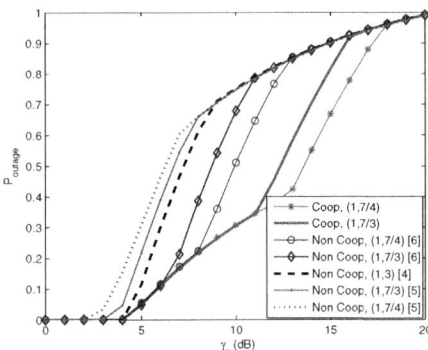

Fig. 6. Outage probability of the received SINR varying with the threshold γ_c(dB)

and 7/4. We can easily conclude from this figure that the proposed cooperative scheme gives a better performance than the three other schemes. The obtained gains compared to [6], [4] and [5] are about respectively $5dB$, $8dB$ and $10dB$.

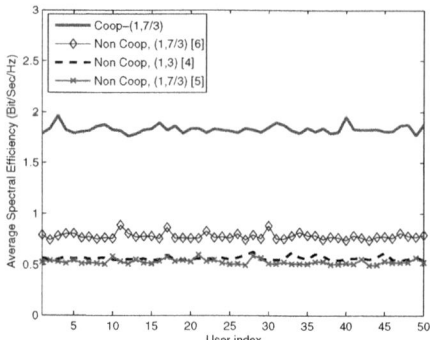

Fig. 7. Average spectral efficiency for the proposed scheme, schemes given in [4], [5] and [6] in the edge of cell 0 with $t_c = 100$ slots, 150 subcarriers (3 groups) and 50 users

We can also presume from the cumulative distribution function (CDF) given in Fig.5 that the cooperative scheme with the FRFs of 1 and 7/3 or 7/4 is more efficient in reducing the amount of the CCI in the cell edge. We notice a performance improvement of about $4dB$, $7dB$ and $9dB$ at a CDF value of 0.4 in comparison respectively with schemes given in [6], [4] and [5].

Fig.6 depicts the outage probability vs. the SINR threshold of acceptable performance γ_c for different schemes. For the proposed scheme, the probability that the received SINR values are below γ_c is smaller than the three other schemes.

In Fig.7, we evaluate the performance of the proposed cooperative scheme in terms of average spectral efficiency (averaged over all simulation times) for the edge of the cell 0. In our scheme, the available 350 subcarriers are divided into seven groups. We evaluate the performance of our scheme with the FRF of 7/3 i.e 3 groups in each cell. The proportional fair scheduling algorithm is adopted for all schemes. We assume that 50 users are distributed uniformly in the edge of the cell 0. As seen in this figure, the proposed scheme with fixed relaying performs much better than the other schemes. It provides an enhancement of about 1 bit/sec/hz, 1.5 bit/sec/hz and 1.75 bit/sec/hz in comparison respectively with schemes presented in [6], [4] and [5].

6   Conclusion

In this paper, we have proposed a fractional frequency reuse scheme in cooperative relaying for multi-cell OFDMA systems with 19-cell structure. The use of the FRF of 7/3 and 7/4 assisted by fixed relays appears to be a promising approach for interference mitigation in the edge of the cell.

References

1. Pischella, M., Belfiore, J.-C.: Achieving a Frequency Reuse Factor of 1 in OFDMA cellular Networks with cooperative Communications. In: IEEE Vehicular Technology Conference, VTC 2008, May 11-14, pp. 653–657 (Spring 2008)
2. Liang, M., Liu, F., Chen, Z., Wang, Y.F., Yang, D.C.: A Novel Frequency Reuse Scheme for OFDMA Based Relay Enhanced Cellular Networks. In: IEEE Vehicular Technology Conference, VTC 2009, April 26-29, pp. 1–5 (Spring 2009)
3. Lei, H., Zhang, X., Yang, D.: A Novel Frequency Reuse Scheme for Multi-Cell OFDMA Systems. In: IEEE Vehicular Technology Conference, VTC 2007, pp. 347–351 (Fall, September-October 2007)
4. Najjar, A., Hamdi, N., Bouallegue, A.: Efficient Frequency Reuse Scheme For Multi-cell OFDMA Systems. In: IEEE Symposium on Computers and Communications, ISCC 2009, Tunisia, July 5-8 (2009)
5. Kim, C.S., Bahk, S., Choi, Y.-J.: Flexible Design of Frequency Reuse Factor in OFDMA Cellular Networks. In: IEEE International Conference on Communication, ICC 2006, vol. 4, pp. 1784–1788 (June 2006)
6. Najjar, A., Hamdi, N., Bouallegue, A.: Fractional Frequency Reuse Scheme with Two and Three regions For Multi-cell OFDMA Systems. In: 17th Telecommunication forum Telfor 2009, Serbia, Belgrade, November 24-26 (2009)
7. Slimane, S.B., Zhouand, B., Li, X.: Delay Optimization in Cooperative Relaying with Cyclic Delay Diversity. In: IEEE International Conference on Communication, ICC 2008, May 19-23, pp. 3553–3557 (2008)
8. Pott, A., Kumar, P., Hellesth, T., Jungnickel, D.: Difference sets, Sequences and their Correlation Properties. Kluwer Academic Publishers, Dordrecht (1998)
9. Nabar, R.U., Bolcskei, H., Kneubuhler, F.W.: Fading relay channels: performance limites and space-time signal design. IEEE Journals on Sel. Areas in Telecommun. 22(6), 1099–1109 (2004)
10. European Cooperative in the Field of Science and Technical Research EURO-COST 231. Urban transmission loss models for mobile radio in the 900 and 1800 MHZ bands (Revision 2) In: The Hague, the Netherlands (September 1991), http://www.lx.it.pt/cost231/final-report.htm
11. Kim, H., Kim, K., Han, Y., Yun, S.: A proportional fair scheduling for multicarrier transmission systems. In: Vehicular Technologie Conference, VTC 2004, September 26-29, vol. 1, pp. 409–413 (Fall 2004)
12. Seo, H., Lee, B.G.: A proportional-fair power allocation scheme for fair and efficient multiuser OFDM systems. In: IEEE Globecom 2004, vol. 6, pp. 3737–3741 (December 2004)

Appendix: Difference Set

let $\Omega = \{0, 1, 2, ..., M\}$ a set.

Definition

Let D_S a subset of Ω which contains N elements and $0 < N < M$. D_s is called a (M,N,K) difference set if the set $\{a - a', a \neq a', a, a' \in \Omega\}$ contains each non zeros element of Ω exactly K-times.

Lemma 1

If D_s is an (M,N,K) difference set in a set Ω, then the set defined as $D'_s = \{D_s + a(mod M), a \in \Omega\}$ is symmetric of D_s.

Lemma 2

Let S_1 and S_2 two different subsets $\in D'_s$, there exist precisely K-elements that are common between S_1 and S_2.

Examples

Let (7,3,1) difference set. If we choose arbitrarily the subset $(1, 2, 4)$ and we apply the lemma 1, we can find the subsets $(2, 3, 5)$, $(3, 4, 6)$, $(4, 5, 7)$, $(5, 6, 1)$, $(6, 7, 2)$ and $(7, 1, 3)$ that satisfy the lemma 2. Indeed, there is exactly a single common element between two any arbitrarily subsets. In the other way and as shown in [5], using the (7,3,1) difference set, the number of shared channels between any two neighboring cells is fixed to 1. Also, with (7,4,2) difference set and by the arbitrary selection of the subset (1,2,3,5), the following subsets $(5, 6, 7, 2)$, $(4, 5, 6, 1)$, $(3, 4, 5, 7)$, $(2, 3, 4, 6)$, $(7, 1, 2, 4)$ and $(6, 7, 1, 3)$ satisfy the property of lemma 2 and can maintain a fixed number of shared channel between any two neighboring cells equals to 2.

Static Inter-Cell Interference Coordination Techniques for LTE Networks: A Fair Performance Assessment

David González G., Mario García-Lozano, Silvia Ruiz, and Joan Olmos*

Universitat Politecnica de Catalunya (UPC),
C/ Esteve Terradas, 7 - 08860 Castelldefels, Spain
{david.gonzalez.gonzalez,mario.garcia-lozano}@upc.edu

Abstract. This paper focuses in the analysis of 100% static and distributed inter-cell interference coordination techniques in the context of LTE networks. Several methods have been modeled and studied with the aim of deriving practical radio planning rules based on the joint effect of operational parameters and thresholds. The investigation places special emphasis on the efficiency vs. fairness tradeoff through a set of metrics that allow not only to evaluate the measurements from different points of view, but also to look at the effectiveness in radio resources usage. Results show that similar levels of spectral efficiency can be achieved by means of a proper and accurate network tuning. Moreover, interesting second order differences appear due to inherent features of each approach. These can be exploited depending on the particular network operator needs.

1 Introduction

In the context of mobile communications, the availability of new services and mobile applications combined with the enhanced terminals capabilities run up the need for higher data rates and to adequate the levels of quality of service. In order to fulfill such expectations, mobile operators are continually optimizing and upgrading their networks. The Long Term Evolution (LTE) of the Universal Mobile Telecommunications System (UMTS) along with WiMAX seems to be the path to follow in a very short term. Both LTE and and WiMAX employ Orthogonal Frequency Division Multiple Access (OFDMA) as the radio access technology for the downlink mainly due to its flexibility for resource allocation and because OFDMA provides intrinsic orthogonality to the users within a cell, which translates into an almost null level of intra-cell interference. Therefore, inter-cell interference is the limiting factor when high reuse levels are intended. In this case, cell-edge users are specially susceptible to the effects of inter-cell interference hence their radio channels are much worse than the ones experienced by the users close to the base station. Thus, unless more resources are assigned to them, fairness among users is jeopardized. Unfortunately, this is at the cost

---

* This work has been funded through the project TEC2008-06817-C02-02 (Spanish Industry Ministry).

A. Vinel et al. (Eds.): MACOM 2010, LNCS 6235, pp. 211–222, 2010.

of reducing the spectral efficiency in the cell area. Clearly, a tradeoff between efficiency and fairness exists.

Within the 3GPP, several alternatives have been proposed [1], inter-cell interference (a) coordination (ICIC), (b) randomization and (c) cancelation. This work focuses on inter-cell interference coordination techniques, particularly in static and distributed approaches, that means configurations that are adjusted during the radio planning process.

Broadly speaking, the main target of any ICIC strategy is to determine the resources (bandwidth and power) available at each cell at any time. Then (and typically), an autonomous scheduler assigns those resources to users. Thus, from the Radio Resource Control perspective, there are two kind of decisions: (a) which resources will be allocated to each cell? and, (b) which resources will be allocated to each user?. The temporality of such decisions is quite different. Whereas the allocation of resources to users ranges in the order of miliseconds, the allocation of resources to cells lasts for much longer periods or may be even fixed. This work deals with the very last case.

Static ICIC schemes are attractive to operators since the complexity of their deployment is very low and there is not need for new extra signaling out of the standard. Besides ICIC is the best option in cells deployments with relatively evenly loaded traffic and some studies show how applications such as VoIP can obtain full benefit from its throughput gains [2].

Static ICIC mostly relies on the fractional reuse concept. This means that users are categorized according to their average Signal-to-Noise-plus-Interference Ratio (SINR), that means basically according to their inter-cell interference, and different reuse factors are applied to them, being higher at regions with more interference, mostly outer regions of the cells. The total system bandwidth is divided into sub-bands which are used by the scheduler accordingly.

Typically, static strategies follow 2 approaches:

- Apply different reuse factors to inner and outer users, being lower in the first case. In this sense, pioneer references can be found in [3, 4, 5] in which reuse factor 1 is applied to users in the central part of the cell, and a factor >1 to the users in the outer part. Some refinements and extensions were done in [6] and a comparison of these proposals in terms of spectral efficiency was presented in [7]. A formal analysis of fractional reuse can be found in [8].
- Reuse factor 1 is applied to the whole band but less power is devoted to inner users to reduce inter-cell interference. The authors in [9, 10] compute the optimum power levels so that the system throughput is maximized. Also, the optimization of static power levels were addressed analytically by the authors in [10].

The contribution of this work is to present a fair comparison among static ICIC schemes highlighting the role of the different elements affecting the efficiency vs. fairness tradeoff. To to do this, we evaluate an extended set of metrics which help us better understand the strengths and weaknesses of each method and we close the analysis providing some recommendations of practical interest based on the findings.

The paper is organized as follows, Section 2 provides the description of the static ICIC schemes we have considered and their corresponding configurations. Next, Section 3 explains the methodology, this is followed by the analysis of results in Section 4. Finally, conclusions and other remarks close the paper in Section 5.

2   Description of ICIC Strategies

In this section, we present a detailed description of the static ICIC strategies we have considered for this study. The choice of the three partiular static ICIC strategies attemps basically to the following reasons: well known problems with classical ICIC schemes based on full frequency reuse (FR) and FR=3 has been reported and explained in [11] and [12]. Moreover, such schemes are actually quite rigids in the sense that do not offer many degrees of freedom for network tunning. On the other hand, some flavors of fractional and soft frequency reuse schemes, as the ones suggested in [13] and [14], appear to be more suitable alternatives to flexibly tune the network. And finally, these schemes also provide a more convenient testing framework to deal with new elements such as classification of users in classes, classification thresholds and bandwidth and power allocations that can be parametrized. These elements are shown for the different ICIC schemes in Table 1.

The section finishes with the list of the performance metrics employed to assess the performance of those strategies.

2.1   Static ICIC

As we commented before, in static ICIC schemes, the resources allocated to each cell do not change over time. They are computed and evaluated during the radio planning process and only long-term readjustments are performed during the operation of the network. The set of sub-carriers and the power levels allocated to them is fixed for each cell. A generic representation of the power profiles corresponding to each scheme is shown in Fig. 1.

These strategies mainly rely on a classification of the users based on the average SINR. Thus, in general, the group to which each user belongs highly depends on the position of the user within the cellular layout. Two possible criteria have been defined because the choice of the threshold to classify users has an immediate impact on the scheduler decisions and so on the system performance, as it is shown later:

1. *Class Proportionality*: SINR thresholds are selected so that each class has the same average number of users.
2. *Bandwidth Proportionality*: The threshold guarantees that the number of users is proportional to its allocated bandwidth.

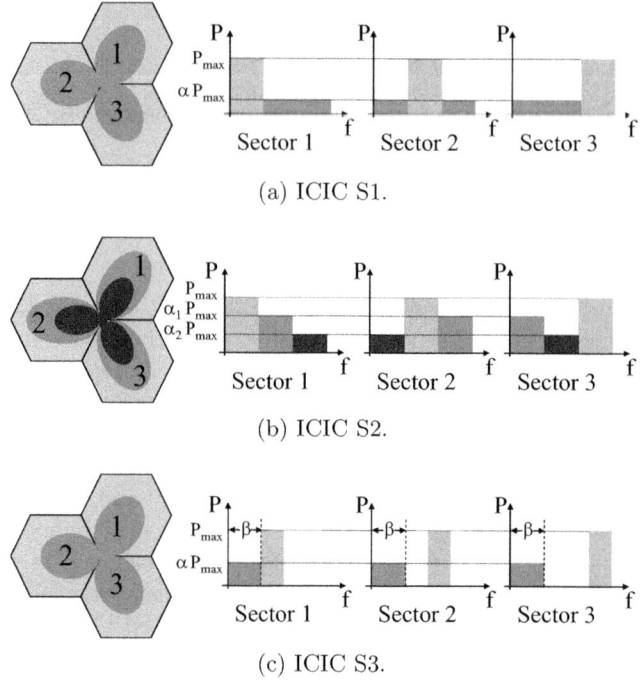

(a) ICIC S1.

(b) ICIC S2.

(c) ICIC S3.

Fig. 1. Classic power profiles in static ICIC schemes

Table 1. Details of static ICIC schemes

ICIC scheme	Sub-bands	Bandwidth (%)	User class
S1	1	33	Inner
	2	66	Outer
S2	1	33	Inner
	2	33	Centrals
	3	33	Outer
S3	1	$\beta \cdot 100$	Inner
	2		
	3	$(1 - \beta)\frac{1}{3} \cdot 100$	Outer
	4		

ICIC S1: In this scheme, we can observe from Fig. 1a that the design is done to guarantee a reuse factor 3 for outer users, while the inner ones have a lower value. In this case, inter-cell interference is also inter-class because of the assignment of bands among different cells. In order to control the amount of interference received by cell-edge users, low power is used in the bands to be used in the central area of the cell. This power is controlled by the parameter α, which is the only tunable factor in this schema. In order to evaluate the impact of a

Table 2. Configurations considered for ICIC S1

Config. Id.	α	Classification criterion	Associated SINR Threshold [dB]
S1.a	0.12		
S1.b	0.37	BW Prop.	0.35
S1.c	0.75		
S1.d	1.00		
S1.e	0.12		
S1.f	0.37	Class Prop.	2.40
S1.g	0.75		
S1.h	1.00		

Table 3. Configurations considered for ICIC S2

Config. Id.	α_1	α_2	Classification criterion	Associated SINR Threshold [dB]
S2.a	1.00	1.00		
S2.b	0.66	0.33	Class Prop.	$\{0.35, 5\}$
S2.c	0.5	0.25		

poorly adjusted value, different cases have been considered in the analysis, these are shown in Table 2.

ICIC S2: This case is a natural extension of the previous scheme in order to study the effect of a more accurate classification of users. In this case we have three classes instead of two. Table 3 shows the different configurations. The analysis of this case is similar to the one for ICIC S1.

ICIC S3: As in ICIC S1, two different classes are also considered, however the novelty here is that inter-class interference is removed completely. In other words, each class has exclusive use of its bandwidth. This is important because the performance in terms of throughput and fairness becomes independent of α since the SINR does not depend on the transmitted power, equal for all cells, as long as the inter-cell interference level is significantly higher than the noise floor. The parameter β controls the width of the band allocated to inner users, hence it also determines the bandwidth available for outer ones. For this reason, in order to be consistent with the classification criteria previously defined it is necessary to evaluate this scheme for additional thresholds of SINR. Table 4 shows the different configurations considered for this strategy.

Table 4. Configurations considered for ICIC S3. $\alpha = 0.5$

Config. Id.	β	Classification criterion	Associated SINR Threshold [dB]
S3.a	0.10	Class Prop.	2.40
S3.b	0.25	Class Prop.	2.40
S3.c	0.25	–	3.95
S3.d	0.25	BW Prop.	7.05
S3.e	0.25	–	12.7
S3.f	0.40	Class Prop.	2.40
S3.g	0.40	BW Prop.	3.95
S3.h	0.49	Class Prop.	2.40

Table 5. Performance metrics

Metric	Units
Spectral efficiency	$[\frac{bps}{Hz \cdot km^2}]$
Average cell rate	[bps]
Average user rate	[bps]
Weakest user rate	[bps]
Fairness (Jain's Index [15]): Assuming that fair means equal throughput, this metric measures if n users are receiving equal treatment. The result ranges from $1/n$ (worst case) to 1 (best case).	N/A
Payload per PRB: Effectiveness in the bandwidth usage.	$[\frac{bits}{PRB}]$
Average transmitted power per cell.	[W]
Payload per Watt: Effectiveness in the power usage.	$[\frac{bits}{W}]$

2.2   Comparison of Strategies: Performance Metrics

In order to evaluate the static ICIC strategies properly, a set of metrics have been taken into account. These allow to look at the performance from several perspectives. Many contributions do not take into account the fairness in their analysis. Moreover, very often a tradeoff oriented analysis is also missing. In this work, the main target is to provide a comprehensive evaluation of these ICIC strategies in which this tradeoff (fairness vs. efficiency) can be clearly understood. Table 5 lists the set of metrics. Note that a Physical Resource Block (PRB) is the minimum bandwidth the scheduler can assign to one single user.

3   Methodology

We consider the downlink of a cellular network composed by 57 cells in a regular tri-sectorial layout. The inter-site distance is 1.5 km and the average number of users per cell is 15. The users are randomly distributed without mobility since the snapshots lifetime is only 18 ms. Experiments were perfomed by means of snapshots which are suitable for comparisson purposes where the most important element is to achieve as many as possible statistically independent realizations.

An OFDMA system level simulation platform has been developed in C++ whose configuration follows the setting established by the LTE standard [16, 17, 18]. Moreover, the link-to-system level interface largely follows the guidelines given by [19]. Specifically, the system has 100 PRB available for the users (18 MHz, 1200 sub-carriers of 15 kHz). Transmission time intervals of 1 ms containing 10 OFDMA symbols are considered. The total available power at each cell is 43 dBm. We studied and quantified the performance of the different strategies by means of a semi-static system level simulator. Statistics were collected from the 21 central cells to avoid border effects. 3GPP urban macro-cellular and ITU Extended Typical Urban (ETU) have been considered as propagation and channel models respectively. 8 dB log-normal shadowing is applied following the model proposed in [20] with a correlation coefficient between cells equal to 0.5. It is important to note that achievable rates were computed taking into account the instantaneous channel conditions and according to the adaptive modulation and coding used in LTE, as specified in [21]. This mapping has been done using the link abstraction model based in mutual information at modulation symbol level [22], which outperforms the classic Effective Exponential SINR model because it is able to predict the BLER with higher accuracy, particularly for higher order modulations, such as 64-QAM.

4   Numerical Results

Table 6 shows the results obtained for the experiments defined in Table 2, 3 and 4. Nevertheless, for comparison purposes, some of the results are shown graphically in Fig. 2.

Figures 2a and 2b show the results corresponding to the first static strategy, S1. In figures 2a and 2b the results obtained for different values of α when the bandwidth and class proportionality criteria were applied are shown. Clearly, when bandwidth proportionality is used, the set of cell-edge users becomes smaller leaving more PRBs than the class proportionality case. Because of this, the values obtained for fairness are slightly better. On the other hand, class proportionality brings a significant efficiency improvement at the expense of a small fairness degradation. Looking at the effect of α, it is evident that an increase of such parameter leads to better values of efficiency no matter which classification criterion has been selected. This means that this parameter can be tuned independently of the number of users of each type.

Fig. 2c shows the results for the static strategy S2. In this scheme, an additional cell area (or class) is introduced in order to obtain a more detailed classification. In this case, class proportionality was applied and the effect of the variations of α_1 and α_2 was studied.

Since users are supposed to be grouped into more classes, fairness is expected to be improved. Nevertheless, S1 outperforms S2 in terms of efficiency since the

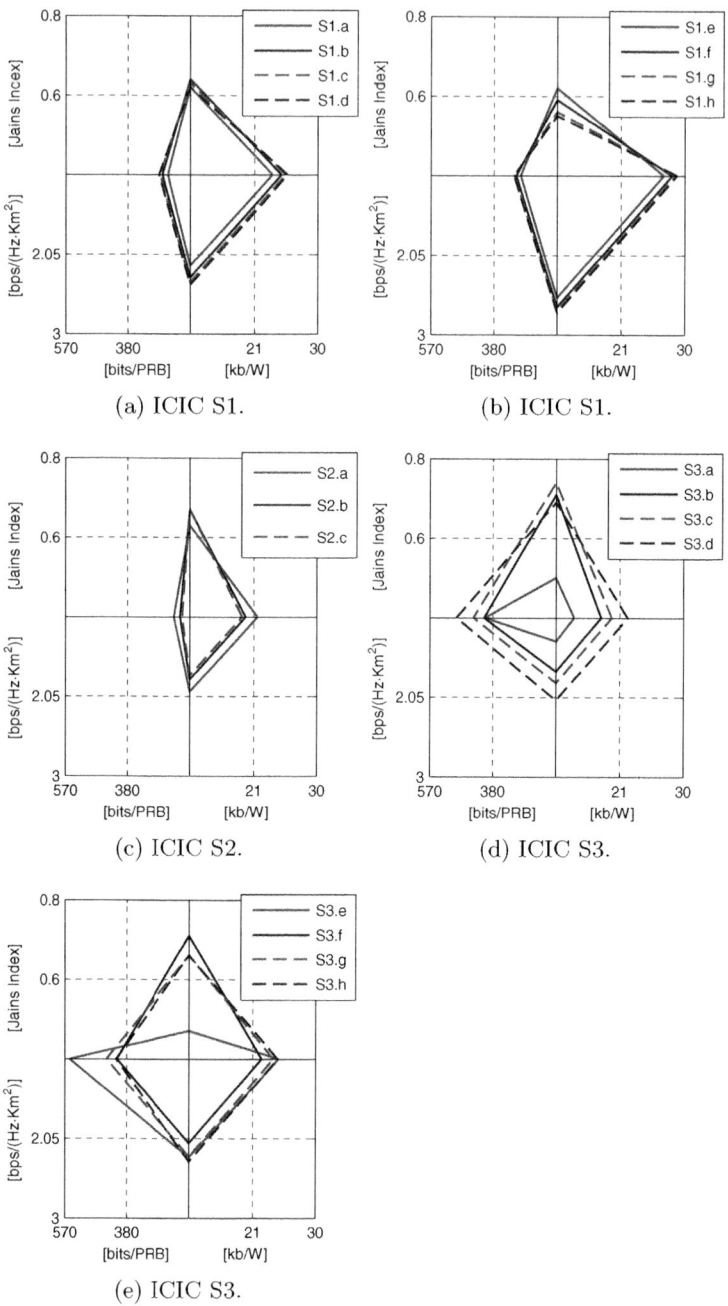

(a) ICIC S1.

(b) ICIC S1.

(c) ICIC S2.

(d) ICIC S3.

(e) ICIC S3.

Fig. 2. Results obtained for the different ICIC approaches

Table 6. Overall comparison

Id.	Spectral efficiency $\left[\frac{\text{bps}}{\text{Hz}\cdot\text{km}^2}\right]$	Cell rate [Mbps]	User rate [Mbps]	Weakest user rate [kbps]	Fairness [J. index]	Payload per PRB $\left[\frac{\text{bits}}{\text{PRB}}\right]$	Power [W]	Payload per Watt $\left[\frac{\text{kb}}{\text{W}}\right]$
S1.a	2.19	25.60	1.91	130	0.62	257.9	19.6	23.5
S1.b	2.33	27.18	2.03	130	0.64	273.8	19.7	24.8
S1.c	2.39	27.92	2.09	100	0.63	281.3	19.8	25.4
S1.d	2.42	28.25	2.11	50	0.62	284.6	19.8	25.7
S1.e	2.56	29.88	2.23	130	0.62	299.1	19.9	27.0
S1.f	2.67	31.21	2.33	120	0.59	312.3	19.9	28.2
S1.g	2.72	31.76	2.37	80	0.56	317.9	19.9	28.7
S1.h	2.74	32.00	2.39	50	0.55	320.2	19.9	28.9
S2.a	2.00	23.44	1.75	50	0.63	238.15	19.64	21.5
S2.b	1.85	21.57	1.61	110	0.67	219.18	19.57	19.84
S2.c	1.80	21.00	1.57	130	0.67	213.36	19.32	19.32
S3.a	1.38	16.12	1.20	30	0.50	403.67	19.91	14.57
S3.b	1.74	20.32	1.52	90	0.71	406.87	19.92	18.36
S3.c	1.88	21.95	1.64	80	0.74	439.19	19.94	19.81
S3.d	2.09	24.46	1.83	70	0.69	490.63	19.91	22.11
S3.e	2.28	26.70	2.00	70	0.47	555.10	19.45	24.71
S3.f	2.11	24.63	1.84	60	0.71	410.77	19.93	22.24
S3.g	2.27	26.48	1.98	50	0.66	441.48	19.94	23.90
S3.h	2.33	27.21	2.03	40	0.66	412.52	19.93	24.57

allocation of a wider band to the set of users enjoying a better radio channel becomes the predominant effect, especially when the SINR threshold is shifted to higher values. On the other hand, when we look at the joint effect of α_1 and α_2, it is clear that the higher the difference in the power assigned to each class, the higher value of fairness is obtained although with a marginal gain of S2.b over S2.c. The effect on efficiency is the opposite. Since the bandwidth assigned to each class is fixed, as we equalize the power allocated to them, we are taking energy previously assigned to the users having worse channel users and moving it to users with better channel conditions.

Finally, the results corresponding to the static scheme S3 are shown in Fig. 2d and 2e. In this case inter-class interference is completely removed. This is done at the expense of a reduction in the available bandwidth at each cell as it can be seen from Fig. 1c. The advantage of doing so is that higher levels of SINR can be achieved within the cell; this opens the door to the use of techniques requiring good channel estimations, such as spatial multiplexing with a MIMO configuration, in a wider area of the cell with possible throughput improvement. Despite having a reduction of 7% in terms of spectral efficiency, the effectiveness in the use of resources is higher than the previous cases; this is evident since the number of PRB and bits per W is significantly greater.

5   Conclusions

A fair comparison among different static ICIC strategies has been presented in this work. In order to assess the performance not only from the efficiency-fairness tradeoff but also from the effectiveness in the resources usage perspective, a set of metrics were employed. Moreover, these results were obtained from different experiments with identical simulation conditions.

The main conclusions are summarized as follows:

- In general, a more accurate classification of users does not imply a better performance. In addition, we have seen that the fairness among classes is more affected by the amount of power allocated to each one.
- Based on the results, it is clear that, when a fine tuning of the parameters is done, a very attractive efficiency-fairness tradeoff can be achieved by using any of the static ICIC strategies. Nevertheless, there are subtle differences among them. Schemes S1 and S2 tend to favorate more the spectral efficiency from the overall network perspective as these schemes employ in general a higher frequency reuse. On the other hand, scheme S3 is an advisable solution when fairness is an important issue, for instance when a high number of users is requesting resources.
- Another practical recommendation, when higher order modulations and advanced techniques requiring higher levels of SINR are expected to be used, is the utilization of fractional reuse approaches such as scheme S3.
- From a practical perspective, static ICIC coordination techniques are a very attractive alternative for operators since there is not complexity associated nor signalling exchange.
- Moreover, it is important to recall that perhaps the most important issue related to static ICIC coordination is the lack of adaptability to the network conditions; these include changing traffic loads, traffic patterns and mobility. Also, fault tolerance is a critic element that could be further investigated.
- Finally, it is of special interest to study the behaviour of such ICIC strategies when channel feedback is prohibitive or not available at all, for instance scenarios with high movility. These among others are future research lines we plan to work on.

References

[1] 3GPP: TR 25.814 v7.1.0 (Release 7) - Physical layer aspects for Evolved Universal Terrestrial Radio Access (UTRA). Technical Report, 3GPP (2006)
[2] Alcatel-Lucent: R1-072568: Voice over IP resource allocation benefiting from Interference Coordination. 3GPP, TSG RAN WG1 Meeting #49: Kobe, Japan (May 2007)

[3] Huawei, G.: 3GPP TSG RAN WG1 Meeting #41 - Soft Frequency Reuse Scheme for UTRAN LTE. Technical Report available as R1-050507, 3GPP (2005)

[4] Ericsson, G.: 3GPP TSG RAN WG1 Meeting #45 - Downlink Inter-cell Interference Co-ordination/Avoidance - Evaluation of Frequency Reuse. Technical Report available as R1-061374, 3GPP (2006)

[5] Alcatel, G.: 3GPP TSG RAN WG1 Meeting #44 - System Simulation Results for Downlink Interference Coordination. Technical Report available as R1-060209, 3GPP (2006)

[6] Zhang, X., He, C., Jiang, L., Xu, J.: Inter-cell Interference Coordination Based on Softer Frequency Reuse in OFDMA Cellular Systems. In: Proc. IEEE Int. Conf. Neural Networks and Signal Processing (ICNNSP 2008), Zhenjiang (China), June 7-11 (2008)

[7] Ruiz, S., Haro, E., González, D., García-Lozano, M., Olmos, J.: Comparison of Different Distributed Scheduling Strategies for Static/Dynamic LTE Scenarios. Technical Report available as TD(09)992, COST, Wien (Austria) (September 28-30, 2100)

[8] Elayoubi, S.E., Haddada, B., Fourestié, B.: Performance Evaluation of Frequency Planning Schemes in OFDMA-based Networks. IEEE Trans. on Wireless Comm. 7(5), 1623–1633 (2008)

[9] Corvino, V., Gesbert, D., Verdone, R.: A Novel Distributed Interference Mitigation Technique Using Power Planning. In: Proc. IEEE Wireless Comm. and Networking Conf (WCNC 2009), Budapest (Hungary), April 5-8 (2009)

[10] Boghe, M., Gross, J., Wolisz, A.: Optimal Power Masking in Soft Frequency Reuse Based OFDMA Networks. In: Proc. European Wireless Conf (EW 2009), Aalborg (Denmark), May 17-20 (2009)

[11] Boudreau, G., Panicker, J., Guo, N., Chang, R., Wang, N., Vrzic, S.: Interference coordination and cancellation for 4G networks. IEEE Communications Magazine 47(4), 74–81 (2009)

[12] Alcatel: R1-050407: Interference Coordination in new OFDM DL air interface. 3GPP, TSG RAN WG1 Meeting #41: Athens, Greece (May 2005)

[13] Nortel: R1-060905: Adaptive Fractional Frequency Reuse. 3GPP, TSG RAN WG1 Meeting #44bis: Athens, Greece (Mar 2006)

[14] Samsung: R1-051341: Flexible Fractional Frequency Reuse Approach. 3GPP, TSG RAN WG1 Meeting #43: Seoul, Korea (November 2005)

[15] Jain, R.: The Art of Computer Systems Performance Analysis, 1st edn. John Wiley & Sons, New Yotk (1991)

[16] 3GPP: TR 36.201 v8.2.0 (Release 8) - LTE Physical Layer - General Description. Technical Report, 3GPP (2008)

[17] 3GPP: TR 36.211 v8.5.0 (Release 8) - Physical Channels and Modulation. Technical Report, 3GPP (2008)

[18] 3GPP: TR 36.212 v8.5.1 (Release 8) - Multiplexing and Channel Coding. Technical Report, 3GPP (2008)

[19] Brueninghaus, K., Astely, D., Salzer, T., Visuri, S., Alexiou, A., Karger, S., Seraji, G.A.: Link performance models for system level simulations of broadband radio access systems. In: IEEE 16th International Symposium on Personal, Indoor and Mobile Radio Communications, PIMRC 2005, vol. 4, pp. 2306–2311 (November 2005)

[20] Fraile, R., Lázaro, O., Cardona, N.: Two Dimensional Shadowing Model. Technical Report available as TD(03)171, COST 273, Prague (Czech Rep.) (September 24-26, 2003)

[21] Olmos, J., Serra, A., Ruiz, S., García-Lozano, M., González, D.: Exponential Effective SIR Metric for LTE Downlink. In: Proc. IEEE Int. Symp. on Personal, Indoor and Mobile Radio Comm. (PIMRC 2009), Tokyo (Japan), September 13-16 (2009)

[22] Zheng, H., Wu, M., Choi, Y., Himayat, N., Zhang, J., Zhang, S.: Link Performance Abstraction for ML Receivers Based on RBIR Metrics. Technical Report C802.16m-08, IEEE (2008)

A Two-Users Transmission Game in OFDM Wireless Networks with Resource Cost

Andrey Garnaev[*] and Anton Toritsyn

Saint Petersburg State University,
Universitetskii prospekt 35, Peterhof, St Petersburg 198504 Russia

Abstract. The goal of this work is to generalize closed form approach suggested by Altman, Avrachenkov and Garnaev for a water-filling problem in game-theoretical frameworks for the case of a general allocation of users. Also we take into account power cost which can produce essential impact on users behaviour. The equilibrium strategies for the extended model are found in closed form. Uniqueness of the equilibrium is proved. Explicit criteria showing when the users prefer to apply all the power they have in disposition, or they would like to reduce it or even to cancel transmission at all are supplied. Algorithm based on a generalization of bisection method is produced and numerical modelling is performed.

Keywords: Resource Cost, Equilibrium Strategies, Resource Allocation, OFDM.

1 Introduction

In wireless networks and DSL access networks the total available power for signal transmission has to be distributed among several resources. In the context of wireless networks, the resources may correspond to frequency bands (e.g. as in OFDM). This spectrum of problems can be considered in game-theoretical multiusers scenario which leads to "Water Filling Game" or "Gaussian Interference Game" [5], [10], [14], where each user perceives the signals of the other users as interference and maximizes a concave function of the noise to interference ratio. A natural approach in the non-cooperative setting is the application of the Iterative Water Filling Algorithm (IWFA) [11]. Recently, in [7] the convergence of IWFA under fairly general conditions was proved. An interested reader can find more references and game theoretical models applied to OFDM/OFDMA resource management in [5], [12], [13], [15] and [16].

We would like to mention that the water filling problem and jamming games with transmission costs have been analyzed in [1]. In [2], [3] a closed form approach to find a Nash equilibrium was developed for so called symmetric water filling game where user are located on the same distance from base station (so, when all the fading channel gains are the same) was suggested. The goal of this work is to extend this approach for a general users allocation case, i.e. for the

---

[*] The work was partly supported by RFBR Grant no.09-01-00334-a.

A. Vinel et al. (Eds.): MACOM 2010, LNCS 6235, pp. 223–234, 2010.

scenario with different fading channel gains. Also we take into account power cost which can produce essential impact on behaviour of users. We supply explicit criteria showing when the users prefer to apply all the power they have in disposition, or they would like to reduce it or even to cancel transmission at all.

Also, note that nowadays energy saving and reduction of electromagnetic pollution become important issues. One approach to these problems is the introduction of taxes on the energy dissipation. The transmission cost can be considered as a variant of such taxation [4], since clearly too big taxes will strongly discourage the users of wireless technology and hinder the progress and too small taxes will lead to a wasteful use of the energy resources and may also lead to reckless use of the radio resources.

The structure of the paper is as follows: In Section 2 the formulation of the problem is give. Also we found in closed form relation between equilibrium strategies and Lagrangian multipliers. This closed form presentation allows us to reduce the problem of finding Nash equilibrium to a problem of solution of two non-linear equations one by one which can be done easily by a variation of bisection method due to obtained monotonous properties on Lagrangian multipliers of the equilibrium strategies. These monotonous properties also allow us easily to establish the uniqueness of the equilibrium. Then, in Section 3 we provide numerical examples which illustrate the theoretical results and in Section 4 a discussion of the results and possible generalization are supplied.

2   The Game

In this section we consider game-theoretical formulation of the situation where two users (transmitters) transmit signals through n sub-carriers taking into account power cost. So, we deal with with multiple access channel situation where multiple uncoordinated transmitters send independent information to a common receiver. It is worth to note that this transmission cost can impact on the user's behavior essentially, namely, for big transmission cost users can reject from transmission at all, and of course for small transmission cost users will employ the network facilities in the full range. A strategy of user j $(j = 1, 2)$ is vector $T^j = (T^j_1, \ldots, T^j_n)$, where $T^j_i \geqslant 0$ and $\sum_{i=1}^n T^j_i \leqslant \bar{T}^j$, where $\bar{T}^j > 0$ is the total signal which user j has to transmit. The payoff to users are quality of service (in our case it is Shannon capacities) minus transmission cost. Thus, the payoffs are given as follows:

$$v^1(T^1, T^2) = \sum_{i=1}^n \ln\left(1 + \frac{\alpha_i T^1_i}{\beta_i^{12} T^2_i + N^0_i}\right) - C^1 \sum_{i=1}^n T^1_i,$$

$$v^2(T^1, T^2) = \sum_{i=1}^n \ln\left(1 + \frac{\beta_i T^2_i}{\alpha_i^{21} T^1_i + N^0_i}\right) - C^2 \sum_{i=1}^n T^2_i,$$

where $N^0_i > 0$ is the uncontrolled noise, $\alpha_i, \beta_i^{12}, \beta_i, \alpha_i^{21}$ are fading sub-carrier gains for sub-carrier i. C^1 and C^2 are transmission cost for users. This is an instance of the Water-Filling or Gaussian Interference Game [5,10,11].

Using notation $g_i^{12} := \alpha_i/\beta_i^{12}$ and $g_i^{21} := \beta_i/\alpha_i^{21}$ where g_i^{12} and g_i^{21} are crosstalk coefficients we can rewrite our problem in the following equivalent form:

$$
\begin{aligned}
v^1(T^1, T^2) &= \sum_{i=1}^{n} \ln\left(1 + \frac{\alpha_i T_i^1}{g_i^{12}\beta_i T_i^2 + N_i^0}\right) - C^1 \sum_{i=1}^{n} T_i^1, \\
v^2(T^1, T^2) &= \sum_{i=1}^{n} \ln\left(1 + \frac{\beta_i T_i^2}{g_i^{21}\alpha_i T_i^1 + N_i^0}\right) - C^2 \sum_{i=1}^{n} T_i^2.
\end{aligned}
\tag{1}
$$

To escape bulkiness in the formulas we restrict ourselves to the symmetric game where crosstalk coefficients for different users coincide, namely, $g_i^{12} = g_i^{21} = g_i$ for $i \in [1, n]$ what relates to the plot when the users are allocated approximately on the same distance from base station. In the discussion section we will explain how obtained result can be extended for the general allocation case. Thus, we will focus on the following payoffs:

$$
v^1(T^1, T^2) = \sum_{i=1}^{n} \ln\left(1 + \frac{\alpha_i T_i^1}{g_i\beta_i T_i^2 + N_i^0}\right) - C^1 \sum_{i=1}^{n} T_i^1,
$$

$$
v^2(T^1, T^2) = \sum_{i=1}^{n} \ln\left(1 + \frac{\beta_i T_i^2}{g_i\alpha_i T_i^1 + N_i^0}\right) - C^2 \sum_{i=1}^{n} T_i^2.
$$

Note that in [2] the case with equal crosstalk coefficients and fading channel gains $g_i = g$, $\alpha_i = \beta_i$, $i \in [1, n]$ was considered.

Our goal is to find a Nash equilibrium for this game. The strategies T^{1*}, T^{2*} are the Nash equilibrium if for any strategies T^1, T^2 the following inequalities hold:

$$
v^1(T^1, T^{2*}) \leqslant v^1(T^{1*}, T^{2*}), v^2(T^{1*}, T^2) \leqslant v^2(T^{1*}, T^{2*}).
$$

Thus, Kuhn–Tucker Theorem from non-linear programming implies that (T^1, T^2) is a Nash equilibrium if and only if there are non-negative ω^1 and ω^2 (Lagrange multipliers) such that

$$
\frac{\alpha_i}{\beta_i g_i T_i^k + \alpha_i T_i^m + N_i^0} - C \begin{cases} = \omega^m & \text{for } T_i^m > 0, \\ \leqslant \omega^m & \text{for } T_i^m = 0, \end{cases}
\tag{2}
$$

where $\{k, m\} = \{1, 2\}$ and

$$
\omega^k \begin{cases} \geqslant 0 & \text{for } \sum_{i=1}^{n} T_i^k = \bar{T}^k, \\ = 0 & \text{for } \sum_{i=1}^{n} T_i^k < \bar{T}^k. \end{cases}
$$

This relation allows to specify in Theorem 1 the structure of equilibrium strategies in closed form as functions on Lagrange multipliers. This closed form presentation will allow us to reduce the problem of finding Nash equilibrium to a problem of solution of two non-linear equation one by one.

Theorem 1. *Let $g_i < 1$. The equilibrium strategies*

$$(T^1, T^2) = (T^1(\omega^1, \omega^2), T^2(\omega^1, \omega^2))$$

as functions on Lagrange multipliers have to have the following form.
(a) Let $\alpha_i/(\omega^1 + C^1) > \beta_i/(\omega^2 + C^2)$. Then

$$T_i^1(\omega^1, \omega^2) = \begin{cases} \dfrac{1}{\alpha_i(1 + g_i)} \\ \quad \times \left(\dfrac{\dfrac{\alpha_i}{\omega^1 + C^1} - \dfrac{g_i\beta_i}{\omega^2 + C^2}}{1 - g_i} - N_i^0 \right), & \dfrac{\dfrac{\beta_i}{\omega^2 + C^2} - \dfrac{g_i\alpha_i}{\omega^1 + C^1}}{1 - g_i} > N_i^0 \\[4ex] \dfrac{1}{\omega^1 + C^1} - \dfrac{N_i^0}{\alpha_i}, & \dfrac{\dfrac{\beta_i}{\omega^2 + C^2} - \dfrac{g_i\alpha_i}{\omega^1 + C^1}}{1 - g_i} \leqslant N_i^0 \\ & \text{and } N_i^0 < \dfrac{\alpha_i}{\omega^1 + C^1} \\[3ex] 0, & \dfrac{\alpha_i}{\omega^1 + C^1} \leqslant N_i^0 \end{cases}$$

$$T_i^2(\omega^1, \omega^2) = \begin{cases} \dfrac{1}{\beta_i(1 + g_i)} \\ \quad \times \left(\dfrac{\dfrac{\beta_i}{\omega^2 + C^2} - \dfrac{g_i\alpha_i}{\omega^1 + C^1}}{1 - g_i} - N_i^0 \right), & \dfrac{\dfrac{\beta_i}{\omega^2 + C^2} - \dfrac{g_i\alpha_i}{\omega^1 + C^1}}{1 - g_i} > N_i^0 \\[4ex] 0, & \dfrac{\dfrac{\beta_i}{\omega^2 + C^2} - \dfrac{g_i\alpha_i}{\omega^1 + C^1}}{1 - g_i} \leqslant N_i^0 \end{cases}$$

(b) Let $\alpha_i/(\omega^1 + C^1) < \beta_i/(\omega^2 + C^2)$. Then

$$T_i^1(\omega^1, \omega^2) = \begin{cases} \dfrac{1}{\alpha_i(1 + g_i)} \\ \quad \times \left(\dfrac{\dfrac{\alpha_i}{\omega^1 + C^1} - \dfrac{g_i\beta_i}{\omega^2 + C^2}}{1 - g_i} - N_i^0 \right), & \dfrac{\dfrac{\alpha_i}{\omega^1 + C^1} - \dfrac{g_i\beta_i}{\omega^2 + C^2}}{1 - g_i} > N_i^0 \\[4ex] 0, & \dfrac{\dfrac{\alpha_i}{\omega^1 + C^1} - \dfrac{g_i\beta_i}{\omega^2 + C^2}}{1 - g_i} \leqslant N_i^0 \end{cases}$$

$$T_i^2(\omega^1, \omega^2) = \begin{cases} \dfrac{1}{\beta_i(1 + g_i)} \\ \quad \times \left(\dfrac{\dfrac{\beta_i}{\omega^2 + C^2} - \dfrac{g_i\alpha_i}{\omega^1 + C^1}}{1 - g_i} - N_i^0 \right), & \dfrac{\dfrac{\alpha_i}{\omega^1 + C^1} - \dfrac{g_i\beta_i}{\omega^2 + C^2}}{1 - g_i} > N_i^0 \\[4ex] \dfrac{1}{\omega^2 + C^2} - \dfrac{N_i^0}{\beta_i}, & \dfrac{\dfrac{\alpha_i}{\omega^1 + C^1} - \dfrac{g_i\beta_i}{\omega^2 + C^2}}{1 - g_i} \leqslant N_i^0 \\ & \text{and } N_i^0 < \dfrac{\beta_i}{\omega^2 + C^2} \\[3ex] 0, & \dfrac{\beta_i}{\omega^2 + C^2} \leqslant N_i^0 \end{cases}$$

(c) Let $\alpha_i/(\omega^1 + C^1) = \beta_i/(\omega^2 + C^2)$. Then

$$T_i^1(\omega^1, \omega^2) = \begin{cases} \dfrac{1}{\alpha_i(1 + g_i)} \left(\dfrac{\alpha_i}{\omega^1 + C^1} - N_i^0 \right), & \dfrac{\alpha_i}{\omega^1 + C^1} > N_i^0 \\ 0, \dfrac{\alpha_i}{\omega^1 + C^1} \leqslant N_i^0 \end{cases}$$

$$T_i^2(\omega^1, \omega^2) = \begin{cases} \dfrac{1}{\beta_i(1 + g_i)} \left(\dfrac{\alpha_i}{\omega^1 + C^1} - N_i^0 \right), & \dfrac{\alpha_i}{\omega^1 + C^1} > N_i^0 \\ 0, & \dfrac{\alpha_i}{\omega^1 + C^1} \leqslant N_i^0. \end{cases}$$

In the next theorem we produce an explicit form for the total power applied by both strategies $T_i^1(\omega^1, \omega^2)$ and $T_{(}^2\omega^1, \omega^2)$ at subcarrier i.

Theorem 2. *The total power applied by both strategies at channel* $i \in [1, n]$ $T_i(\omega^1, \omega^2) = T_i^1(\omega^1, \omega^2) + T_i^2(\omega^1, \omega^2)$ *has the following form:*
(a) Let $\alpha_i/(\omega^1 + C^1) > \beta_i/(\omega^2 + C^2)$. Then

$$T_i(\omega^1, \omega^2) = \begin{cases} \dfrac{1}{1 + g_i} \left(\dfrac{\dfrac{\beta_i - \alpha_i g_i}{\beta_i(\omega^1 + C^1)} + \dfrac{\alpha_i - \beta_i g_i}{\alpha_i(\omega^2 + C^2)}}{1 - g_i} \right. \\ \qquad \left. - \dfrac{N_i^0}{\alpha_i} - \dfrac{N_i^0}{\beta_i} \right), & \dfrac{\dfrac{\beta_i}{\omega^2 + C^2} - \dfrac{g_i \alpha_i}{\omega^1 + C^1}}{1 - g_i} > N_i^0 \\ \dfrac{1}{\omega^1 + C^1} - \dfrac{N_i^0}{\alpha_i}, & \dfrac{\dfrac{\beta_i}{\omega^2 + C^2} - \dfrac{g_i \alpha_i}{\omega^1 + C^1}}{1 - g_i} \leqslant N_i^0 \\ & \text{and } N_i^0 < \dfrac{\alpha_i}{\omega^1 + C^1} \\ 0, \dfrac{\alpha_i}{\omega^1 + C^1} \leqslant N_i^0. \end{cases}$$

(b) Let $\alpha_i/(\omega^1 + C^1) < \beta_i/(\omega^2 + C^2)$. Then

$$T_i(\omega^1, \omega^2) = \begin{cases} \dfrac{1}{1 + g_i} \left(\dfrac{\dfrac{\beta_i - \alpha_i g_i}{\beta_i(\omega^1 + C^1)} + \dfrac{\alpha_i - \beta_i g_i}{\alpha_i(\omega^2 + C^2)}}{1 - g_i} \right. \\ \qquad \left. - \dfrac{N_i^0}{\alpha_i} - \dfrac{N_i^0}{\beta_i} \right), & \dfrac{\dfrac{\alpha_i}{\omega^1 + C^1} - \dfrac{g_i \beta_i}{\omega^2 + C^2}}{1 - g_i} > N_i^0 \\ \dfrac{1}{\omega^2 + C^2} - \dfrac{N_i^0}{\beta_i}, & \dfrac{\dfrac{\alpha_i}{\omega^1 + C^1} - \dfrac{g_i \beta_i}{\omega^2 + C^2}}{1 - g_i} \leqslant N_i^0 \\ & \text{and } N_i^0 < \dfrac{\beta_i}{\omega^2 + C^2} \\ 0, & \dfrac{\beta_i}{\omega^2 + C^2} \leqslant N_i^0. \end{cases}$$

(c) Let $\alpha_i/(\omega^1 + C^1) = \beta_i/(\omega^2 + C^2)$. Then

$$T_i(\omega^1, \omega^2) = \begin{cases} \dfrac{1}{1 + g_i} \left(\dfrac{2}{\omega^1 + C^1} - \dfrac{N_i^0}{\alpha_i} - \dfrac{N_i^0}{\beta_i} \right), & \dfrac{\alpha_i}{\omega^1 + C^1} > N_i^0 \\ 0, & \dfrac{\alpha_i}{\omega^1 + C^1} \leqslant N_i^0. \end{cases}$$

Denote by $H^k(\omega^1,\omega^2)$ and $H(\omega^1,\omega^2)$ the total power applied by strategy $T^k(\omega^1,\omega^2)$ of user k separately and both these strategies jointly, i.e.

$$H^1(\omega^1,\omega^2) = \sum_{i=1}^{n} T_i^1(\omega^1,\omega^2),$$

$$H^2(\omega^1,\omega^2) = \sum_{i=1}^{n} T_i^2(\omega^1,\omega^2),$$

$$H(\omega^1,\omega^2) = H^1(\omega^1,\omega^2) + H^2(\omega^1,\omega^2).$$

From explicit formulas of Theorems 1 and 2 we have the following result describing some important properties of equilibrium.

Theorem 3. *Let $g_i < 1$, $i \in [1,n]$. The equilibrium strategies have the following monotonous and continuous properties:*

1. *$T^1(\omega^1,\omega^2)$, $T^2(\omega^1,\omega^2)$, $T(\omega^1,\omega^2)$ and $H^1(\omega^1,\omega^2)$, $H^2(\omega^1,\omega^2)$, $H(\omega^1,\omega^2)$ are continuous on ω^1 and ω^2,*
2. *$T^1(\omega^1,\omega^2)$ and $H^1(\omega^1,\omega^2)$ are decreasing on ω^1 and increasing on ω^2, they are strictly monotonous while they are positive and they become equal zero for enough big ω^1,*
3. *$T^2(\omega^1,\omega^2)$ and $H^2(\omega^1,\omega^2)$ are increasing on ω^1 and decreasing on ω^2, they are strictly monotonous while they are positive and they become equal zero for enough big ω^2,*
4. *if $g_i \leqslant \min\{\alpha_i/\beta_i, \beta_i/\alpha_i\}$, $i \in [1,n]$ then $T(\omega^1,\omega^2)$ and $H(\omega^1,\omega^2)$ are decreasing on ω^1 and ω^2 and they are strictly decreasing while they are positive.*

Our next goal is to find the Lagrange multipliers. By Kuhn-Tucker Theorem each couple of Lagrange multipliers (ω^1,ω^2) satisfying the following conditions (3)-(5) supplies a Nash equilibrium:

$$H^1(\omega^1,\omega^2) \leqslant \bar{T}^1, \quad H^2(\omega^1,\omega^2) \leqslant \bar{T}^2 \tag{3}$$

such that

$$\text{if } H^1(\omega^1,\omega^2) < \bar{T}^1 \text{ then } \omega^1 = 0 \tag{4}$$

and

$$\text{if } H^2(\omega^1,\omega^2) < \bar{T}^2 \text{ then } \omega^2 = 0. \tag{5}$$

In the next three theorems we show that such couple of Lagrangian multipliers exists and is unique. Theorem 4 deals with the situation where power costs are big, Theorem 5 considers the plot with small positive power costs and Theorem 6 deals the case where power costs are equal to zero.

Theorem 4. *Let $C_1 C_2 > 0$ and $g_i < \min\{\alpha_i/\beta_i, \beta_i/\alpha_i\}$ for any i. If*

$$H(0,0) \leqslant \bar{T}^1 + \bar{T}^2 \tag{6}$$

then the game has unique Nash equilibrium (T^1, T^2), where

(a) If
$$H^1(0,0) \leqslant \bar{T}^1 \text{ and } H^2(0,0) \leqslant \bar{T}^2 \tag{7}$$

then $(T^1, T^2) = (T^1(0,0), T^2(0,0))$.

(b) If
$$H^1(0,0) > \bar{T}^1 \text{ and } H^2(0,0) < \bar{T}^2 \tag{8}$$

then $(T^1, T^2) = (T^1(\omega_{10}^*, 0), T^2(\omega_{10}^*, 0))$, where ω_{10}^* is unique solution of equation

$$H^1(\omega_{10}^*, 0) = \bar{T}^1. \tag{9}$$

(c) If
$$H^1(0,0) < \bar{T}^1 \text{ and } H^2(0,0) > \bar{T}^2 \tag{10}$$

then $(T^1, T^2) = (T^1(0, \omega_{01}^*), T^2(0, \omega_{01}^*))$, where ω_{01}^* is unique solution of equation

$$H^2(0, \omega_{01}^*) = \bar{T}^2. \tag{11}$$

(Note that by (6) the inequality $H^1(0,0) > \bar{T}^1$ and $H^2(0,0) > \bar{T}^2$ cannot hold. Thus, Theorem 4 covers all possible cases.)

Proof. By Theorem 3 $H(\omega^1, \omega^2)$ is decreasing on ω^1 and ω^2. Thus, $\omega^1 \omega^2 > 0$ implies that $H(\omega^1, \omega^2) < H(0,0) \leq \bar{T}^1 + \bar{T}^2$. So, either $H^1(\omega^1, \omega^2) < \bar{T}^1$ or $H^2(\omega^1, \omega^2) < \bar{T}^2$. Then (3)-(5) yield that one of the Lagrangian multipliers (either ω^1 or ω^2) has to be equal zero. Hence, only the following three situations of relation between them arise: (a) $\omega^1 = \omega^2 = 0$, (b) $\omega^1 > 0, \omega^2 = 0$ and (c) $\omega^1 = 0, \omega^2 > 0$

(a) The case $\omega^1 = \omega^2 = 0$ holds if and only if $H^1(0,0) \leqslant \bar{T}^1$ and $H^2(0,0) \leqslant \bar{T}^2$ and (a) follows

(b) The case $\omega^1 > 0, \omega^2 = 0$ holds if and only if $H^1(\omega^1, 0) = \bar{T}^1$ and $H^2(\omega^1, 0) \leqslant \bar{T}^2$. Since $H^1(\omega^1, 0)$ is decreasing then the last two relations hold if and only if $\omega^1 = \omega_{10}^*$, and $H^1(0,0) > \bar{T}^1$ and $H^2(0,0) < \bar{T}^2$.

(c) follows from (b) by symmetry.

Theorem 5. Let $C_1 C_2 > 0$ and $g_i < \min\{\alpha_i/\beta_i, \beta_i/\alpha_i\}$ for any i. Let

$$H(0,0) > \bar{T}^1 + \bar{T}^2. \tag{12}$$

Then the game has unique Nash equilibrium (T^1, T^2).

(a) Let
$$H^1(0, \bar{\omega}^2) > \bar{T}^1 \text{ and } H^1(\bar{\omega}^1, 0) < \bar{T}^1 \tag{13}$$

or what is equivalent to

$$H^2(0, \bar{\omega}^2) < \bar{T}^2 \text{ and } H^2(\bar{\omega}^1, 0) > \bar{T}^2. \tag{14}$$

where $\bar{\omega}^1$ and $\bar{\omega}^2$ are such that

$$H(\bar{\omega}^1, 0) = \bar{T}^1 + \bar{T}^2 \text{ and } H(0, \bar{\omega}^2) = \bar{T}^1 + \bar{T}^2 \tag{15}$$

(since H is decreasing by both parameters such $\bar\omega^1$ and $\bar\omega^2$ exist). Then

$$(T^1, T^2) = (T^1(\omega^*, \omega^2(\omega^*)), T^2(\omega^*, \omega^2(\omega^*)))$$

where $\omega^2(\omega)$ is decreasing continuous function on $[0, \bar\omega^1]$ such that

$$H(\omega, \omega^2(\omega)) = \bar T^1 + \bar T^2 \text{ for } \omega \in [0, \bar\omega^1] \tag{16}$$

and ω^* is the unique root of the equation

$$H^1(\omega^*, \omega^2(\omega^*)) = \bar T^1 \tag{17}$$

or what is equivalent to

$$H^2(\omega^*, \omega^2(\omega^*)) = \bar T^2. \tag{18}$$

(b) Let

$$H^1(\bar\omega^1, 0) \geq \bar T^1 \text{ and } H^1(0, \bar\omega^2) \geq \bar T^1 \tag{19}$$

or what is equivalent to

$$H^2(\bar\omega^1, 0) \leq \bar T^2 \text{ and } H^2(0, \bar\omega^2) \leq \bar T^2. \tag{20}$$

Then $(T^1, T^2) = (T^1(0, \bar\omega^*_{01}), T^2(0, \bar\omega^*_{01}))$.

(c) Let

$$H^1(\bar\omega^1, 0) \leq \bar T^1 \text{ and } H^1(0, \bar\omega^2) \leq \bar T^1. \tag{21}$$

or what is equivalent to

$$H^2(\bar\omega^1, 0) \geq \bar T^2 \text{ and } H^2(0, \bar\omega^2) \geq \bar T^2. \tag{22}$$

Then $(T^1, T^2) = (T^1(\bar\omega^*_{10}, 0), T^2(\bar\omega^*_{10}), 0)$.

Proof. We are looking ω^1 and ω^2 satisfying (3)-(5). Since H is decreasing then by (12) ω^1 and ω^2 cannot be equal to zero simultaneously. Thus, only the following three situations of relation between them arise: (a) $\omega^1 > 0, \omega^2 > 0$, (b) $\omega^1 > 0, \omega^2 = 0$ and (c) $\omega^1 = 0, \omega^2 > 0$.

First note that since H is decreasing by (12) $\omega^2(\omega)$ exists for $\omega \in [0, \bar\omega^1]$. Also, $\omega^2(0) = \bar\omega^2$ and $\omega^2(\bar\omega^1) = 0$ with $\bar\omega^1$ and $\bar\omega^2$ given by (15). Besides, $\omega^2(\omega)$ is continuous and decreasing in $[0, \bar\omega^1]$.

(a) $\omega^1 > 0, \omega^2 > 0$ holds if and only if $H^1(\omega^1, \omega^2) = \bar T^1$ and $H^2(\omega^1, , \omega^2) = \bar T^2$. Thus, $H(\omega^1, \omega^2) = \bar T^1 + \bar T^2$ and $\omega^2 = \omega^2(\omega^1)$ with $\omega^1 \in [0, \bar\omega^1]$. Since $\omega^2(\omega^1)$ is decreasing and $H^1(\omega^1, \omega^2)$ is decreasing on ω^1 and increasing on ω^2 then $\bar H^1(\omega^1) = H^1(\omega^1, \omega^2(\omega^1))$ is decreasing for $\omega^1 \in [0, \bar\omega^1]$ from $H^1(0, \bar\omega^2)$ for $\omega^1 = 0$ to $H^1(\bar\omega^1, 0)$ for $\omega^1 = \bar\omega^1$. The equation $\bar H^1(\omega^1) = \bar T^1$ has the unique root $\omega^1 = \omega^*$ if and only if $H^1(0, \bar\omega^2) > \bar T^1$ and $H^1(\bar\omega^1, 0) < \bar T^1$. Since instead of $\bar H^1(\omega^1)$ we could consider $\bar H^2(\omega^1) = H^2(\omega^1, \omega^2(\omega^1))$ which is increasing from $H^2(0, \bar\omega^2)$ for $\omega^1 = 0$ to $H^2(\bar\omega^1, 0)$ for $\omega^1 = \bar\omega^1$ the case (a) follows.

(b) $\omega^1 > 0, \omega^2 = 0$ holds if and only if $H^1(\omega^1, 0) = \bar{T}^1$ and $H^2(\omega^1, 0) \leq \bar{T}^2$.
Thus, $\omega^1 = \omega_{10}^*$. Also, $H(\omega_{10}^*, 0) = H^1(\omega_{10}^*, 0) + H^2(\omega_{10}^*, 0) \leq \bar{T}^1 + \bar{T}^2$ and
since $H(\omega, 0)$ is decreasing then $\omega_{10}^* \geq \bar{\omega}^1$. Thus, since $H^1(\omega, 0)$ is decreasing
then $H^1(\bar{\omega}^1, 0) \geq H^1(\bar{\omega}_{10}^*, 0) = \bar{T}^1$. Besides, if also $H^1(0, \bar{\omega}^2) \geq \bar{T}^1$ then neither
equation $H^1(\omega, \omega^2(\omega)) = \bar{T}^1$ or $H^2(\omega, \omega^2(\omega)) = \bar{T}^2$ can have positive root and
(b) follows.

(c) can be proved analogously to (b).

Finally note that the situation where there is no transmission costs is essentially simpler and it is described by the following theorem.

Theorem 6. *Let* $C_1 = C_2 = 0$ *and* $g_i < \min\{\alpha_i/\beta_i, \beta_i/\alpha_i\}$ *for any* i. *Then the game has unique Nash equilibrium* (T^1, T^2) *and*

$$(T^1, T^2) = (T^1(\omega^*, \omega^2(\omega^*)), T^2(\omega^*, \omega^2(\omega^*)))$$

with ω^* *and* $\omega^2(\omega^*)$ *given by (17) and (16).*

3   Numerical Results

Theorems 4 and 5 reduce the problem of finding Nash equilibrium to a problem of solving either nonlinear equations (9), (11), (15) and or two nonlinear equations (16) and (17) with monotonous structures. That is why they can be solved easy by bisection method. In this section we supply a numerical example of finding equilibrium strategies for different power costs based on this algorithm.

We assume that there are five channels ($n = 5$) and the background noise is permanent for all them ($N_i^0 = 0.1$, $i \in [1, 5]$). Let the fading channel gains are $g = [0.9, 0.8, 0.7, 0.6, 0.5]$, $\alpha_i = \beta_i = 1$ for $i \in [1, 5]$ and the total signal user 1

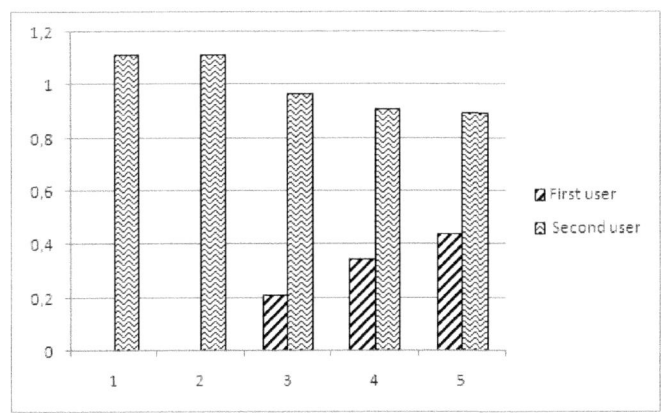

Fig. 1. The equilibrium strategies for $C = 0.1$

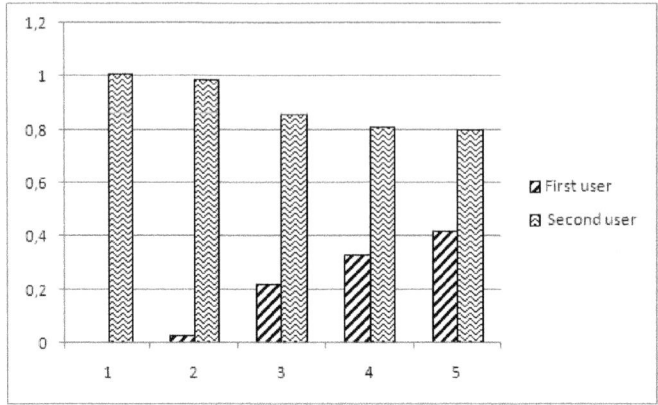

Fig. 2. The equilibrium strategies for $C = 0.9$

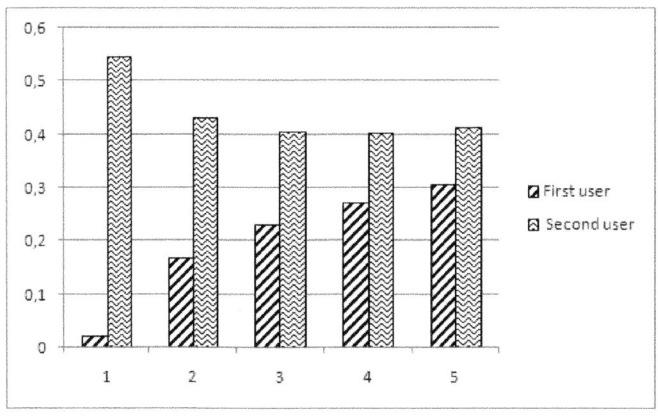

Fig. 3. The equilibrium strategies for $C = 1.5$

(2) has to transmit is $\bar{T}^1 = 1$ ($\bar{T}^2 = 5$). On Figures 1, 2 and 3 the equilibrium strategies of the users are given for the power costs $C^1 = C^2 = C = 0.1, 0.9, 1.5$ respectively.

For the low power cost $C = 0.1$ (Figure 1) both users transmit all the signal they intend to ($\sum_{i=1}^{5} T_i^1 = 1$ and $\sum_{i=1}^{5} T_i^2 = 5$). In this situation user 2 employs all the five channels while user 1 employs only three channels with highest quality.

If the power cost grows up to $C = 0.9$ (Figure 2) user 2 decreases the transmitted signal ($\sum_{i=1}^{5} T_i^2 < 5$) and no longer send the total signal (namely, he send $\sum_{i=1}^{5} T_i^2 \approx 4.47$) and user 1 employs more channels (namely, four channels) while his transmitted signal remains $\sum_{i=1}^{5} T_i^1 = 1$.

With further increasing of the power cost up to $C = 1.5$ (Figure 3) user 1 becomes to employ all the five channels ($\sum_{i=1}^{5} T_i^1 = 1$), while the signal transmitted by user 2 goes on to decrease ($\sum_{i=1}^{5} T_i^2 \approx 2.2$).

4   Discussion

In this paper we generalized closed form approach suggested in [2] for a water-filling problem in game-theoretical frameworks for general non-symmetrical allocation of users and also taking into account power cost which can produce essential impact on users behaviour. The equilibrium strategies for the extended model are found in closed form. Uniqueness of the equilibrium is proved. We show which impact can produce power costs on user's behaviour. Explicit criteria showing when the users prefer to apply all the total power they have in disposition ((13) and (12)), or they would like its reducing ((8) and (10) for (6) and (19) and (22) for (12)) or even they inclined to cancel transmission at all (6) and (7) are supplied. Algorithm based on a generalization of bisection method is produced which allows to perform numerical modelling. Finally note that suggested approach easily to apply for finding Nash equilibrium for the general payoffs (1). In this situation we have to assume that $g_i^{12}, g_i^{21} < \min\{\alpha_i/\beta_i, \beta_i/\alpha_i\}$. Then the monotonous and continuous properties announced in Theorem 3 still will be hold which allows to get equilibrium and prove its uniqueness as it is done in Theorems 4 and 5. Our future work involves investigating the problem when more than two users present in network, especially the case where the number of users is large.

References

1. Altman, E., Avrachenkov, K.E., Garnaev, A.: A Jamming game in wireless networks with transmission cost. In: Chahed, T., Tuffin, B. (eds.) NET-COOP 2007. LNCS, vol. 4465, pp. 1–12. Springer, Heidelberg (2007)
2. Altman, E., Avrachenkov, K., Garnaev, A.: Closed form solutions for water-filling problems in optimization and game frameworks. In: Proc. of ValueTools 2007 (2007)
3. Altman, E., Avrachenkov, K., Garnaev, A.: Closed form solutions for water-filling problem in optimization and game frameworks. Telecommunication Systems Journal (forthcomming, 2010)
4. Altman, E., Avrachenkov, K., Garnaev, A.: Taxation for Green Communication. In: WiOpt 2010, Avignon, France (2010)
5. Lai, L., El Gamal, H.: The water-filling game in fading multiple access channels. IEEE Trans. Information Theory (2005) (submitted), http://www.ece.osu.edu/~helgamal/
6. Lin, C.R., Gerla, M.: Adaptive clustering for mobile wireless networks. IEEE JSAC 15(7), 1265–1275 (1997)
7. Luo, Z.-Q., Pang, J.-S.: Analysis of iterative waterfilling algorithm for multiuser power control in digital subscriber lines. EURASIP Journal on Applied Signal Processing (2006)
8. Popescu, O., Rose, C.: Water filling may not good neighbors make. In: Proc. of GLOBECOM 2003, vol. 3, pp. 1766–1770 (2003)
9. Popescu, D.C., Popescu, O., Rose, C.: Interference avoidance versus iterative water filling in multiaccess vector channels. In: Proc. of IEEE VTC 2004, vol. 3, pp. 2058–2062 (2004)

10. Yu, W.: Competition and cooperation in multi-user communication environements. PhD Thesis, Stanford University (2002)
11. Yu, W., Ginis, G., Cioffi, J.M.: Distributed multiuser power control for digital subscriber lines. IEEE JSAC 20, 1105–1115 (2002)
12. Scutari, G., Palomar, D.P., Barbarossa, S.: Optimal linear precoding strategies for wideband non-cooperative systems based on game-theory - Part I: Nash equilibria. IEEE Trans. on Signal Processing (2007) (to appear)
13. Scutari, G., Palomar, D.P., Barbarossa, S.: Optimal Linear Precoding Strategies for Wideband Non-Cooperative Systems Based on Game Theory-Part II: Algorithms 56(3), 1250–1267 (2008)
14. Shum, K., Leung, K.-K., Sung, C.W.: Convergence of iterative waterfilling algorithm for Gaussian interference channels. IEEE Journal on Sel. Areas in Comm. 25(6), 1091–1100 (2007)
15. Han, Z., Ji, Z., Liu, K.J.R.: Fair Multiuser Channel Allocation for OFDMA Networks Using Nash Bargaining Solutions and Coalitions. IEEE Trans. on Comm. 53(8), 1366–1376 (2005)
16. Han, Z., Ji, Z., Liu, K.J.R.: Non-Cooperative Resource Competition Game by Virtual Referee in Multi-Cell OFDMA Networks. IEEE Journal on Sel. Areas in Comm. 25(6), 1079–1089 (2007)

QoS-Driven Radio Resource Allocation for OFDMA Networks Based on a Game Theoretical Approach

Claudio Sacchi and Fabrizio Granelli

University of Trento
Department of Information Engineering and Computer Science (DISI)
Via Sommarive 14, I-38050, Povo (Trento), Italy
{sacchi,granelli}@disi.unitn.it

Abstract. In this paper, we propose a cooperative strategy for OFDMA radio resource allocation based on game theory. The main novelty with respect to state-of-the-art is that the utility function considers the application-oriented Mean Opinion Score (MOS) rather than the gross data rate attributed to each user flow. In such a scenario, data flows compete in cooperative way in order to maximize the perceived Quality-of-Service (QoS). Experimental results show that the MOS achievable by the proposed resource allocation strategy is better than that one provided both by uncoordinated strategies based on water-filling and by cooperative strategies based on pure data rate maximization.

Keywords: OFDMA, Radio Resource Allocation, Game theory, QoS measurements.

1 Introduction

Orthogonal Frequency Division Multiplexing (OFDM) is a very promising solution to design high performance physical layer in digital radio communications. Some well-known wireless standards adopt OFDM as radio interface, namely: IEEE 802.11a, IEEE 802.11g, and IEEE 802.16e (WiMAX). One of the best features of OFDM is the increased flexibility in radio resource management with respect to single-carrier systems. Radio resource management (RRM) is made adaptive with respect to channel conditions with the objective of providing controlled QoS levels [1]. In a single-user OFDM system, RRM essentially consists of dynamically adapting modulation constellation and transmit power on each subcarrier. In a multi-user OFDM context, there is a need for a multiple access scheme to allocate subcarriers, bit loading and transmit power to the various users.

The problem of optimal allocation of power and subcarriers in OFDM and OFDMA systems has been widely dealt by literature. The tradeoff between efficiency (i.e.: maximum attainable data rate) and fairness (i.e.: equity in radio resource assignment) is one of the most challenging aspect of OFDMA RRM. State-of-the-art solutions look suboptimal from this viewpoint. Efficiency trends to privilege users with good channel conditions (that are generally closer to the base-station – BS).

A. Vinel et al. (Eds.): MACOM 2010, LNCS 6235, pp. 235–246, 2010.
© Springer-Verlag Berlin Heidelberg 2010

Fairness is based on criteria like max-min that do not consider the notion that different users might have different requirements [1]. In some recent works, some solutions have been proposed for OFDMA RRM based on "negotiation" strategies inspired to everyday life. RRM resembles to a marketplace where transmitting users can exchange goods (i.e.: power and/or subcarriers) and negotiate transactions so that people can be satisfied through bargaining. In a single-cell multi-user OFDMA system, the BS acts as the market and the distributed users can negotiate via the BS to cooperate in making decision on the subcarrier usage [2]. This fact fully motivates the application of game theory [3] to this specific problem. Game theory can achieve fairness while maximizing the overall system rate [2]. The Nash Bargaining Solution (NBS) is considered in [2] together with coalition in order to find an optimal agreement among negotiating users. Other approaches considering the use of game theory in OFDMA RRM have been proposed by Han, Ji and Liu in [4] and by Noh in [5]. In [4] a non-cooperative resource competition game is proposed in opposition to the cooperative approach of [2]. Non-cooperative game can solve the issue of some undesirable Nash equilibrium that sometimes affects cooperative strategies. In [5], an iterative resource control algorithm for distributed OFDMA systems using an auction game is proposed. Users and BS control transmitting power and bidding to maximize their utility. These controls are aimed at maximizing system capacity and fairness. In this paper, we propose a novel approach for OFDMA RRM that intrinsically maximizes the fairness in terms of user's satisfaction. Saul and Auer in [6] show the advantage to optimize cross-layer between application and MAC layer by considering user-centric metrics linked to the perceived quality-of-service (like e.g. Mean Opinion Square – MOS), rather than network-centric metrics like bit-rate or bit-error-rate. Therefore, we propose a cooperative game theoretic approach, where user data flows cooperate to achieve the best possible perceived QoS. In particular, the considered utility function is the maximization of the minimum MOS. In such a way, the RRM should increase fairness in order to allow users to be as much as possible satisfied with respect to their specific QoS requirements.

The rest of the paper is organized as follows: section 2 will describe the proposed OFDMA RRM algorithm, section 3 will present some selected experimental results, and finally section 4 will draw paper conclusions.

2   The Proposed OFDMA RRM Strategy

2.1   Problem Statement

Let's consider an OFDMA transmission system sharing among K users a fixed bandwidth B spanned around a transmission frequency f_c. The available bandwidth is partitioned into N subcarriers, each one of bandwidth B/N [Hz]. As usually done in literature (see e.g. [1]) the channel is assumed flat over each assigned subcarrier; therefore the signal-to-noise ratio measured by user k on its subcarrier n is given by:

$$\gamma_{n,k} = \frac{p_{n,k} g_{n,k}}{\sigma^2} \tag{1}$$

being $p_{n,k}$ the power allocated to the user k on the subcarrier n, $g_{n,k}$ the channel power gain, and σ^2 the Guassian noise variance. The objective of RRM in OFDMA systems is to maximize the system data rate with a constraint on bit-error-rate (BER). The general problem can be summarized as follows [1]:

$$\begin{cases} \max_{c_{k,n},p_{k,n}} \dfrac{B}{N} \sum_{n=0}^{N-1} \sum_{k=1}^{K} c_{n,k} \log_2 \left(1 + \dfrac{\alpha_{gap} p_{n,k} g_{n,k}}{\sigma^2} \right) \\ \sum_{k=1}^{K} c_{k,n} = 1 \,\forall n, \quad \sum_{n=1}^{N} c_{k,n} p_{k,n} \le P_{tot} \quad \forall k \end{cases} \tag{2}$$

where $c_{n,k} \in \{0,1\}$ is the subcarrier allocation coefficient, and α_{gap} is the signal-to-noise ratio gap [1], expressed as a function of the desired BER. In the presence of constraints on the total available power, the resource allocation strategy maximizing the total system throughput is *water-filling* [1]. Water-filling tends to maximize power allocation on those subcarriers having the highest signal-to-noise ratio $\gamma_{n,k}$, while, on the other hand, to penalize those subcarriers having lowest SNR values by minimizing power allocation on them. This strategy is simple to be implemented, but it privileges users with highest channel gains and penalizes users experimenting worse channel response. Therefore, strategies to counteract this issue and support "fairness" in radio resource allocation should be established [1]. A thorough overview of rate maximization scheduling schemes has been presented in [7]. Proportional fairness has been proposed in [7] as feasible sub-optimal approach for uplink OFDMA. This algorithm requires the knowledge of the target rate and the available transmit power of each user, together with the knowledge of $\gamma_{n,k}$ values. Substantially, the algorithm compares the normalized ratios of each user and assigns more carriers to users whose proportional rates are the least met. It is shown in [7] that, although decreasing the total data rate, proportional fairness makes the rate distribution significantly fairer as compared to other schemes based on global data-rate maximization.

2.2 The MOS-Based Utility Function

The key concept underlying the proposed approach is to consider as utility function of the cooperative game the Mean Opinion Score (MOS) of users competing for the access in a mixed traffic configuration (audio, video, best-effort data download). In the classical OFDMA RRM problem, expressed in (1), the cost function that is maximized is the total system data-rate. Although fairness is introduced in the RRM (see e.g. [1] and [7]), the optimization with respect to "gross" data rate may not match with the QoS requirements of users transmitting differentiated data streams. Therefore, in this paper, we propose a cost function directly quantifying the QoS effectively perceived by the competing users. Each user is represented by a data flow, with the related quality target. The MOS is expressed by a real number ranging from 1 (very dissatisfied user) to 4.5 (very satisfied user). The user's satisfaction threshold is commonly set to 4. In the present paper, we consider three different classes of users (i.e.: players): "video streamers", "audio streamers" and "best effort", the latter indicating users requiring file transfer. For audio and file transfer applications, a suitable expression of MOS can be given as follows [6]:

$$MOS_{audio,BE} = a \log \left[bR(1 - PEP) \right]$$ (3)

R being the transmission rate and PEP the packet-error-probability. The constants a, b are computed by fixing the MOS at a given rate value R, in the absence of packet errors (PEP=0). Considering the QoS requirements of WiMax applications expressed in [8], we can regard as fully satisfactory data rates (corresponding to MOS=4.5): r_b=200 Kb/s for audio streamers, and r_b=2 Mb/s for video streamers and best-effort users. As far as video streaming is concerned, the following MOS model has been considered [9]:

$$MOS_{video} = \begin{cases} 1.0 & PSNR \leq PSNR_{1.0} \\ d \log PSNR + e & PSNR_{1.0} < PSNR < PSNR_{4.5} \\ 4.5 & PSNR \geq PSNR_{4.5} \end{cases}$$ (4)

The parameters $PSNR_{1.0}$ and $PSNR_{4.5}$ denote the Peak-Signal-to-Noise-Ratio (PSNR) at which the perceived QoS drops to "not acceptable" (MOS=1.0) and exceeds "very satisfied" (MOS=4.5), respectively. The constants d and e are derived according to this. The simple model that relates PSNR with the data-rate of the video stream proposed in [9] has been considered also in this paper:

$$PSNR_{dB} = u + v \sqrt{\frac{R}{w}} \left(1 - \frac{w}{R} \right)$$ (5)

The parameters u, v, and w characterize a specific video sequence (see in Fig.1 three sample frames corresponding to different MOS values).

2.3   The Cooperative RRM Algorithm Based on Game Theoretical Approach

Game theory is mainly applied in Economics in order to provide a reliable model for company competition [3]. Nevertheless, other applications in the field of information science and networking are considered by literature [10]. Game theory represents a formal method that analyzes conflict situations, searching for competitive and/or

(a)   (b)   (c)

Fig. 1. Sample frame of "Foreman" sequence corresponding to three different MOS: (a): *MOS=3*, (b): *MOS=3.5*, and (c): *MOS=4.5*

cooperative solutions obtained by means of specific models. A game substantially consists of three elements:

- A set of players;
- A set of strategies or actions available for each player;
- A utility function that rewards the player for its strategy combination.

In the OFDMA case, the users competing for access share limited radio resources (i.e.: a finite subcarrier set). Each user tries to allocate the "best" subcarriers, i.e.: those subcarriers that are not severely attenuated by the frequency selective fading. In such a way, user conflicts are unavoidable. The aim of the cooperative game is to manage these conflicts in order to allow users to achieve a satisfactory agreement. The equilibrium point of a cooperative game is represented by the so-called Nash equilibrium [3]. If all players reach Nash equilibrium, none of them can improve their performance by modifying its own strategy. Nash equilibrium is substantially a sort of "optimal agreement" that is not necessarily the optimal solution of the game.

Players are represented by data flows, with the corresponding resource requirements. User profiles in terms of data-rates and BER are derived according to [8] and shown in Tab.1.

Table 1. BER and data-rate requirements for different user typologies

	Video streamers	Audio streamers	Best Effort users
BER $\leq$	10^{-6}	10^{-4}	$2 \cdot 10^{-4}$
Rate [kbps] for MOS=4	384	24.6	56
Rate [kbps] for MOS=4.5	2000	200	2000

In order to enable negotiation of available resources, the game is organized like a "championship" in several rounds, with challenges or negotiations between pairs of users. Negotiation is targeted to identify the partitioning of available resources maximizing the joint utility function. The following pseudo code provides additional information on the proposed algorithm:

```
%Algorithm initialization
•Assign N₁=floor(N/K) subcarriers to each users
•Perform water-filling over the assigned subcarriers
•Computation of utility function: U(1)=min(MOS(P(0),C(0)))
 %minimum MOS after initial assignment: P(0), and C(0) are
 the power and subcarrier allocation matrices computed by
 initial water-filling
%Competition among user pairs
for i=1:K*(K-1) %for the no. of user pairs
•The subcarrier sets of the selected users pair i are
 merged;
%Min-MOS based user pair negotiation
•Subcarrier indexes j are ordered in decreasing order with
 respect to the ratio gⱼ₁(i)/gⱼ₂(i)
•for j=1:N(i) %N(i)=cardinality of subcarrier set related
 to pair i
```

- ▪ Subcarriers from 1 to *j* are associated to user 1 of pair *i* and water-filling is performed on these subcarriers
- ▪ Subcarriers from *j*+1 to *N(i)* are attributed to user 2 of pair *i* and water-filling is performed on these subcarriers
- ▪ Computation of utility function: $U(i) = min(MOS_1(i), MOS_2(i))$
- ▪ *Return* index *j* that maximizes *U(i)*
  End **%end of user pair negotiation**
- • *Return* power allocation matrix *P(i)* and bit allocation matrix *C(i)* updated after the challenge (negotiation) *i*
- • Computation of utility function $U(i+1) = min(MOS(P(i), C(i)))$ for all users
- • If *U(i+1)*>4   %all users are satisfied
  break
  End
End **%end of competition**

The user pair negotiation is similar to that one proposed in [2], but using a different utility function (the minimum MOS achieved by the two negotiating users). On the other hand, the multi-user negotiation methodology is different. In [2], random coalitions of user pairs and the Hungarian algorithm for optimal coalition selection are considered. The first methodology is clearly sub-optimal; the second one may be computationally expensive. Moreover, the challenge is interrupted when the minimum MOS, computed for all users, trespasses the satisfaction threshold. Resulting complexity is in the order of $O(KN \log N)$, while exhaustive search would lead to $O(K^N)$. The sub-optimal RRM algorithm based on proportional fairness of [7] has a complexity of order $O(KN)$. Therefore, the proposed RRM algorithm looks computationally affordable.

The most significant novelty with respect to state-of-the-art yielded by the proposed approach is represented by the refereed competition for reciprocal maximization of perceived quality of service, instead of a competition for maximizing the gross data-rate as shown e.g. in [2]. Practically speaking, the proposed game aims at partitioning the available resources in order to allocate enough capacity to each data flow to achieve a satisfactory MOS level. In the next section, the proposed scheme is numerically validated, demonstrating that this kind of competition can turn onto a substantial performance improvement, impacting on measurable perceived QoS.

3   Experimental Results

Intensive simulation trials have been performed in MATLAB environment in order to test the effectiveness of the proposed RRM methodology. *K*=30 users of mixed typology (10 video streamers, 10 audio streamers and 10 best-effort users) are sharing a set of 256 subcarriers distributed over a 4MHz bandwidth. The frequency-selective Stanford University Interim (SUI) channel model of type 5 has been considered for simulations [11]. Fig.2 shows the minimum MOS vs. SNR obtained by the different RRM algorithms assessed in this paper, i.e.: a) the proposed approach based on game theory and perceived QoS, b) an approach based on game theory and rate maximization

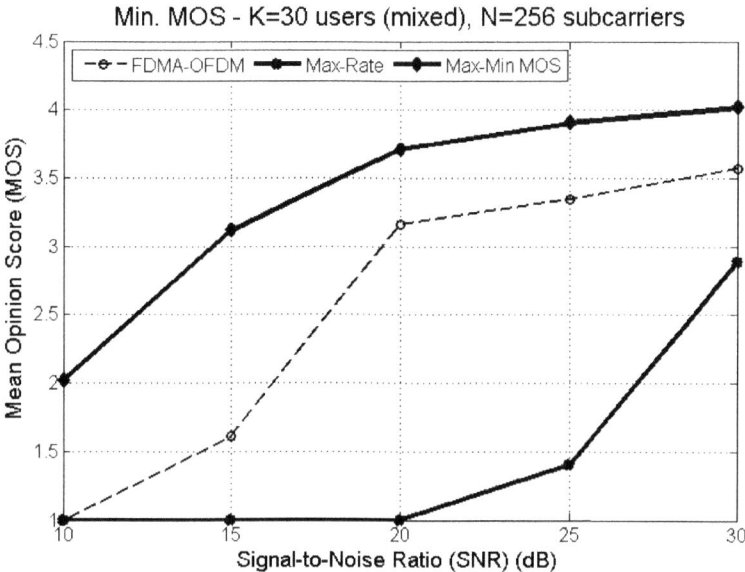

Fig. 2. Minimum MOS achieved by the different RRM strategies considered in the paper: cooperative min-MOS maximization, cooperative rate maximization, not-cooperative FDMA-OFDM with water-filling (channel: SUI-5, bandwidth 4 MHz)

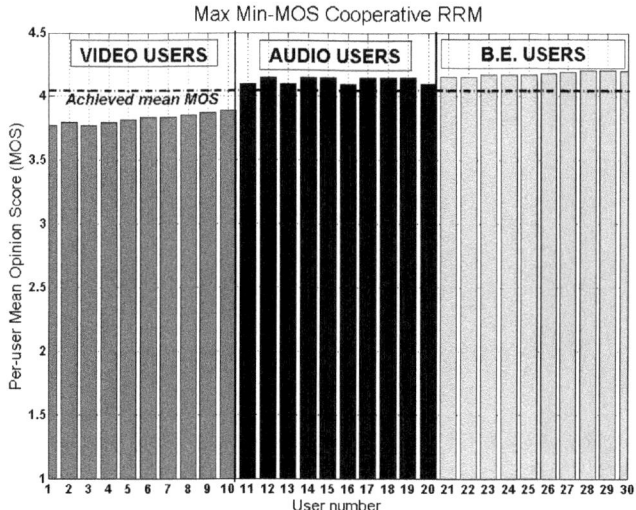

Fig. 3. MOS achieved by cooperative RRM based on minimum-MOS maximization (bar plot), K=30 users, SNR=20dB

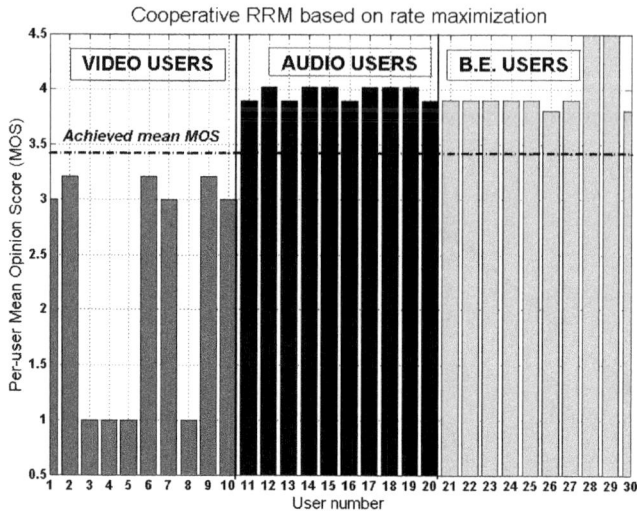

Fig. 4. MOS achieved by cooperative RRM based on rate maximization (bar plot), K=30 users, SNR=20dB

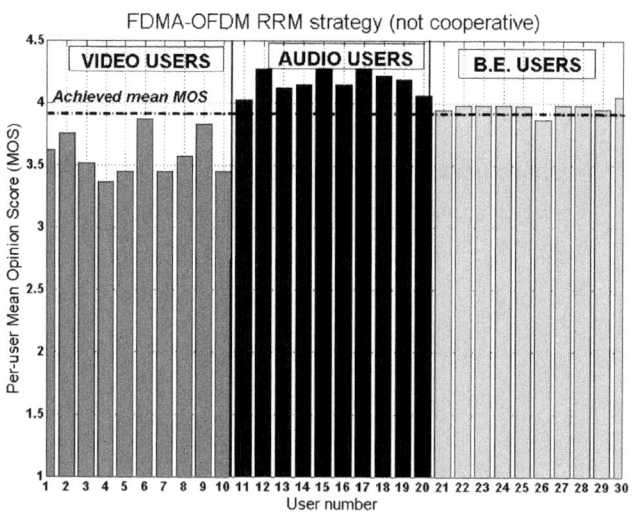

Fig. 5. MOS achieved by RRM based on not-cooperative FDMA-OFDM (bar plot), K=30 users, SNR=20dB

(similar to that one proposed in [2]) and c) an OFDM-FDMA approach with static allocation of users' subcarriers and waterfilling inside each user's group with fairness constraint [12]. The improvement of MOS index achieved by the proposed RRM strategy is evident both with respect both to OFDM-FDMA strategy and to game theoretical rate-maximization algorithm.

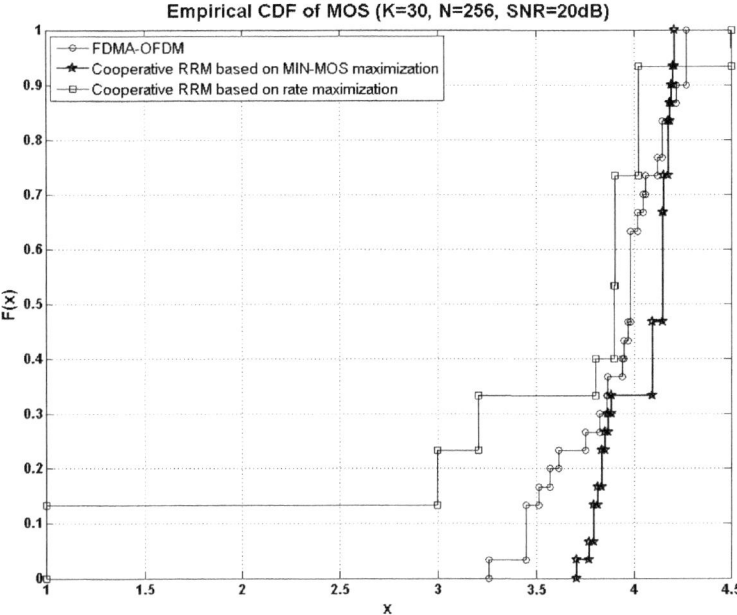

Fig. 6. Empirical CDF of MOS related to different RRM strategies proposed in this paper, obtained using MOS values drawn in bar plots of Figs.3-4 and 5

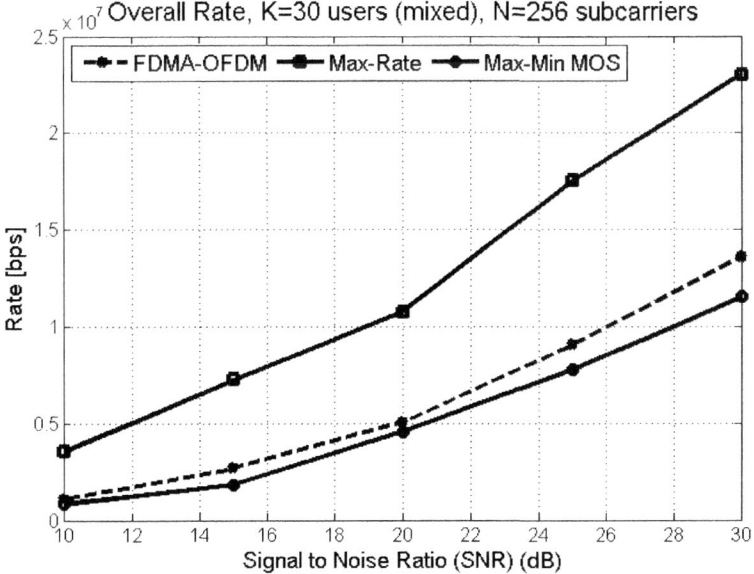

Fig. 7. Overall data rate achieved by the different RRM strategies considered in the paper: theoretical game-based min-MOS maximization, theoretical game rate maximization, static FDMA-OFDM with water-filling (channel: SUI-5, bandwidth 4 MHz)

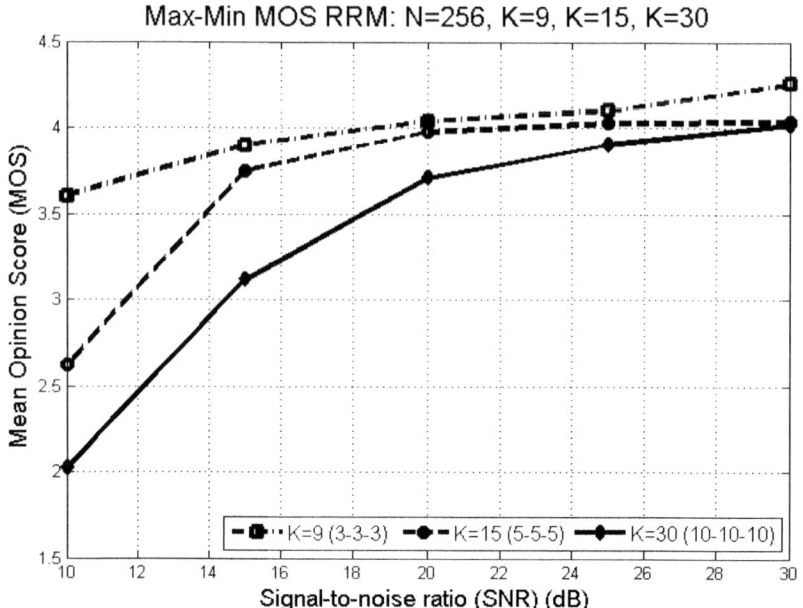

Fig. 8. MOS achieved by the proposed cooperative Max-min MOS RRM strategy vs. SNR and user number (channel: SUI-5, bandwidth 4 MHz)

In Figs.3-5, MOS bar plots related to the different RRM strategies for SNR=20dB clearly underline that the proposed algorithm provides a satisfactory QoS both for audio streamers and best-effort users, while video streamers are in any case very close to the satisfaction threshold. On the other hand, game theoretical rate maximization RRM severely penalizes some classes of users while satisfying other ones. Finally, Fig.5 shows that OFDM-FDMA RRM is more in favor of audio streamers (that require the lowest rate), while all best-effort users are close around the satisfaction threshold, and video streamers are clearly below the satisfaction threshold. Bar plot results shown in Figs.3-5 are condensed in empirical Cumulative Distribution Functions (CDFs) of MOS drawn in Fig.6. One can note from graphs of Fig.6 that the statistical spread of MOS values is very large when considering cooperative RRM based on rate maximization. On the other hand, the proposed cooperative RRM strategy, based on minimum-MOS maximization, is characterized by the smallest statistical spread of MOS values. This is a further confirmation of the fairness introduced by users' cooperation targeted at reciprocally enhancing perceived QoS.

It is interesting to note that the MOS index is not directly related with the overall system data-rate – which is the usual indicator of effectiveness of RRM strategies in wireless networks (see Fig.7). This happens as max-rate strategies merely maximize the overall system data rate, without considering the impact on the effective QoS perceived by the users.

Other interesting results have been shown in Fig. 8, where the impact of the number of users on the proposed RRM algorithm is shown. One can note that for high transmission SNR, the achievable utility values are substantially invariant with respect to the number of users, as all the available subcarriers are characterized by a satisfactory signal-to-noise ratio. On the other hand, for lower SNR and larger user number, the role of the negotiation becomes relevant. This is expectable, because the competing users have to share radio resources that cannot provide to the overall community a satisfactory QoS. If SNR drops below 15dB, larger user communities (i.e. $K=15$ and $K=30$) cannot reach a satisfactory QoS.

4 Conclusion and Future Works

In this paper, a cooperative RRM strategy, based on game theory, has been studied with the clear objective of maximizing the perceived QoS of OFDMA users. The proposed approach, explicitly including users' perceived quality (measured in terms of Mean Opinion Score – MOS), may ensure the right balance between efficiency and fairness, resulting in a perceived QoS increase with respect to state-of-the-art non-cooperative RRM strategies. Future works may concern with the impact on performance of channel estimation (in this paper, we assumed that CSI is ideally known) and novel cooperative strategies involving all competing users together, rather than to make them negotiating in couples.

Acknowledgements

Authors wish to thank Dr. Fabrizio Vicari of University of Trento (Italy) for his valuable help in collecting paper results. This work has been partially supported by the Italian Ministry of University and Scientific Research, under the framework of SALICE (COFIN 2007RFTYY7_002) and WORLD (project code: COFIN 2007R 989S) research projects.

References

1. Sadr, S., Anpalagan, A., Raahemifar, K.: Radio Resource Allocation Algorithms for the Downlink of Multiuser OFDM Communication Systems. IEEE Communications Survey & Tutorials 11(3), 92–106 (2009) (3rd quart.)
2. Han, Z., Ji, Z., Liu, K.J.R.: Fair Multiuser Channel Allocation for OFDMA Networks Using Nash Bargaining Solutions and Coalitions. IEEE Trans. on Comm. 53(8), 1366–1376 (2005)
3. Fudenberg, D., Tirole, J.: Game Theory. MIT Press, Cambridge (1991)
4. Han, Z., Ji, Z., Liu, K.J.R.: Non-Cooperative Resource Competition Game by Virtual Referee in Multi-Cell OFDMA Networks. IEEE Journal on Sel. Areas in Comm. 25(6), 1079–1089 (2007)
5. Noh, W.: A Distributed Resource Control for Fairness in OFDMA Systems: English-Auction Game with Imperfect Information. In: Proc. of IEEE GLOBECOM 2008 Conf., New Orleans (LA), November 30-December 4, pp. 1–6 (2008)

6. Saul, A., Auer, G.: Multiuser Resource Allocation Maximizing Perceived Quality. EURA-SIP Journal on Wireless Communications and Networking, Article ID 341869 2009, 1–15 (2009)
7. Ma, Y., Kim, D.I.: Rate-Maximization Scheduling Schemes for Uplink OFDMA. IEEE Trans. on Wireless Communications 8(6), 3193–3205 (2009)
8. WiMAX forum, Can WiMAX address your applications? (2005)
9. Choi, L.U., Irvlac, M.T., Steinbach, E., Nossek, J.A.: Sequence-level models for distortion-rate behaviour of compressed video. In: Proc. of IEEE ICIP 2005 Conf., Genoa (Italy), September 11-14, vol. 2, pp. 486–489 (2005)
10. Jorswieck, E.A., Larsson, E.G., Luise, M., Poor, H.V.: Game theory in signal processing and communications. IEEE Signal Processing Magazine, 17–18 (September 2009)
11. IEEE 802.16 Broadband Wireless Access Working Group. Channel Models for Fixed Wireless Applications (2001), http://wirelessman.org
12. Bohge, M., Gross, J., Wolisz, A.: The Potential of Dynamic Power and Sub-Carrier Assignments in Multi-User OFDM-FDMA Cells. In: Proc. of IEEE GLOBECOM 2005 Conf., St. Louis (MO), November 28-December 2, pp. 2932–2936 (2005)

Using Agent-Oriented Simulation System AGNES for Evaluation of Sensor Networks

Dmitry Podkorytov, Alexey Rodionov, Olga Sokolova,
and Anastasia Yurgenson

Institute of Computational Mathematics and Mathematical Geophysics
SB RAS, Pr. Lavrentieva, 6, Novosibirsk, 630090, Russia
d.podkorytov@gmail.com, alrod@sscc.ru, {olga,nastja}@rav.sscc.ru

Abstract. Agent-oriented simulation became popular now for performance evaluation of distributed systems, in possession of elements with cognizant behavior in particular. The agent-oriented simulation package named AGNES (AGent NEtwork Simulator) is presented as a tool for simulation of sensor networks. The multi-relay selection technique for one-to-many signal delivering is described as example of its work.

Keywords: Agent-oriented simulation, JADE, sensor networks, multi-relay scheme.

1 Introduction

Design and analyze of wireless sensor networks (WSN) include optimal energy consuming as one of its main goals. Different multi-relay schemes are specially designed or can be adopted for similar problems of WSN [1]. Before applying different schemes for special network its performance is to be evaluated. Computer simulation is an acknowledged method for this task [2]. Many commercial or free simulation tools can be used, but few of them have special facilities for modeling intelligent behavior in distributed computing system. Agent-oriented simulation (AOS) is more suitable for this problem [3, 4]. One of the main advantages of AOS tools is a possibility of scaled-down simulation. Simulated parts of the system under evaluation can be combined with real ones [5]. Another advantage is that program agents are intended for intelligent collective achieving of some goal and thus AOS systems can be used for simulation optimization [6].

In this paper we present new AOS system AGNES and example of its usage for simulation message delivery in a sensor network. AGNES is oriented on simulation of large info-telecommunication networks.

2 Agent-Oriented Simulation System AGNES

Multi-agent approach is very popular and widely used in different areas including simulation and modeling. Main advantages of this method are: simple realization (only information about elements is needed, not whole model), scalability

A. Vinel et al. (Eds.): MACOM 2010, LNCS 6235, pp. 247–250, 2010.

(quantity of agents doesn't effect to system functionality), state of distribution (agents can work on different networks nodes and communicate with each other through communication network). Many different multi-agent simulation and modeling systems for different domains are available for: life simulation - Framsticks , social process simulation - Brahms, ecology simulation - Echo, politics simulation - PS-I and etc. General platforms are accessible, some of packages are free (Repast, Swarm, Mason), and some are commercial (AnyLogic). Most of these tools are effective, usable and flexible ABM (agent-based modeling) or ABS (agent-based simulation) systems. But they lack for good fault-tolerant possibilities and they are not oriented to simulation of large networks with local management. AGNES (AGent NEtwork Simulator) has been designed specially for achieving these benefits. It contains mechanism for providing high level of fault-tolerance and its specialization is simulation of large networks.

AGNES is implemented basing on JADE (Java Agent Development Framework) that is a well-known program environment for developing program agents in Java [7]. In one's turn, JADE is based on the FIPA (Foundation for Intelligent Physical Agents) standards that provide developer with means for:

- agent creation, registration and migration;
- announcement of agent's functions;
- inter-agents communications;
- graphical supports at debugging and deployment phases;
- etc.

Agents forms two groups in AGNES: control (CA) and functional (FA) agents.
    CAs control the simulation process and solve the following tasks:

- model initialization and starting;
- clustering of agents: all agents are divided between "clusters" according to their needs in resources and current settings of simulation environment. Special control agent (manager, MA) is created for each cluster.
- run-time control of simulation process and data collection.

CAs are organized in a network similar to sensor one. CA can be passive or active one (PA or AA). After self-organizing of CAs' network, each agent starts a group of FAs that are under its supervision. "Passive" agents (PA) executes orders of "active" ones (AA) and inform them about emergency situations. When running, CAs cooperate and can change their role, depending on current situation.

    Special CAs are designed for data collection and event synchronization.

    FAs have to be designed for simulating behavior of modeled system components and can be organized in problem-oriented libraries.

    Except to the problem-oriented tasks each agent must execute some functions that provide the simulation process itself (periodical backup, synchronization, routing of messages between agents etc.). Model-depended functions are realized by the following procedures:

Sleep() – suspends simulation;
Wakeup() – resumes simulation;

Start (array InitParameters) – serves for model initialization and start;

CreateBackup(agent Receiver) – creates a reserve copy of data needed for the agent recovery. Parameter of procedure is an address of agent that is assigned for storing this data;

RestoreBackup(array BackupParameters) – reconstructs an agent from its backup data;

GetLog() – gets data about the state of simulation.

The following system functions of AGNES's agent are model-independent:

SaveBackup(agent Sender, time BackupMoment, array BackupData) – stores backup data for "Sender" at time BackupMoment;

SendBackup(agent Receiver, time BackupMoment, agent BackupAgent) – transfers backup data;

SetTime(time ModelTime) – sets simulation time for the current agent;

GetAgentTime() – gets agent's simulation time;

SendLog() – sends a broadcasts message with agent's current state;

Ping() – checks an working state of an agent.

3 Using AGNES for Simulating Multi-relay Selection in Sensor Networks

Model of sensors networks contains a large number of similar elements with autonomous behavior. In this case usage of intellectual agents for sensor network simulation seems as a good solution. Conceptually agents and sensors are elements with similar nature: they are independent, stand-alone and don't contain information about whole their environment. Creation of a number of agents in necessary quantity is possible.

We consider the Problem of Multi-Relay Selection in sensor networks as an example for AGNES' performance capabilities for simulation of sensor networks. Sending a package of information (message) to other nodes is one of essential functions of any node in sensor network. Techniques that are used for routing in network systems can be classified as unicast and broadcast. Both techniques are of bad performance when information needs to be sent to a group of selected users (destinations). This situation has lead to the development of network protocols called multicast routing protocols [8]. The problem of message delivery to a selected group of nodes (multicast group) is the problem of multicast routing and multi-relay selection.

As a model for sensor network we use a geometrical graph that consists of a number of nodes (source, relays and destinations), and two nodes are connected only if packets can be transmitted between them (nodes are equipped with wireless interfaces). Nodes which are not able to communicate directly use other intermediate nodes as relays. The main problem in such delivery is minimizing a number of relays. This task has special significance in sensor network because each retransmission requires consumption of some quantity of a limited energy store. In [9] authors describe the problem of computing minimal cost multicast tree in network. This problem is well known as the Steiner tree problem for graphs, and it is NP-complete even for the case when every connection has the same cost. The problem of multicast tree is reformulated to the problem of minimizing the set of relays (and the selection of this set). This problem is NP-complete also and some heuristics are proposed for approximate solution. For the simulation example we use Guha and Khuller algorithm [10] for finding suboptimal connected dominating set.

For simulation we choose square area 200x200 with 100 sensors placed randomly, radio distance is 40. Probability of successful message delivery for any transmitting is 0.8. For this example we obtain that average maximum number of transmitters is about 13, while the time needed for delivering a message to all sensors is about 18.

4   Conclusion

Multi-agent approach is useful for solving multi-relay selection problem in sensor networks. The AGNES simulation package shows good usability for simulating sensor networks. Been written in Java, it can be installed and used under all modern operation systems. Been based on free agent development framework JADE, it can be used with low or no cost.

References

1. Khajehnouri, N., Sayed, A.H.: Distributed MMSE relay strategies for wireless sensor networks. IEEE Trans. Signal Processing 55, 3336–3348 (2007)
2. Antoine-Santoni, T., Santucci, J., De Gentili, E., Costa, B.: Discrete Event Modeling and Simulation of Wireless Sensor Network Performance. Simulation 84(2-3), 103–121 (2008)
3. Wooldridge, M.: An Introduction to MultiAgent Systems. John Wiley & Sons Ltd., Chichester (2002)
4. Shoham, Y., Leyton-Brown, K.: Multiagent Systems: Algorithmic, Game-Theoretic, and Logical Foundations. Cambridge University Press, Cambridge (2008)
5. Kotenko, I.V., Ulanov, A.V.: The Software Environment for multi-agent Simulation of Defense Mechanisms against DDoS Attacks. In: The International Conference on Intelligent Agents, Web Technologies and Internet Commerce, IAWTIC 2005, Vienna, Austria, pp. 283–289 (2005)
6. Andradottir, S.: A Review of Simulation Optimization Techniques. In: Simulation. Conference 1998, Proceedings of the Winter, vol. 1, pp. 151–158 (1998)
7. http://jade.tilab.com/
8. Oliveira, C.A.S., Pardalos, P.M., Resende, M.G.C.: Optimization Problems in Multicast Tree Construction. In: Handbook of Optimization in Telecommunications, pp. 701–731. Springer Science + Business Media, Heidelberg
9. Ruiz, P.M., Gomez-Skarmeta, A.F.: Approximating optimal multicast trees in wireless multihop networks. In: Proceedings of 10th IEEE Symposium on Computers and Communications, ISCC 2005, pp. 686–691 (2005)
10. Guha, S., Khuller, S.: Approximation Algorithms for Connected Dominating Sets. Algorithmica 20(4), 374–387 (1998)

Multiple Metrics in MANET with End-to-End QoS Support for Unicast and Multicast Traffic

Evgeny Khorov and Alexander Safonov

Institute for Information Transmission Problems
of the Russian Academy of Science,
B.Karetny lane 19, 127994 Moscow, Russia
{khorov,safa}@iitp.ru

Abstract. The paper proposes an approach of using multiple metrics in a wireless multihop network, when one of the metrics called *optimizable* reflects consuming network resources, and other metrics called *restrictive* reflect traffic QoS requirements. Compared to popular Hop Count and Air Time Link metrics, a set of metrics is proposed, increasing the network capacity measured as the number of unicast voice calls with tolerable quality. The metrics are further used in a proposed multicast tree construction algorithm.

Keywords: wireless network, routing, metric, QoS.

1 Introduction

In both wired and wireless networks, a well-known problem of routing has received a lot of attention from academia and standardization bodies. Whatever route criterion a routing protocol implies, an optimizable function called *link metric* is used to weight the links in the network graph and ultimately choose the best route between the source and destination nodes. The criterion usually represents the amount of network resources consumed to deliver a packet.

Quality of Service (QoS) requirements imposed by many applications make the problem more complicated. To satisfy the applications, the route criterion reflecting consuming network resources is amended by the list of QoS restrictions. For example, three parameters determine the quality of voice received through a network: the average packet delivery time, the jitter, i.e. the variation of the delivery time, and the packet delivery ratio. International Telecommunication Union (ITU) recommends an empirically obtained formula for so-called R-factor mapping a combination of these parameters to a perceptional voice quality [1]. So, R-factor determines the boundary values of them which shall not be crossed for a chosen voice quality.

In this paper, we propose an approach of using multiple metrics simultaneously, with one of the metrics which we call *optimizable*, reflecting consuming network resources, and other metrics which we call *restrictive*, reflecting QoS

A. Vinel et al. (Eds.): MACOM 2010, LNCS 6235, pp. 251–262, 2010.

requirements. If a route length crosses a threshold in at least one of the restrictive metrics, the route shall not be chosen for packet delivery, to escape network resources waste. So, the best route is chosen in an optimizable metric, in the class of routes allowed by restrictive metrics. The approach is applicable for both unicast and multicast traffic, as shown in the paper.

The rest of the paper is organized as follows. In Section 2, we introduce terminology used in the paper and define QoS routing problem formally. Section 3 overviews the simplest and thus popular metrics for wireless networks. A family of metrics is proposed in Section 4, which may be used as optimizable and/or restrictive. In Section 5, we compare proposed and overviewed metrics with a simulation model. Section 6 proposes an algorithm to construct a multicast tree, using multiple metrics, and estimates the algorithm complexity.

2   Preliminaries

The efficiency criterion to compare various routing protocols may be stated, in general, as the total value of packets delivered during some time interval: a protocol providing bigger total value is more efficient. The value of a delivered packet depends on type q of the packet, the amount of consumed network resources, and the end-to-end packet delivery time or other factors imposed by the Quality of Service (QoS) requirements. It also may be negative, when the packet cannot be delivered with appropriate quality of service, e.g. when the packet delivery time reaches some threshold.

Let W be the current total value of packets already delivered in a network. When a packet p of type q is delivered over route l, W is increased by $\delta W(p, q, l)$. The function opposite in sign to the mathematical expectation of $\delta W(p, q, l)$

$$\omega_q(l) = -E_p[\delta W(p, q, l)] \tag{1}$$

is called *the route metric*. It serves to evaluate the quality of route l for delivering a packet of type q. In a particular case when all packets are of the same type, the route metric is a one-variable function of the route: $\omega_q(l) \equiv \omega(l)$.

When routing a packet, the goal is to find such a route $l_q \in L$ that

$$l_q = \arg_l \left[\min_{l \in L} \{\omega_q(l)\} \right], \tag{2}$$

where L is the set of all possible routes in the network for this packet.

In a network of peers with distributed decision making, the value of the metric of a route is not evaluated directly. Instead, another function is introduced representing the contribution of corresponding links to the route metric in question. This function is called *the link metric*.

The physical interpretation of a link metric is usually clearly connected with the routing efficiency criterion. For example, Airtime Link metric (see Section 3) introduced in IEEE 802.11s standard draft [2] represents the channel occupation time required to transmit a packet over the link, including possible retransmissions. The routing efficiency criterion behind this link metric is the total amount of channel resources consumed by all (re-)transmissions of the packet over the route. As we discuss further, this criterion is in a way general, but not connected with any QoS requirement, so the default routing protocol of IEEE 802.11s uses this single criterion for all packet types.

Aiming at providing in a network several levels of service for packets of different types, one needs to introduce several routing criteria clearly reflecting the corresponding QoS expectations. In general, QoS requirements for packet type q may be written as an k_q-dimensional vector $\overrightarrow{\Omega_q}$ of upper bounds of k_q parameters. To estimate the actual value of each parameter, a corresponding link metric $w_q^{(i)}, i = \overline{1, k_q}$, may be used. Further in the paper, we refer to this metrics as to *restrictive*, in contrast to $w_q(l)$ which we refer to as *optimizable*. Then, denoting optimizable metric $w_q^{(0)}(l) \equiv w_q(l)$, routing problem (2) turns to be a bounded problem which may be written as follows:

$$l_q = \arg \left[\min_{\substack{l \in L \\ w_q^{(i)}(l) \leq \Omega_q^{(i)}, i=\overline{1,k_q}}} \left\{ w_q^{(0)}(l) \right\} \right]. \qquad (3)$$

The issues of link quality estimation and routing information dissemination, which shall be resolved to find solutions of problems (2) and (3), are out of scope of this paper. So, we consider an abstract proactive link state hop-by-hop routing protocol.

Both problems (2) and (3) are also valid for *multicast* route selection, if we put L be the set of all multicast trees covering the source node and destination nodes, and assume that individual transmissions with the same acknowledgment policy as for unicast traffic are used to deliver multicast packets over the tree. Among various multicast tree construction protocols, the common idea is to build a tree of minimal weight, also known as Steiner tree. Obviously, Steiner tree is the solution of (2). In this paper, we discuss a possible solution of (3).

3   "Classical" Link Metrics

This section overviews simple and thus popular metrics for multihop wireless networks. The simplest metric is called Hop Count. For any route, the metric value equals the number of links the route consists of. The solution of (2), when Hop Count is used, is the path containing minimal possible number of links. Also, the routing efficiency criterion may be interpreted as the number of nodes involved in the packet delivery process.

Thanks to its simplicity, Hop Count is defined as default metric in numerous routing protocols, e.g. AODV [3], OLSR [4], and ZRP [5]. Though, it is known to choose the worst paths in wireless networks, e.g. see [6]. Hop Count metric does not take into account the fact that links data rate and error rate varies a lot in wireless networks. The result of minimizing the number of nodes on the route is that the longest links with lowest signal to noise ratio, and consequently lowest data rate and longest transmission time, are always used. High collision probability provoked by long transmissions increases the number of retries which, in turn, increases the packet service time contributing to the end-to-end packet delivery time. Simulation results of comparing Hop Count with other metrics discussed in this paper are presented in Section 5.

A link metric which directly accounts for lossy links is called Expected Transmission count (ETX) [6]. The ETX of a route is the sum of the ETX for each link in the route. The metric finds paths with the fewest expected number of (re-)transmissions required to deliver a packet all the way to its destination. ETX is calculated based on statistics of already transmitted packets on each wireless link and is proved to find paths with higher throughput, under assumption that a single rate is used on all the links of the network.

Removing this assumption, IEEE 802.11s standard draft [2] introduces Air Time Link (ATL) metric which contains the expected transmission count as a factor and accounts for multi-rate links:

$$\mu_A(i, j) = \left(O + \frac{P}{r_{ij}} \right) \frac{1}{1 - e_{ij}}, \tag{4}$$

where O and P are constants representing channel access overhead and standard packet size respectively, r_{ij} is the link (i, j) data rate, and e_{ij} is the probability of transmission error. The route metric is calculated as the sum of the corresponding link metrics.

The physical meaning of ATL metric is as follows: its value equals the time interval when the channel is busy with transmitting the packet over the link. Routing efficiency criterion is the amount of channel resources consumed by this transmission. Implicit assumption that the channel cost is the same on all links is obviously oversimplified, as a transmission in a multihop wireless network only occupies the channel in some neighborhood around the transmitter. Among the nodes in the network, the number of neighbor nodes varies as well as the number of their transmissions in a time unit. Consequently, the channel cost for a node depends on the number of active nodes in its vicinity. Metrics proposed in this paper does take into account active nodes in a transmitter neighborhood, overperforming ATL metric in terms of a number of criteria, as shown in Section 5.

Although the metrics discussed in this section were developed for unicast routing, they are also used in various multicast routing protocols such as MAODV, MOLSR, ODMRP and others. A multicast tree constructed by any of these protocols is simply the union of corresponding unicast paths to all multicast destinations and it is out of line with problem (2). Still, if any of discussed above metrics is used for proactive unicast routing, e.g. by means of OLSR, a multicast tree of minimal weigh may be constructed over known network graph, which would indeed be the solution of (2).

4   Proposed Metrics

4.1   B (Busy)

As mentioned above, the channel cost in different parts of network varies. To enhance ATL metric by taking this fact into account, one need to consider as

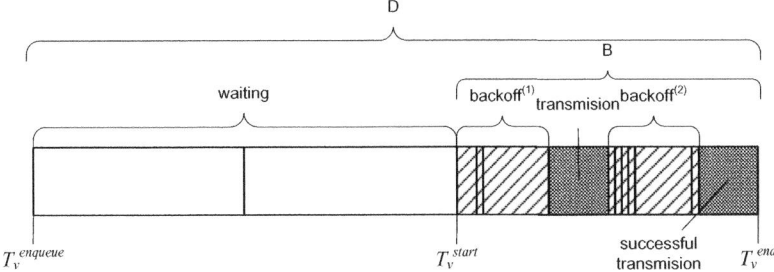

Fig. 1. Packet service time, B, and expected packet delay on a node, D

the link metric the average *packet service time* on the link instead of the channel busy time. The service time of a packet consists of intervals when the packet is actually (re-)transmitted, that is the channel busy time $\mu_A(i, j)$, and intervals when the node counts down its backoff timer, which total length is $backoff_{ij}$, as illustrated in Fig. 1.

The link metric, which we refer to as B (Busy), may be written as follows:

$$\mu_B(i, j) = E\left[backoff_{ij}\right] + \mu_A(i, j). \tag{5}$$

The route metric value equals the sum of corresponding link metric values. Metric B differs from ATL metric by term $backoff_{ij}$. Some papers, e.g. [7], claim this term to be negligible as backoff slots are very small compared to packet duration time. However, this is only true when a single node in the neighborhood is the transmitter and all backoff slots are of minimal length, or, in Bianchi's analytical model terms, all virtual slots are empty [8]. But in the case when several nodes in the transmission range of each other have packet to transmit, the mean duration of a backoff slot grows and may reach packet duration time. So, $backoff_{ij}$ may be even sufficiently greater than $\mu_A(i, j)$.

The value of link metric (5) is easily estimated by statistical data collected by nodes and requires no additional information exchange between nodes. Denote the moment when packet v is enqueued by $T_v^{enqueue}$, and the moments when its service starts and is completed by T_v^{start} and T_v^{end} respectively, see Fig. 1. If a packet is pushed in an empty queue, its service starts immediately: $T_v^{start} = T_v^{enqueue}$. Otherwise, the packet service starts when the previous packet transmission is over: $T_v^{start} = T_{v-1}^{end}$. The packet service is completed when the an ACK is received or the retry threshold is reached.

4.2   D (Delay)

Let us define as link metric D (Delay) the average packet delay on a transmitter. Then, the route metric calculated as the sum of corresponding link metrics represents the end-to-end packet delivery time.

D consists of packet waiting in the queue and packet service time, as illustrated in Fig. 1. The waiting interval starts when the packet is enqueued and ends when the packet service starts. So, for D we write:

$$\mu_D(i,j) = E[waiting_i] + E[backoff_{ij}] + \mu_A(i,j). \tag{6}$$

Waiting interval length only depends on the transmitter, i, while the second and the third terms in (6) depend on the number of retries and hence on the receiver, j, too. Metric D finds the route with the smallest packet delivery time.

4.3   P (Packet Loss Ratio)

In wireless networks, packets may be lost due to the following reasons: the retry threshold reached, node buffer overflown, lifetime expired. Let p_{ij} be the the probability that node i finally succeeds in packet transmission to neighbor j.

Assuming that packets are only lost when the retry threshold, R, is reached,

$$p_{ij} = 1 - e_{ij}^{R+1}, \tag{7}$$

where e_{ij} is the probability of a transmission failure.

If path l consists of, say, two links (i,j) and (j,k), packet delivery ratio over the path equals $p_{ij} \cdot p_{jk}$, i.e. route metric is the product of link metrics, but not the sum. To make it additive, define link metric P as follows:

$$\mu_P(i,j) = -ln(p_{ij}). \tag{8}$$

Metric P chooses a route with the highest packet delivery ratio. However, the metric is in a way selfish and does not take account the amount of consuming network resources. Thus, as shown in Section 5, this metric cannot be used as optimizable.

5   Simulation results

5.1   Simulation Setup

To compare the metrics proposed in Section 4 and the metrics overviewed in Section 5, we use simulation tool NS3 [9] with IEEE 802.11s module developed by IITP RAS [11].

Instead of default routing protocol HWMP, we use an abstract link state proactive hop-by-hop routing protocol broadcasting topology info with the refresh interval equal to 1 s.

As traffic source, we use a UDP application which generates packets of size PS. The interval between packets is randomized in $(0.9 \cdot PI, 1.1 \cdot PI)$. The UDP application runs in 2 configurations: "Voice" ($PS = 20$ bytes, $PI = 0.02$ s) and "Data" ($PS = 1024$ bytes, PI is variable).

To analyze proposed metrics we consider two scenarios. In both scenarios, we analyze how voice traffic is delivered via a multihop wireless network. We define the availability of voice service, VA, as the probability that R-factor exceeds 50, according to ITU recommendation [1], and consider VA as the ultimate criterion of routing efficiency.

5.2   Scenario "Circle"

In this scenario, we consider the topology shown in Fig 2. Nodes connected with a line are neighbors and can exchange packets directly.

Other pairs of nodes do not sense transmission of each other.

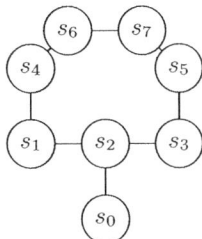

Fig. 2. Topology for scenario "Circle"

Application "Data" at node s_0 generates packets destined to node s_2, with access category AC_BE. Another application "Voice" at node s_4 generates packets destined to node s_3. Two routes exist between nodes s_4 and s_3.

In this scenario, we analyze the dependence of VA for "Voice" traffic on "Data" traffic rate. In the results shown in Fig. 3, line "3" corresponds to the case when static 3-hops route s_4, s_1, s_2, s_3 is used, and the line "4" corresponds to the static 4-hops route s_4, s_6, s_7, s_5, s_3. Other lines correspond to the cases when metrics ATL, B, D, and P are used.

To explain the VA results, let us consider the curves of the packet delivery time, packed loss ratio and route length. When the load on link (s_0, s_2) is low, any route gives high voice availability. However, the 3-hop route is preferred because it ensures lower network resources consumption. The collision probability of voice and data packets increases with the load on link (s_0, s_2), and 4-hops route becomes the best choice.

As to Hop Count metric, it always finds the 3-hops route, except for the case when the load on (s_0, s_2) is very high. In this case, the topology control frames often come into collisions, so the chosen path is unstable, switching between 3-hops and 4-hops routes from time to time.

By taking the backoff time into account, metrics B and D appear more sensitive to "Data" load growth than ATL, so the 3-hop path switches to 4-hop path just in time, providing better VA. Unlike ATL, these metrics grow with both the number of retries and the average length of a virtual slot, as explained in Section 4. ATL metric only prefers the 4-hops route when (s_0, s_2) load is close to maximum.

B and D metrics show almost the same results because $\mu_D(i, j)$ differs from $\mu_B(i, j)$ by a significant value only if there are packets in the queue of node i during long time interval. It does not happen in this scenario.

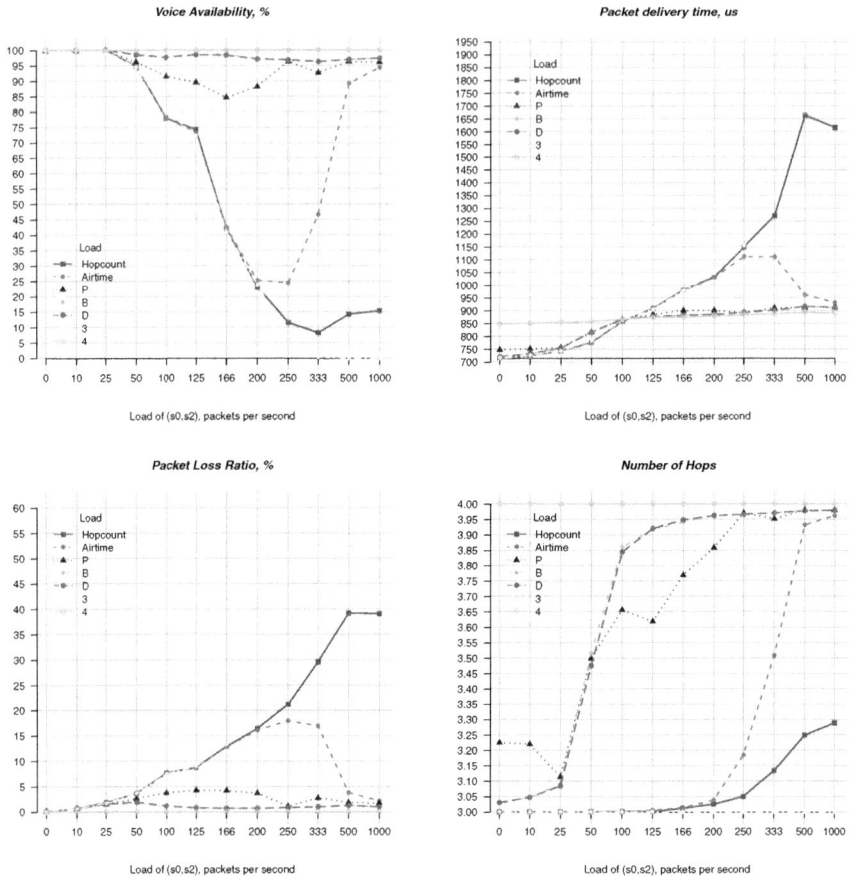

Fig. 3. Voice availability, packet delivery time, packet loss ratio and average route length in scenario "Circle"

5.3   Scenario "Grid"

In this scenario, we consider a network of NxN grid topology. Pairs of source-destination are chosen randomly. Let σ be the network load measured as the average number of "Voice" flows $F = \sigma N^2$.

Voice availability and packet loss ratio for the case when N=4 are shown in Fig. 4. For any σ, B and D metrics gives the voice unavailability about twice lower than ATL.

Let us measure the network capacity as the number of unicast voice calls with tolerable quality. Consider the voice quality as tolerable if VA is greater than a threshold. For a reasonable threshold, say, 90% or 95%, B and D metrics provides higher network capacity than ATL metric by about 30%. D metric behaves slightly better, because it feels queue size and keep off bottlenecks.

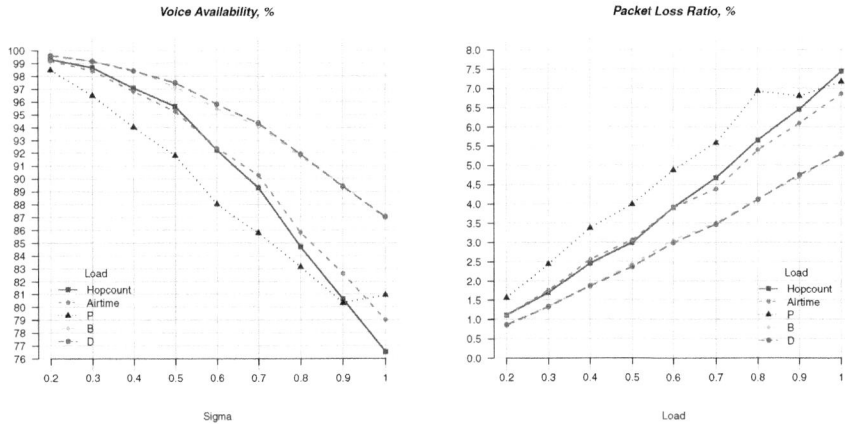

Fig. 4. Voice availability, packet loss ratio in scenario "Grid"

Despite P metric is designed to select the routes with lowest packet loss ratio, it provides the worst VA. Metric P is selfish and finds long routes with low transmission error probability. This policy results in high consumption of network resources.

As VA depends on packet delivery ratio, metric P may be used as a restrictive metric. E.g., a route is found by an optimizable metric, say, D, and then is inspected by metric P whether the packet delivery ratio is high enough. If it is not, the route shall not be used for packet delivery, to prevent network resources waste.

6   Metric Usage in Multicast Routing

In this section, we consider the problem of multicast routing from source node s to set of destination nodes DS. A lot of algorithms are proposed in literature to construct multicast trees. The solution of problem (2) (see Section 2) is the tree of minimal weight measured in metric ω_q, if we assume L be the set of all multicast trees covering the source node and destination nodes. The minimal weight tree is known as Steiner tree, and (2) is known as Steiner problem.

We propose to use multiple metrics as discussed in previous sections not only for unicast routing, but also for multicast routing. In this case, problem (3) replaces (2). It differs from Steiner problem by the vector of restrictions, $\Omega_q^{(\cdot)}$, on the tree depth measured in restrictive metrics $\omega_q^{(\cdot)}$, while the tree weight is still measured in optimizable metric $\omega_q^{(0)}$, as in Steiner problem.

Problem (3) may also be defined as follows: find the tree of minimal weight in metric $\omega_q^{(0)}$ in the class of trees which depth in metrics $\omega_q^{(\cdot)}$ does not exceed the corresponding upper bound $\Omega_q^{(\cdot)}$. In other words, when restrictions $\Omega_q^{(\cdot)}$ are weak, problem (3) is reduced to Steiner problem. Consequently, problem (3) is NP-complete, as Steiner problem is known to be NP-complete [10].

Further in this section, we specify problem (3) for the case of voice traffic and propose a heuristic algorithm to construct a tree which is the solution of the defined problem. As we consider voice packets only, for brevity and without loss of generality, we omit index q further in this section.

In Section 5, we show that the decisive restricting parameter of the received voice quality is the end-to-end packet delivery ratio. So, excluding the average packet delivery time and the jitter from consideration, we consider vector $\Omega^{(\cdot)}$ consisting of the only component $\Omega^{(1)} = P_{max}$ which is the upper bound of the end-to-end packet delivery ratio measured by metric P defined in Section 4 and denoted as $\omega^{(1)}$ further.

As the optimizable metric we propose to use metric B which reflects consuming network resources well, as shown in Section 5.

It may happen that no tree satisfies restriction $\Omega^{(1)}$, that is there is at least one destination node $d^* \in DS$ such that $\omega^{(1)}(s, d^*) > P_{max}$. To address this case in our algorithm, we propose to reject packets destined to d^*, which seems to be rational as such packets cannot be delivered to d^* with appropriate QoS anyway. Additionally, such packet drops reduce network resources consumed by the multicast flow.

Further, we propose a heuristic algorithm to solve the defined problem by constructing on the network graph, (V, E), the desired tree, $\mathbf{T} \equiv (V_T, E_T)$. Initially, $V_T = \{s\}$ and $E_T = \emptyset$. Let l_{uv}^T, $l_{uv}^{(0)}$ and $l_{uv}^{(1)}$ be the only path from u to v in the tree $\mathbf{T}$, the shortest path in metric $\omega^{(0)}$ and the shortest path in metric $\omega^{(1)}$ respectively. Let $\widetilde{DS}$ be the set destination nodes not covered by $\mathbf{T}$ yet. Initially, $\widetilde{DS} = DS$.

The tree construction algorithm is the following.

1. By means of Dijkstra's algorithm, find two sets of the shortest paths $l^{(0)}$ and $l^{(1)}$ in metrics $\omega^{(0)}$ and $\omega^{(1)}$ respectively from every node in the network to every node in set $\widetilde{DS}$.
2. For every $d \in \widetilde{DS}$, if $\omega^{(1)}(l_{sd}^{(1)}) > \Omega^{(1)}$, i.e. the length of path $l^{(1)}$ from s to d is above the threshold, remove d from $\widetilde{DS}$.
3. If $\widetilde{DS} = \{\emptyset\}$, the algorithm stops and $\mathbf{T}$ cannot be constructed with necessary QoS restrictions. Otherwise, add routes to the nodes from $\widetilde{DS}$ to $\mathbf{T}$: while $\widetilde{DS} \neq \{\emptyset\}$ do
   (a) choose node $d_{next} \in \widetilde{DS}$ with the longest path from tree $\mathbf{T}$, $l_{\mathbf{T}d_{next}}^{(1)}$.

$$d_{next} = \arg \left[\max_d \left[\min_{d \in \widetilde{DS}} \left(\min_{v \in V_T} \left\{ \omega^{(1)}(l_{vd}^{(1)}) \right\} \right) \right] \right], \qquad (9)$$

   where v is is the start node in the shortest route from the tree to d, $l_{vd_{next}}^{(1)}$ is the shortest route from v to d.
   (b) let $\Upsilon^{(k)}(\mathbf{T}, d_{next})$ be $\{v \in V_T | \omega^{(1)}(l_{sv}^T \oplus l_{vd_{next}}^{(k)}) \leq \Omega^{(1)}\}$ and $\lambda^{(k)}(\mathbf{T}, d_{next})$ be function

$$\lambda^{(k)}(\mathbf{T}, d_{next}) = \arg \left[\min_l \min_{v \in \Upsilon^{(k)}(\mathbf{T}, d_{next})} \left\{ \omega^{(0)}(l_{vd_{next}}^{(k)}) \right\} \right]; \qquad (10)$$

if $\Upsilon^{(0)}(\mathbf{T}, d_{next}) \neq \emptyset$, then route $l = \lambda^{(0)}(\mathbf{T}, d_{next})$, else $l = \lambda^{(1)}(\mathbf{T}, d_{next})$;

(c) add all the nodes and links of route l to $\mathbf{T}$, and exclude d_{next} and all the nodes of route l_{new} from $\widetilde{DS}$ (if any).

If there is a set $\widetilde{DS}$ of several arguments d that comes to minimization of some function $f(d)$ the expression $\arg\min_d\{f(d)\}$ returns a random value of $d \in \widetilde{DS}$.

An upper bound of the running time of this algorithm can be expressed as a function of $|V_T|$ and $|DS|$ using the Big-O notation. The running time of the first step is $O(|V_T|log(|V_T|)|DS|)$. The running time of choosing each node d_{next} is $O(|V_T||DS|)$ and it takes $O(|V_T|)$ operations to find the route to d_{next} if we store the value $\omega^{(1)}(l_{sv}^T)$ after adding node v to the tree. So, the running time of the algorithm is $O\left(|V_T|log(|V_T|)|DS| + |V_T||DS| + |V_T|\right) \sim O\left[|DS||V_T|log(|V_T|)\right]$.

The running time of proposed algorithm has polynomial growth with the grow of the network size and can be used in practical application.

7   Conclusions and Further Investigation

In this paper, we have proposed metrics, B and D, which provide significant growth of network capacity measured in the number of voice calls with tolerable voice quality as compared to simple and thus popular Hop Count or Airtime Link metrics defined as default in numerous routing protocols, as it is proved by simulation results.

We also propose an approach of multiple metrics usage in multihop ad hoc networks with end-to-end QoS support for unicast and multicast traffic. Authors are going to perform extensive simulations to evaluate this approach in the nearest future with NS3 simulation tool [9].

References

1. ITU-T. Recommendation G.107. The E-Model - A Computational Model In Use In Transmission Planning (2005)

2. IEEE P802.11s/D4.0. Draft STANDARD for Information Technology Telecommunications and information exchange between systems. Local and metropolitan area networks Specific requirements Part 11: Wireless LAN Medium Access Control (MAC) and Physical Layer (PHY) specifications Amendment: Mesh Networking (2009)

3. Perkins, C., Belding-Royer, E., Das, S.: Ad Hoc On-Demand Distance Vector Routing Protocol, IETF MANET Working Group (2003), http://www.ietf.org/rfc/rfc3561.txt

4. Jacquet, P., Muhlethaler, P., Clausen, T., Laouiti, A., Qayyum, A., Viennot, L.: Optimized Link State Routing Protocol. In: Proc. of IEEE INMIC 2001, Lahore, Pakistan, pp. 62–68 (2001)

5. Haas, Z.J., Pearlman, M.R., Samar, P.: The Zone Routing Protocol (ZRP) for Ad Hoc Networks. In: IETF MANET Working Group (2002), http://www.ietf.org/proceedings/55/I-D/draft-ietf-manet-zone-zrp-04.txt

6. De Couto, D.S.J., Aguayo, D., Bicket, J., Morris, R.: A high-throughput path metric for multi-hop wireless routing. In: Proc. of MobiCom 2003, USA, pp. 134–146 (2003)
7. Draves, R., Padhye, J., Zill, B.: Routing in multi-radio, multi-hop wireless mesh networks. In: Proc. of MobiCom 2004, USA, pp. 114–128 (2004)
8. Bianchi, G.: Performance Analysis of the IEEE 802.11 Distributed Coordination Function. IEEE Journal on Selected Areas in Communications 18(3), 535–547 (2000)
9. The ns-3 network simulator, http://www.nsnam.org/
10. Karp, R.M.: Complexity of Computer Computations. Reducibility Among Combinatorial Problems (1972)
11. Andreev, K., Boyko, P.: Simulation Study of VoIP Performance in IEEE 802.11 Wireless Mesh Networks. In: Proc. of MACOM 2010, Barcelona (2010)

Performance of MAC Protocols in Beaconing Mobile Ad-Hoc Multibroadcast Networks

Cristina Rico Garcia, Andreas Lehner, Patrick Robertson, and Thomas Strang

German Aerospace Center (DLR), Institute for Communications and Navigation

Abstract. The investigation of infrastructureless safety applications in different transportation systems is a hot research topic. The nodes in the network are designed to advertise to the rest of the nodes information-about the current traffic situation by means of short beacon messages containing speed, direction, positions and other relevant safety information. The scheduling should be organized by the MAC layer so that the transmitted messages arrive successfully as soon as possible at the receiver. The networks that support these kind of applications are Mobile Ad-hoc Multibroadcast Networks (MAMNETs). In this paper we present the challenges the MAC layer for MAMNETs should overcome. We discuss the most important performance metrics of the MAC layers in order to obtain a system independent analysis and show a survey of the factors that may influence the behavior of the MAC layers.

1 Introduction

A family of new transportation related safety systems is being extensively investigated in the last years. These new safety systems share the same basic principles: The nodes in the network which can be any kind of vehicle, pedestrians or any moving entity, must be constantly aware of the situation in their surroundings in order to avoid dangerous constellations. To do so, they do not rely on previously installed infrastructure. Instead, they periodically broadcast a short beacon containing important safety information like position, speed, direction, etc.

We call the kind of networks that support this type of safety systems beaconing Mobile Ad-hoc Multibroadcast Networks (B-MAMNETs). The term multibroadcast stresses that, in contrast to traditional broadcast networks, where a fixed number of nodes are broadcast transmitters and the rest of the nodes in the network are receivers, in multibroadcast networks all nodes are broadcast transmitters and receivers.

Typical MAMNETs are given in : Car-to-Car Communications (Car2Car) [6] for automotive transportation; Automatic Dependent Surveillance Broadcast systems (ADS-B) [2] for air transportation; Railway Collision Avoidance Systems (RCAS) [13] for railway transportation; Automatic Identification Systems (AIS) [3] for maritime transportation, as well as various military applications. For the future, we can foresee many promising applications in the natural disaster management communication systems area when communication cannot rely

A. Vinel et al. (Eds.): MACOM 2010, LNCS 6235, pp. 263–274, 2010.

on infrastructure any more, like monitoring of rescue teams in earthquake or tsunami scenarios.

The performance of these systems and therefore the safety enhancement they can offer, depends directly on the successful reception of the beacons. Here, the Medium Access Control (MAC) layer plays an important role.

Traditionally, the IEEE 802.11 protocol is used for mobile ad-hoc networks (MANETs). However, the fact that the safety systems are working in a multi-broadcast beaconing MAMNET network, introduces several challenges in the design of the MAC layer. First of all, the multibroadcast communication mode prevents the usage of any acknowledgment (ACK) packet, as well as the RTS/CTS protocol, both used in the IEEE 802.11 standard. Secondly, the vehicles in the network are moving quickly, so that the topology is changing continuously and the MAC layer should be able to "follow" these changes. And thirdly, since we are dealing with a safety system the delay of the messages must be minimized.

Consequently, the basic assumptions underlying the IEEE 802.11 that were responsible for its success are no longer present for MAMNETs. Therefore, the performance of IEEE 802.11p [12], i.e., the IEEE 802.11 version for MAMNETs, as well as other protocols for MAMNETs must be investigated in detail from an application independent point of view. In this way, the performance of the protocols can be easily mapped to any present or future system.

The first contribution of this paper is the identification of the factors that influence the behavior of the MAC layers in MAMNETs. The second contribution is the establishment of the performance criteria that should be followed in MAMNETs. These criteria must be application independent and useable for any future MAMNET system. We will introduce the Update Delay as the most relevant performance criterion in MAMNETs, since it indicates the capability of the nodes to react to dangerous situations.

The remainder of this paper is organized as follows: Section 2 shows the evolution of wireless networks and the new challenges introduced. In section 3 we discuss influence factors of the MAC layers in MAMNETs. In section 4 the performance criteria of MAMNETs will be analyzed. Finally, section 5 provides the summary.

2   General Network Classification

The design of a MAC layer should be adapted for the kind of network where it will be used. In the literature there are different surveys of MAC layers that can be used for wireless networks [8], for ad-hoc networks [10] and for mobile ad-hoc networks [16].

This section gives an overview about different types of wireless networks and introduces the mobile ad-hoc multibroadcast networks. Figure 1 depicts our classification of the different wireless networks.

Closed wireless network vs. open wireless network: A closed wireless network is defined as being a wireless network where all the nodes are in range of each other. This means, the nodes are able to receive a message from any another

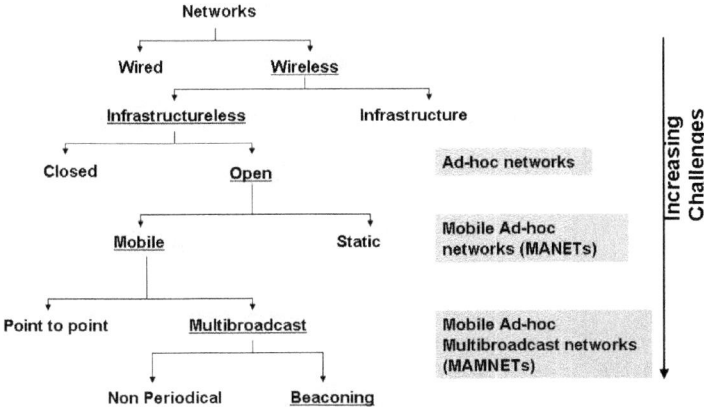

Fig. 1. Network classification

node in the network, no matter how far away it is. On the other hand, in an open wireless network, the network area extends further than the range of the nodes and therefore nodes which are far away are not able to receive each other. In an open network, the hidden terminal problem arises, which is a major design issue for MAC layers. Note a MAC layer always defines no more than a single-hop network since multi-hop is a task of upper layers. Therefore, a single-hoe network might be closed or open.

Ad-hoc network: An Ad-hoc network is a wireless network working in a decentralized mode. This means, there is no infrastructure in charge of coordinating the traffic of the network. The nodes may enter or leave the network suddenly and they must organize themselves in a distributed way so that collisions among messages are avoided.

Mobile Ad-hoc network: A MANET is an ad-hoc network where the nodes can move. The main challenge of this kind of network is that there is no knowledge of how many nodes are in the network and where they are. Therefore, the MAC layer should be able to "follow" the changes experienced in the network.

Multibroadcast Mobile Ad-hoc network: A MAMNET is a MANET where each node is a broadcast station. The multibroadcast communication mode prevents the usage of any acknowledgment (ACK) packet, as well as the RTS/CTS protocol. First, because the broadcasting node does not know from how many nodes ACKs should be received as there is no knowledge about the state of the network. Second, because, the ACKs sent by all receivers would produce collisions among them. Beaconing MAMNETs are a special case of MAMNETs where the messages are short beacons broadcasted with a fixed periodicity. The length of an ACK and a beacon does not differ substantially, therefore, the relative MAC overhead is too costly.

Obviously, MAMNETs introduce new challenges when designing a suitable MAC protocol. Various MAC layer protocols have been proposed for MAMNETs [9], [15], [5], [4]. However, since there is neither a consensus about the necessary metrics in these kinds of networks nor about the MAC performance measurement criteria, there is no way of comparing the performance of the different MAC layers. In the next section, the performance criteria for the analysis of MAC layers in MAMNETs will be discussed.

3   Factors Influencing MAC Performance in MAMNETs

In this section the influencing factors that determine the performance of the layers in MAMNETs are analyzed. In order to study the performance of a MAC layer for a specific system, the influencing factor of the system should be extracted and mapped to the performance criteria curves of a MAC layer, e.g speed and range to node dynamics.

Protocol parameters: The parameters of MAC layer protocols clearly have an influence on the behavior of a MAC layer protocol. Examples of these parameters are the backoff windows in the IEEE 802.11p or the SI parameter in the SOTDMA protocol [9].

Frame duration: An important parameter in MAMNETs is the frame length. A frame duration is defined as the *average time between two consecutive transmissions sent by a node*. It defines the minimum time the system needs to update the information coming from all the nodes in the network. Typical frame duration values range from 0.5 s to 2 min [6], [3].

Closed/open network: A closed or open network has an influence on the MAC layer performance for all those protocols which have to listen to the medium since in open networks the hidden terminal problem arises. The hidden terminal problem is responsible for strong degradation of the performance of MAC layers.

3.1   Net Channel Load

It is widely believed that the net channel load is the most important influencing factor in the performance of a MAC layer [1]. Its relevance has been recognized by the C2C-CC community as well [5]. The net channel load or offered traffic is defined as the *fraction of a frame occupied by data, i.e. payload*. A net channel load of 100% would be given when there are so many nodes in the network that if they were perfectly synchronized so that they would transmit the data packets one after each other, they would produce neither collisions nor free spaces between data packets. Figure 2 shows a configuration where the number of data packets gives a 100% net channel load. The net channel load is related to the amount of data transmitted within a frame and not the order in which they are transmitted, so that the same total number of packets transmitted as in Figure 2,

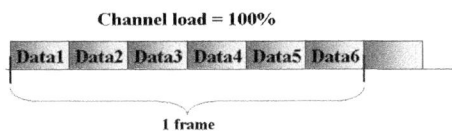

Fig. 2. An example of a frame with 100% Channel Load

but all at the same time will produce 100% channel load, too. Please, note that the offered traffic can be larger than 100% as shown in Figure 8.

It must be noted that in the calculation of the channel load, only the data bits in the packet should be counted. I.e. neither MAC layer header, nor the backoff time or other waiting times introduced by the MAC layer should be taken into account as part of the offered traffic. This ensures fairness when comparing the performance of different MAC layers.

A MAC layer that requires a larger MAC header could probably deliver a better performance than a MAC layer that does not introduce any header to the data, since it has more valuable information for the MAC layer. However, in order to support the same number of nodes in the network, in the first case the system would need a higher bandwidth than in the second case. The MAC layers must always be compared assuming the same system where only the MAC layer is changed.

The C2C-CC White Paper on Network Design Limits [5] recognizes the channel load as the most relevant network design limit and gives the following definition: *The fraction of time that the channel is sensed busy.* The problem with this definition is that when the header bits are being transmitted, the channel will be sensed as busy, and as stated above, the header should not be computed in the channel load.

Several publications use the number of nodes instead of the channel load in the network as an influencing factor in the analysis of MAC layer performance [7]. However, this is a complicated parameter to manage as influencing factor. Since different system parameter constellations with the same number of nodes deliver different performances and on the other side, different system parameter constellations with different nodes can deliver the same performance. To overcome this problem, the system parameters can be mapped to the a net channel load in the following way:

$$Load(\%) = \frac{Datapackets/frame}{Load_{max}}, \tag{1}$$

where $Load_{max}$ tells us how many packets are in the system under 100% net channel load. The calculation of this factor is performed in a different way depending on whether the network is closed or open.

Net channel load in a closed network: The 100% net channel load in a closed network is given by

$$Load_{max_{closed}} = \frac{frame}{Data_{time}}, \tag{2}$$

where the *frame* is in seconds and the $Data_{time}$ is the length in seconds of a packet calculated as $Data_{time} = Data_{bits} \cdot tx_{rate}$. The transmission rate tx_{rate} must be calculated computing the whole available bandwidth for the MAC layer.

A MAC layer that separates its bandwidth in two channels, for example one for MAC control data and another one for data will normally have a better performance in the data channel than a protocol that only uses the data channel without any control channel. Note that in both cases the data transmission rate is the same although the total available bandwidth is larger in the first case. Therefore, again for the sake of fairness in the comparison of different MAC layers the transmission rate should be derived from the available bandwidth for the complete MAC layer and not only from the data bandwidth.

Net channel load in a open network: The 100% net channel load in a open network is given by

$$Load_{max_{open}} = Area_{network} \cdot density_{max}, \tag{3}$$

where the $density_{max} = \frac{Load_{max_{closed}}}{Area_{range}}$. The $Area_{range}$ is the area around a node where its transmissions are received, i.e the communication range area.

Payload/header length: The MAC layer header usually has a fixed length independent of the payload length. A system with a high payload/header rate will be able to attain a bigger channel load than a system with a low payload/header rate. Although MAMNETs systems transmit in general very short messages of around 200 bits, other payload/header rates should be taken into account as influencing factor.

3.2   Topological Network Dynamic

We define the topological network dynamic δ as the quotient of maximum node velocity v_{max} and minimal communication range $Range_{min}$. I.e. $\delta = \frac{v_{max}}{Range_{min}}$. The influence factor δ indicates how instable the network can be.

Table 1. Topological network dynamic factor for different systems

Transport system	Min. comm. range	Maximum velocity	Topological network dynamic
Ships - AIS	40 km	60 km/h	1,5 h^{-1}
Airplanes - ADSB	56 km	1000 km/h	16 h^{-1}
Trains - RCAS	5 km	200 km/h	40 h^{-1}
Cars - C2C	1 km	140 km/h	140 h^{-1}

traffic model: The traffic model plays an important role. It indicates how the nodes are moving altogether. A network may have a high topological network dynamic factor but if the traffic model is a uniform and parallel movement of the nodes, then the influence of the factor will be lower than if the traffic model is a random and independent movement of the nodes.

3.3   Radio Channel

The last influencing factor listed in this paper is the radio channel. The MAC layers should be tested under the conditions of different path loss radio channels, since the influence of the radio channel on the performance of the MAC layer is remarkable. Important radio channels are for example the ideal channel, i.e, the signal is not attenuated in the communication range and afterwards it is zero and the logarithmic path loss channels. In order to study the influence of the interference of nodes beyond the communication range, a logarithmic path loss channel can be compared to a logarithmic-ideal channel, i.e., a channel with a logarithmic profile in the communication range but with infinite attenuation afterwards.

The Bit Error Rate (BER) might as well have an influence in the performance of the MAC layer for MAC protocols if these need to decode the receiver messages in order to take decisions.

4   Performance Criteria and Metrics for the Analysis of MAC Layers in MAMNETs

In this section the performance criteria to measure the behavior of the MAC protocols will be discussed. The analysis of the MAC layers should be accomplished in such a way that performance criteria are studied for the different influencing factors listed in the previous section.

4.1   Update Delay

In classical ad-hoc networks, the *transmission delay* is an important performance criterion, specially for real time applications. However it has low practical relevance in MAMNETs.

Transmission Process in ad-hoc Networks: The transmitter wants to send a packet. This packet can be short or long, data, video, sound, etc. The packet is delivered in the buffer queue, and waits there until the MAC layer allows its transmission. Additionally to the delay in the buffer, the delay due to the distance between sender and receiver should be added. Then, the transmitter waits for an acknowledgement from the receiver. In case no acknowledgement is received, the transmitter resends the packet again until it receives a valid acknowledgement. The total transmission delay TD is:

$$TD = R \cdot (Q + t_c + t_{ack}) \tag{4}$$

where R is the number of transmission per single packet necessary for successful reception, Q is the time spent in the buffer queue and other MAC layer waiting time, t_c is the transmission time due to the distance between transmitter and receiver and t_{ack} is the time the transmitter should wait for an acknowledgement before trying a retransmission.

Transmission Process in MAMNETs: The transmitter is continuously getting information from its onboard localization unit and wants to broadcast a packet every period of time T. This packet is a short beacon, which contains updated safety information like the node position, speed, etc. The packet is delivered in the buffer queue. If time T goes by before the MAC layer has ordered transmission of this packet, a new beacon arrives in the queue and replaces the old undelivered beacon since the status information it contains is outdated. The transmitter does not wait for an acknowledgement. On the one hand, in a multi-broadcast network acknowledgements are not feasible and on the other hand, there is no sense in retransmitting outdated information, in case no acknowledgement would be received – in particular when the transmitter has more recent status information to send. The total transmission delay TD is:

$$TD = (Q + t_c) \qquad (5)$$

Obviously the transmission delay in a MAMNET is negligible for the MAC functionality compared to the transmission delay in a standard ad-hoc network and has no practical relevance in the MAC design.

The moving nodes in a MAMNET take decisions and react according to the received update status beacons. The larger the time between status update beacons, the lower the available reaction time. Figure 3 shows the surveillance strategy of the RCAS system. When the distance between the two trains is below BC, the trains must brake, if they do not want to risk an accident. The status update beacons indicate to the trains the distance the other trains are. In the case no status update beacon is received between TA and BC, the trains will not be aware of the traffic situation and will not be able to brake in time. Obviously, the time between the reception of consecutive status update beacons is a fundamental performance criterion in MAMNETs.

For example, in a MAMNET system with a beacon rate of 1 Hz (each node broadcasts a status update beacon every second), the failure in the reception of one status update beacon implies two seconds of update delay. For a node moving at 160 km/h, this means 90 m less to react in time before an accident might occur.

We define the **Update Delay** as the time between the reception of two consecutive status update beacons coming from the same node. The Update Delay is the most relevant performance criterion in the MAC layer of the MAMNETs. Since the time between the transmission of two consecutive beacons, around 1 Hz depending on the application, is called *frame*, the Update Delay can be as well defined as *the number of frames between two consecutive status update*

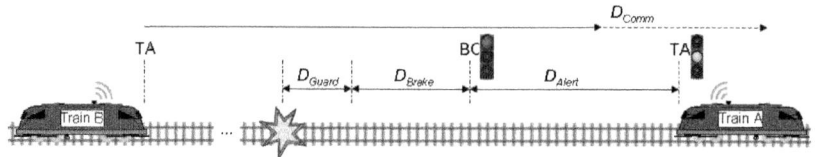

Fig. 3. RCAS surveillance strategy

beacons coming from the same node. Some authors use related criteria in the analysis of MAC layers in MAMNETs: *mean beacon transmission delay* [14] or *Maximum distance between two successful communications* [15]. It is necessary to establish a general metric for the Update Delay criterion. This metric must give a complete description of the behavior of the Update Delay and should be application independent, i.e, should not depend on a concrete number of nodes in the MAMNET or a particular beacon rate. Both proposed metrics in [14] and [15] are not complete descriptions of the behavior of the Update Delay in the network, nor application independent.

We propose the *Complementary Cumulative Distribution function (CCDF)* of the Update Delay as the main metric for the performance of MAC layers in MAMNETs. This metric represents for each point the probability that the update delay UD is larger than the number of frames tf in the abscissa.

$$CCDF(ud) = P(UD > ud = tf) \qquad (6)$$

Figure 4 shows an example of the CCDF of the Update Delay.

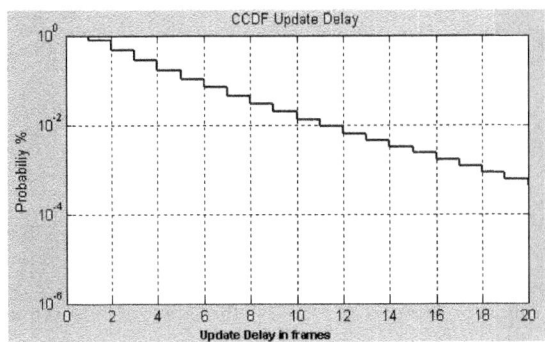

Fig. 4. Update delay metric: CCDF

The abscissa of the curve can be easily translated into seconds for a concrete system in the following way: $t = ud \cdot tf$ where t is the update delay in seconds, ud is the update delay in frames and tf is the duration of a frame in seconds. For example for the Car2Car communication system $tf = 0.5$ since the cars send an update beacon twice per second. This translation can be directly done when the propagation channel is an ideal channel. On the other case, the CCDF should be computed for different frame lengths.

Figures 5 and Figure 6 shows two examples of update delay CCDFs.

4.2   Throughput

The throughput is defined as the fraction of the capacity of the channel that is used. It indicates the efficiency of the MAC layer. The maximum throughput is achieved when the number of correctly delivered packets equals the number of

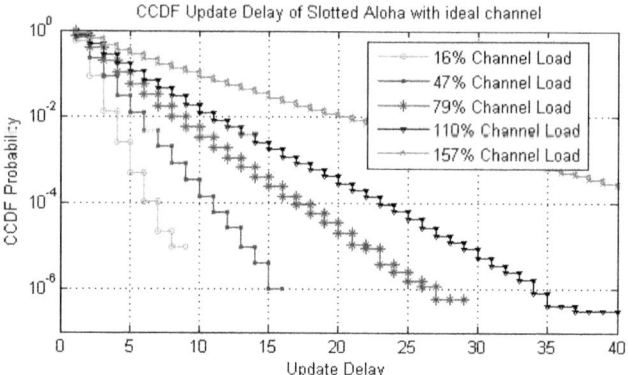

Fig. 5. CCDF of Slotted Aloha for different net channel loads

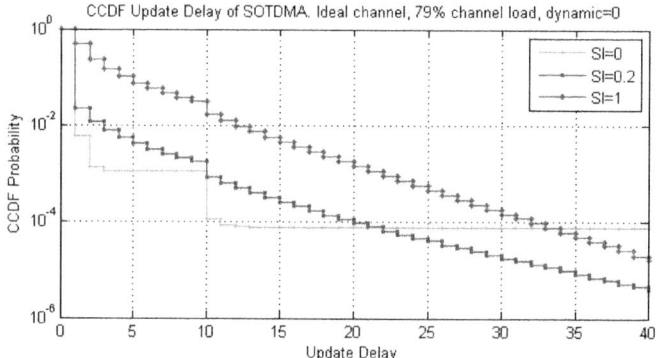

Fig. 6. CCDF of SOTDMA for different SI factors

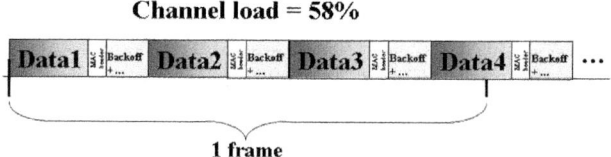

Fig. 7. A full channel that does not reach 100% throughput

packets for a 100% net channel load. Commonly, the MAC layers are not able to reach 100% throughput; there are two main reasons: Firstly, the MAC layer may add a header and waiting times so that the length of the packet is bigger than the real payload. Figure 7 shows this case. Secondly, due to collisions the number of correctly delivered packets cannot reach the number of packets for a 100% net channel load.

In Figure 8 the throughput of slotted Aloha and SOTDMA under the same influence factors can be observed.

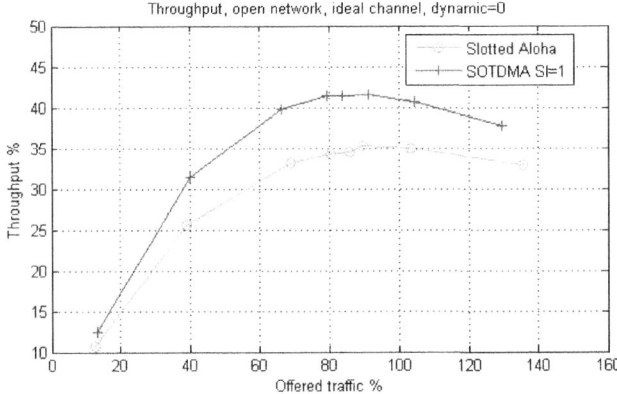

Fig. 8. Comparison of the throughput of Slotted Aloha and SOTDMA

4.3   Collision Rate

The collision rate is defined as the colliding fraction of the transmitted messages
[11]. There is a direct relation between the collision rate CR and the throughput
TH when the channel load is 100%: $CR_{ChannelLoad100\%} + TH_{ChannelLoad100\%} = 1$.

Note that in a radiobroadcast network, every time a node transmits a message,
it is received by many nodes in its surroundings. Therefore, to compute the
number of delivered messages, the total amount of received messages should be
divided by the total amount of nodes that should receive it, i.e. the nodes within
the communication range.

Figure 9 shows the influence of the propagation channel on the collision rate
of slotted aloha.

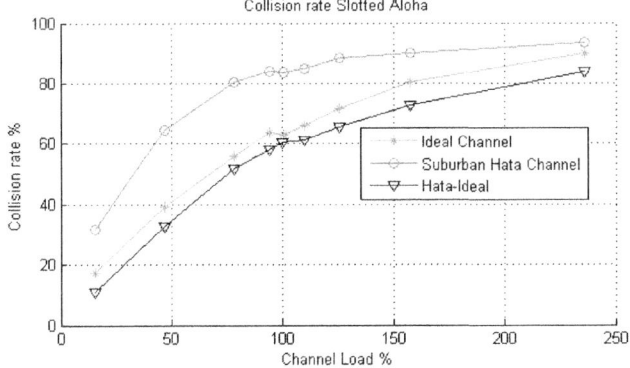

Fig. 9. Collision rate of Slotted Aloha for different propagation channels

5 Conclusions

This paper has presented a road map to analyse the performance of the MAC layers in mobile ad-hoc multibroadcast networks. The particularities and challenges of this kind of networks have been introduced. We have identified the factors that influence the behaviour of the MAMNETs MAC layer protocols and finally the most relevant MAC performance metrics in MAMNETs have been studied.

References

1. Abramson, N.: Computer Communication Networks. Prentice-Hall Inc., Englewood Cliffs (1973)
2. ADS-B Home Page, http://adsb.tc.faa.gov/ads-b.html
3. AIS official website, http://www.navcen.uscg.gov/enav/ais/default.html
4. Rico García, C., Lehner, A., Strang, T.: Comb: Cell based orientation aware manet broadcast mac layer. In: IEEE GLOBECOM 2008 (2008)
5. Brakemeier, A.: Network design limits and vanet performance. White Paper V0.5 (2008-11), Car 2 Car Communication Consortium (November 2008)
6. Car2Car communication consortium, http://www.car-to-car.org/
7. Eichler, S.: Performance evaluation of the IEEE 802.11p WAVE communication standard. In: Proceedings of the 1st IEEE International Symposium on Wireless Vehicular Communications (WiVeC) (September 2007)
8. Gummalla, A.C., Limb, J.O.: Wireless medium access control protocols. IEEE Communications Surveys & Tutorials 3(2), 2–15 (2009)
9. ITU. Itu-r m.1371* technical characteristics for a universal shipborne automatic identification system using time division multiple access in the vhf maritime mobile band (2001)
10. Kumar, S., Raghavan, V.S., Deng, J.: Medium access control protocols for ad hoc wireless networks: A survey. Ad Hoc Networks 4(3), 326–358 (2006)
11. Turletti, T., Romdhani, L., Ni, Q.: Adaptive edcf: Enhanced service differentiation for IEEE 802.11 wireless ad hoc networks. In: IEEE W3reless Communications and Networking Conference (WCNC 2003), March 16-20 (2003)
12. Status of Project IEEE 802.11 Task Group, http://www.ieee802.org/11/reports/tgp-update.html
13. Strang, T., Meyer zu Hörste, M., Gu, X.: A railway collision avoidance system exploiting ad-hoc inter-vehicle communications and galileo. In: 13th World Congress and Exhibition on Intelligent Transportation Systems and Services (ITS 2006) (October 2006)
14. Vinel, A., Koucheryavy, Y., Andreev, S., Staehle, D.: Estimation of a successful beacon reception probability in vehicular ad-hoc networks. In: IWCMC 2009: Proceedings of the 2009 International Conference on Wireless Communications and Mobile Computing, pp. 416–420. ACM, New York (2009)
15. Imai, J., Suzuki, N., Tadokoro, Y., Ito, K., Itoh, N.: A new approach for evaluation of vehicle safety communications with decentralized tdma-based mac protocol. In: IEEE Intelligent Vehicle Symposium (2008)
16. Zhai, H., Wang, J., Chen, X., Fang, Y.: Medium access control in mobile ad hoc networks: challenges and solutions: Research articles. Wirel. Commun. Mob. Comput. 6(2), 151–170 (2006)

Quality of Service Oriented Analysis of Cross-Layer Design in Wireless Ad Hoc Networks

Ulrike Korger[1], Christian Hartmann[1],
Katsutoshi Kusume[2], and Joerg Widmer[2]

[1] Institute of Communication Networks
Technische Universität München
Arcisstr. 21, 80290 Munich, Germany
[2] DOCOMO Euro-Labs
Landsbergerstr. 312, 80687 Munich, Germany

Abstract. In wireless ad hoc networks, cross-layer design aims at reducing multiple access interference and thus obtaining a higher spatial reuse. In order to identify the most suitable solution for future wireless systems, we compared the aggregate throughput achieved by two types of cross-layer designs. While the first approach suppresses the interference by power control at the transmitter side, the second cancels interference at the receiver with a technique called multiuser detection. It turned out that applying power control as basis for cross-layer design in very high traffic density scenarios can result in even worse performance than applying no cross-layer design at all. The gains achieved by the multiuser detection based cross-layer design were shown to be based on the interaction between physical and MAC layer, and not only on the more advanced physical layer technique.

However, to investigate the provided quality of service, also other criteria have to be taken into account. Thus, in this paper, we investigate delay and fairness for both cross-layer designs. We show that power control based cross-layer design leads not only to lower aggregate throughput, but also to a lower offered quality of service.

1 Introduction

Due to the lack of any central entity in ad hoc networks, medium access must be organized in a fully distributed manner. Thus, in order to allow for spatial reuse, a joint design of both, physical and MAC layers, naturally plays an important role. In the literature, several approaches exist that combine different physical layer techniques with an appropriate MAC layer to a cross-layer design (CLD).

One conventional technique on the physical layer is power control. Combined with specific MAC protocols, many authors as well adopt this technique to the distributed nature of ad hoc networks in order to suppress Multiple Access Interference (MAI) [9], [3], [11].

A. Vinel et al. (Eds.): MACOM 2010, LNCS 6235, pp. 275–286, 2010.
© Springer-Verlag Berlin Heidelberg 2010

An alternative physical layer technique on the receiver side is the so called MultiUser Detection (MUD) receiver [13]. With this technique, the receiver detects streams and cancels out unwanted interference. Although in general the complexity of multiuser detection increases exponentially with the number of branches detected [12], there exist also detectors with reduced complexity that achieve similar performance [7]. Exploiting the gains offered by multiuser detection by an appropriate MAC layer is the aim of newly presented CLDs for ad hoc networks [2], [1], [8].

Both types of cross-layer solutions, the power control based as well as the MUD based, aim at the same goals, namely improving the spatial reuse in ad hoc networks. Still, the strategies applied differ completely. The performance of the two physical layer techniques, power control and multiuser detection, is well known in the literature. However, to the best of our knowledge, up to this point no numerical comparison between the different types of CLDs applied in ad hoc networks was investigated yet.

In order to get insight into the Quality of Service (QoS) offered by the CLDs, we present detailed throughput investigations in [6]. There we compare the so called Progressive BackOff Algorithm (PBOA) approach [11], a good representative for power control based CLD, to the MUD-MAC CLD that was presented in [8]. It turns out that MUD-MAC offers larger gains in terms of aggregate throughput than PBOA. Moreover, these gains are not solely based on an advanced physical layer technique, but also on the better interaction between physical and MAC layers of the proposed CLD.

However, investigating the service provided by a CLD only in terms of aggregate throughput can still lead to unsatisfactory solutions. Namely, a solution that only serves the best users will offer good results in terms of aggregate throughput. Most of the users will though suffer unacceptable delays and thus are excluded from the channel access. On the other hand, if the protocol tries to always equally serve all users, system resources might be wasted by serving weak users, leading to a poor spectral efficiency. Thus, a protocol should ideally offer a good trade-off between overall system performance and fair medium access with acceptable packet delays.

In [6] we show that the MUD based CLD leads to large benefits in overall system throughput for random ad hoc scenarios. Within this paper we present the related delay and fairness results. We show that the MUD based CLD outperforms the power control based CLD not only in terms of throughput, but also offers the better mean packet delays and is fairer.

The outline of the paper is as follows. In Section 2 we summarize the PBOA algorithm that serves as a comparison scheme for our MUD-MAC protocol. The MUD-MAC CLD is described in Section 3. We explain the applied delay and fairness measures in Section 4. Throughput, fairness, and delay results are presented in Section 5 for a random network with medium traffic density. Section 6 draws the conclusions.

2   The PBOA Algorithm

The authors of [11] present a power control based CLD, called Progressive Back-off Algorithm (PBOA). The design proposed assumes a certain time slotted structure, called *frame* that is depicted in Fig. 1. The first part of the frame is related to a contention phase and consists of several pairs of minislots. Each minislot is divided into the transmission of an RTS and a CTS signal. The second part of the frame is used for the transmission of data. Notice that no additional acknowledgement is assumed by the authors of [11]. Before the data is transmitted, the different terminals, willing to transmit, start contending for channel access. I.e., at the beginning of the contention phase each potential transmitter simultaneously transmits its RTS signal. Fig. 2 illustrates this (First Minislot). T_1 to T_4 thereby represent simultaneous transmissions during the contention phase.

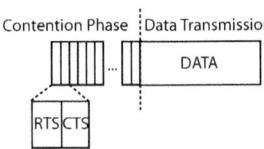

Fig. 1. General frame structure of the PBOA algorithm

If the receiver announced can decode the RTS, it replies with a CTS, also with maximum power. Depending on its receive Signal-to-Interference-and-Noise-Ratio (SINR) and its actual SINR requirement, it includes an factor into the CTS that tells its associated transmitter how much to power down in the next RTS minislot of the contention phase.

An exemplary behavior is depicted in Fig. 2 in the middle (Second Minislot), where the transmission power of T_4 starts to decrease. The successive power reduction goes on in consecutive minislots, unless a minimum for the acceptable transmission power is reached. Afterwards, the receiver of T_4 will abstain from transmitting further CTS messages. Its associated transmitter however will proceed transmitting RTS signals with the minimum transmission power until the contention phase ends. This enables other receivers to still correctly estimate the interference expected during data transmission.

If a transmitter is not successful during one minislot, it will stay contending during the consecutive slot with a so called win probability p, or it will go to backoff and turn into a potential receiving node until the end of the frame with the probability of $1 - p$.

In Fig. 2, T_1, T_2 and T_3 are not successful during the first minislot of the contention phase. While T_2 looses and goes into backoff, T_1 and T_3 try to succeed again during the second minislot. Notice, however, that T_3 chooses a different receiver, namely the receiver of the second packet in its transmission queue. This is proposed by the authors of PBOA, in order to increase the probability that RTS messages reach the intended receivers.

By progressively reducing transmission powers and number of potential transmitters (backoff), other transmitters are able to reach their intended receivers. This is illustrated in Fig. 2 in the third minislot, where T_1 and T_3 can reach their receiver due to the reduced interference.

After the contention phase all successful transmitters send their data to the intended receivers with the minimum transmission power they agreed on. An additional acknowledge is not required, since the channel is assumed to stay constant for the duration of the whole frame [11].

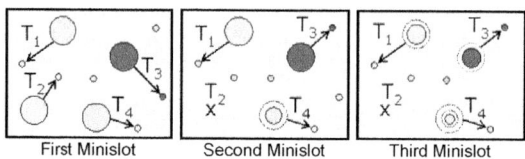

Fig. 2. Power adaptation and backoff during the contention phase of the PBOA protocol

3   The MUD-MAC Protocol

In the following, we summarize a multiuser detection based CLD, the so called MUD-MAC protocol [8]. Similar to PBOA, MUD-MAC requires a time-slotted structure, referred to as *block* in the following. Each data frame is subdivided into N blocks. The block structure of MUD-MAC is depicted in Fig. 3.

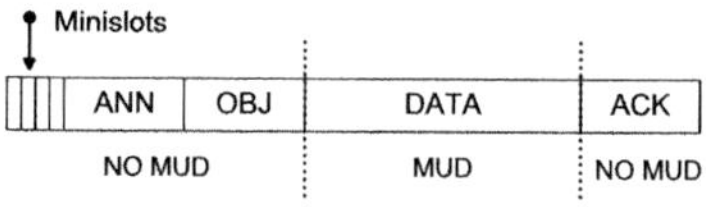

Fig. 3. One block of the MUD-MAC protocol

Each block consists of several control signals, namely announcement (ANN), objection (OBJ), and acknowledge (ACK), and a slot for data transmission (DATA). Notice that during the control signaling slots, no multiuser detection capabilities are required.

Unlike PBOA, transmitters should not start their control signaling simultaneously. Instead, each transmitter randomly chooses one minislot and abstains from a planned transmission if it senses another transmitter signaling in an earlier slot. This kind of contention resolution mostly avoids collisions during the ANN phase. The successful transmitter announces the planned transmission to its associated receiver. It includes a signature used during the data phase into the ANN signal. This signature is required, since a spread spectrum multiple access scheme, e.g., CDMA or IDMA, is considered. Also the number of consecutive blocks N, used for conveying a packet, is included into the ANN. Notice that a transmission lasting multiple blocks is announced only once per packet. A new transmission can be started in each new ANN slot, resulting in a maximum of N parallel transmissions.

With the help of the ANN signals, channel estimation can be performed by the associated receiver as well as receivers that are already involved in ongoing transmissions. During the OBJ phase, the latter ones have the opportunity to object to the planned transmission. This happens, if they cannot handle the additional interference, e.g., if they have no more free MUD branches.

If no OBJ can be sensed, the transmitter starts transmitting the first of N blocks. The size of the blocks is thereby chosen such that the channel coherence time is larger than the time required for the transmission of all N blocks. If the transmission is successful, the receiver acknowledges the reception of multiple blocks once at the end of the transmission.

4   Delay and Fairness Measures

In order to get insight into the QoS offered by a CLD, in addition to the system throughput, delay and fairness have to be carefully investigated.

4.1   Delay

Within this paper, we measure the delay as the delay per packet that nodes experience while transmitting. According to [5], besides traffic that has no delay restrictions, there exist real-time streaming services with very high delay demands (150 ms-250 ms) and non-real time services that are interactive. The latter require at least delays that are lower than 2 s. However, for e.g., web browsing, as service contained in this group, a maximum delay of 0.5 seconds would be desirable [5]. Thus, we restrict the maximum delay $\Delta_{\max}$ a packet can tolerate to 1s. If the delay exceeds this limit, the packet is removed from the packet queue and lost.

We define the mean packet delay $\overline{\Delta_{\mathrm{p}_k}}$ of the received packets each node k experiences as the sum of the packet delays $\Delta_{\mathrm{p}_{k,i}}$ of all successfully transmitted packets i over the number of successfully transmitted packets N_k for this node, respectively:

$$\overline{\Delta_{\mathrm{p}_k}} = \frac{\sum_{i=1}^{N_k} \Delta_{\mathrm{p}_k,i}}{N_k}. \tag{1}$$

In order to take fairness into consideration as well, we subsequently evaluate the median of these mean packet delays per node. Unlike a mean, the median is insensible to outliers. It is the value separating the higher half of the realizations from the lower half. In case of unfair medium access, single nodes that are frequently granted medium access can significantly decrease the overall mean delay. However, the median will not be strongly influenced by these nodes.

Notice, however, that $\overline{\Delta_{\mathrm{p}_k}}$ corresponds only to the packets *successfully transmitted*. In case a CLD looses many packets, e.g., by silencing some of the users, the remaining ones can be transmitted faster and thus $\overline{\Delta_{\mathrm{p}_k}}$ per node and accordingly the median delay for the *transmitted* packets is lower. Thus, as a second measure, we calculate an overall mean delay $\overline{\Delta_{\text{pOut}}}$ that includes also a maximum delay $\Delta_{\max}$ for the number of packets lost N_{Out_k} for the K nodes as follows

$$\overline{\Delta_{\text{pOut}}} = \frac{\sum_{k-1}^{K} \left(\sum_{i=1}^{N_k} \Delta_{\text{p}_{k,i}} + N_{\text{Out}_k} \Delta_{\max} \right)}{\sum_{k=1}^{K} (N_k + N_{\text{Out}_k})}. \tag{2}$$

4.2   Fairness

In order to get insight into the fairness behavior of the CLDs, we evaluate the variance of both, the mean packet delay values $\overline{\Delta_{\text{p}_k}}$ for different nodes, and the one for the average throughput per node. It can be stated that the lower the variance of these values is, the fairer is the access to the medium.

Another measure for the fairness of medium access is the so called Jain's fairness index [4]. This index is defined for K nodes as

$$F_J(w) = \frac{(\sum_{k=1}^{K} g_k(w))^2}{K \sum_{k=1}^{K} g_k^2(w)} \text{ with } 0 < F_J(w) \le 1, \tag{3}$$

where w reflects a sliding window with a size of multiple packets, and $g_k(w)$ reflects the fraction of the overall medium access, a node k achieved within this window. The window is stepwise increased over the pattern of medium accesses, thereby reflecting the change from short-term to long-term fairness.

In case of perfect fair channel access, all $g_k(w)$ equal $\frac{1}{K}$ and Jain's fairness index is equal to 1. A scheme is fairer if its Jain's fairness index is closer to 1 and vice versa.

5   Simulation Results

The following section presents simulation results that compare the two CLDs in terms of their achieved QoS. The QoS is thereby investigated in terms of aggregate throughput, delay, and fairness, offered by both, PBOA, and MUD-MAC. Additionally, we compare the two CLDs to the 802.11 protocol.

All simulations are performed using a self-developed simulator written in c++. The general system parameters used are listed in Tab. 1. We assume that the inter arrival times of the packets are Poisson distributed. The channel is modeled with a modified free space path loss model, and line-of-sight is assumed. At the moment, fading influences are not included into the channel model. Notice, however, that both approaches stick to a certain time slotted structure. Since the duration of a frame (N consecutive blocks) of MUD-MAC as well as the frame duration of PBOA are similar and both assume that the channel stays constant for the transmission of the complete frame, we do not expect that the results of the *comparison* are strongly influenced. Including a block-fading channel model is expected to reduce the performance of both schemes, MUD-MAC, as well as PBOA, in a similar way, due to the likewise assumptions.

We model the probability that a packet is corrupted by the error probability formula derived in the additive white gaussian noise channel [10]. As modulation

Table 1. Simulation parameters

	MUD-MAC	PBOA	802.11
Control sig. bit rate	1 Mbit/s		
Data bit rate	2 Mbit/s		
Packet size	8192 bit		
Number of minislots	4	15	-
Transmission Power	100 mW		
Decoding sensitivity	- 81 dBm		
Carrier sensing sensitivity	- 91 dBm		
Carrier frequency	2 GHz		
Bandwidth	22 MHz		
Path loss exponent	3		
modulation scheme data	QPSK		
modulation scheme control signals	BPSK		

alphabet, we assume BPSK for the control packets, and QPSK for the data transmission. For a more detailed description of the channel model, please refer to [8].

Further on, a non global unique address room with one byte per address is assumed for all compared schemes. The overhead for the two CLDs includes all overhead contained in the 802.11 control packets. Only the bits for synchronization and transmission durations are not required for the two CLDs, since they are frame-level synchronous.

In [6] we show the throughput for a high traffic density random network where all nodes are within sensing range (50 nodes uniformly distributed on a square area of 50m × 50m). The aggregate throughput for the power control based PBOA CLD is quite low. We also present the throughput results in a medium traffic density scenario (50 nodes uniformly distributed on a square area of 500m × 500m) where not all nodes are within sensing range. The power control based CLD is worse than the MUD based CLD. However, it achieves at least good gains compared to 802.11. The latter results are depicted in Fig. 4. The MUD-MAC

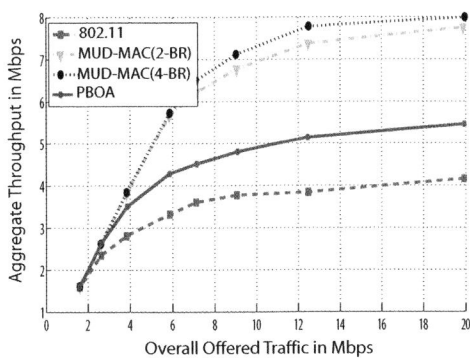

Fig. 4. Aggregate Throughput over offered traffic for the 802.11, the MUD-MAC and the PBOA protocol with 50 nodes in a 500m × 500m random network

CLD with both, four and two decoder branches, clearly outperforms the other CLD as well as 802.11. We investigate the related delay and fairness results for this medium traffic scenario in the following for PBOA, 802.11, and MUD-MAC with two and four decoder branches.

Notice that we do not assume that transmitters can switch to the next receiver awaiting the transmission of a packet in their queue, as proposed from the authors of PBOA. A pure First In Fist Out (FIFO) packet queueing is simulated instead. This makes the results comparable for all simulated schemes. Further on, a win probability p of 0.7 turned out by means of simulations to be the best choice for the PBOA protocol. The power adaptation assumes a minimum receive SINR of 14 dB, corresponding to a packet error probability of 10^{-2}.

5.1 Delay

We start our investigations with the median of the mean packet delays $\overline{\Delta_{p_k}}$ of all nodes over the aggregate delivered traffic, depicted in Fig. 5. MUD-MAC with both, two and four branches, clearly outperforms the PBOA CLD. In order to investigate the performance further, we look at real-time streaming services with very high delay requirements of 200 ms (150 ms - 250 ms [5]) - marked with a dash line in Fig.5. Even here, the multiuser detection based CLD offers about 6 Mbps (6.37 Mbps with four branches, 5.92 Mbps with two branches) delivered traffic. This respectively corresponds to a gain of 136% (four branches), and 119% (two branches), over 802.11 (2.70 Mbps).

The power control based cross-layer solution cannot offer that huge amount of throughput for applications with that stringent delay requirements. However, with 3.47 Mbps aggregate delivered traffic it can achieve a gain of 28% over the 802.11 protocol. Still, MUD-MAC with four branches provides 83% more throughput for highly time critical applications than PBOA.

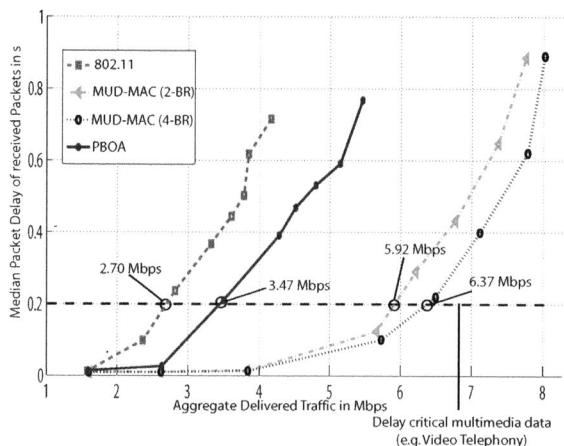

Fig. 5. Median of mean packet delay $\overline{\Delta_{p_k}}$ over delivered traffic for the 802.11, the MUD-MAC and the PBOA protocol with 50 nodes in a 500m × 500m random network

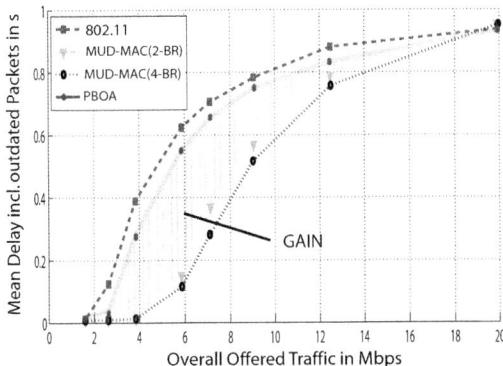

Fig. 6. Mean delay including outdated packets over offered traffic for the 802.11, the MUD-MAC and the PBOA protocol with 50 nodes in a 500m × 500m random network

As already stated, $\overline{\Delta_{p_k}}$ only includes delays for the delivered traffic. The overall delay including outdated packets ($\overline{\Delta_{p_{Out}}}$) that corresponds to the results of Fig. 5 is shown in Fig. 6 over the offered traffic. The shaded area marks the achievements of the multiuser detection based scheme with two branches over the power control based solution. Notice that all schemes tend to approximate the maximum delay Δ_{max} of 1s for increasing overall offered traffic values.

5.2   Fairness

To get insight into the fairness of the investigated CLDs, the variance of the throughput per node over the offered traffic and the mean packet delay per node over the delivered traffic are plotted in Fig. 7 and 8, respectively. The variance of

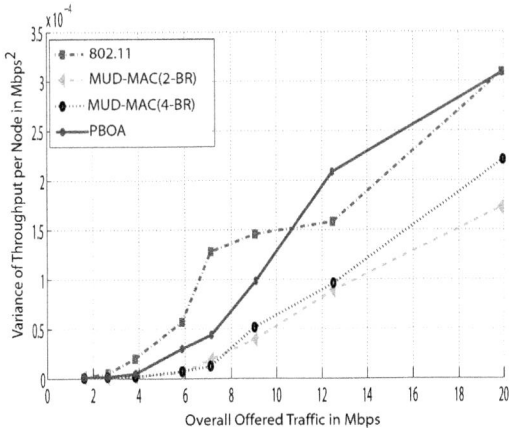

Fig. 7. Variance of throughput per node over offered traffic for the 802.11, the MUD-MAC and the PBOA protocol with 50 nodes in a 500m × 500m random network

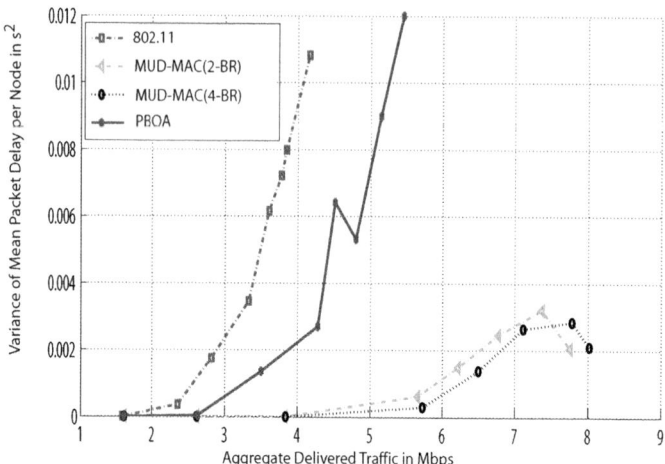

Fig. 8. Variance of mean packet delay per node over delivered traffic for the 802.11, the MUD-MAC and the PBOA protocol with 50 nodes in a 500m × 500m random network

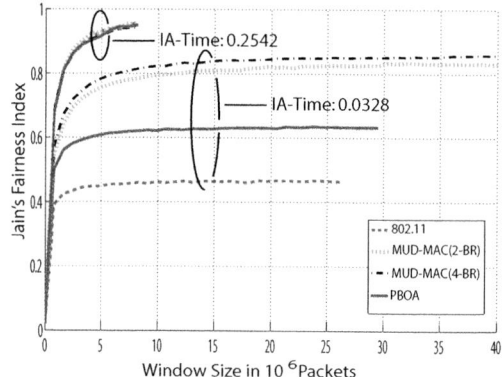

Fig. 9. Jain's Fairness Index for different inter arrival times for the 802.11, the MUD-MAC and the PBOA protocol with 50 nodes in a 500m × 500m random network

the throughput per node over the offered traffic shows an almost linear behavior for the two CLDs as well as 802.11. However, the variance of the mean packet delay per node over the delivered traffic rapidly increases for values near to the saturated traffic of the 802.11 and the PBOA protocol. We define the saturated traffic as the aggregate throughput value that the protocols achieve if the inter arrival time approaches zero. This value turns out by simulations to be 7.99 Mbps for the MUD-MAC CLD, 5.38 Mbps for the PBOA CLD and 4.15 Mbps for the 802.11 protocol (cf. Fig. 4). This indicates that both, the 802.11 as well as the power control based PBOA CLD achieve high delivered traffic only by sacrificing fairness. Opposite to this trends, the multiuser detection based CLD shows no rapid increase of the variance of the mean packet delay per node, fairly realizing the improved overall spectral efficiency.

The tendency that 802.11 and the power control based CLD start to get unfair for high traffic can also be observed by the Jain's fairness index described in Sec. 4. It is plotted for two inter arrival times in Fig. 9. While for an inter arrival time of 0.2542 all schemes achieve similar good fairness values, for high traffic density (inter arrival time of 0.0328), the 802.11 protocol cannot even treat 50% of the users fair. PBOA is considerably fairer and handles 63% of the users equally. However, MUD-MAC with both, two and four branches, shows the best fairness trends and can achieve a fair behavior for more than 83% (two branches) and 85% (four branches) of the users.

6   Conclusions

Within this paper, we compared the quality of service offered by two cross-layer designs that aim at reducing multiple access interference. While the first one (PBOA) suppresses interference by means of power control, the second one (MUD-MAC) applies multiuser detection at the receiver for interference cancelation.

We investigated two QoS measures, namely fairness and delay, and simulated a medium traffic density scenario with 50 nodes randomly distributed in a square area of 500m × 500m. It turned out that the multiuser detection based cross-layer design offers huge benefits for both, fairness as well as delay, compared to the power control based cross-layer design. This held also for low complexity multiuser detectors with only two branches.

Acknowledgment

The authors would like to thank I. Aad for many productive discussions.

References

1. Casari, P., Levorato, M., Zorzi, M.: DSMA: an Access Method for MIMO Ad Hoc Networks Based on Distributed Scheduling. In: Proc. ACM/IEEE International Wireless Communications and Mobile Computing Conference (IWCMC 2006), Vancouver, Canada (July 2006)
2. Eisenberg, Y., Conner, K., Sherman, M., Niedzwiecki, J., Brothers, R.: MUD Enabled Media Access Control for High Capacity, Low-Latency Spread Spectrum Communications. In: Proc. IEEE Military Communications Conference (MILCOM 2007), Orlando, FL (October 2007)
3. Ho, I.W.H., Liew, S.C.: Impact of Power Control on Performance of IEEE 802.11 Wireless Networks. IEEE Trans. on Mobile Computing 6(11), 1245–1258 (2007)
4. Jain, R., Chiu, D., Hawe, W.: A quantitative measure of fairness and discrimination for resource allocation in shared computer systems. Tech. Rep. TR-301, Digital Equipment Corporation (September 1984)

5. Janevski, T.: Traffic Analysis and Design of Wireless IP Networks. Artech House Inc. (2003)
6. Korger, U., Kusume, K., Hartmann, C., Widmer, J.: Power Control versus Multiuser Detection based Cross-Layer Design in Ad Hoc Networks. In: IEEE International Symposium on Personal, Indoor and Mobile Radio Communications, PIMRC 2010 (submitted 2010)
7. Kühn, V., Böhnke, R., Kammeyer, K.: Multi-user detection in multicarrier-CDMA systems. e & i Elektrotechnik und Informationstechnik 119, 395–402 (2002)
8. Kusume, K., Vilzmann, R., Müller, A., Hartmann, C., Bauch, G.: Medium Access in Spread Spectrum Ad Hoc Networks with Multiuser Detection. EURASIP Journal on Advances in Signal Processing - Special Issue on Cross-Layer Design for the Physical, MAC, and Link Layer in Wireless Systems 2009 (2009)
9. Muqattash, A., Krunz, M.: POWMAC: A Single-Channel Power-Control Protocol for Throughput Enhancement in Wireless Ad Hoc Networks. IEEE Journal on Selected Areas in Communications 23(5), 1067–1084 (2005)
10. Proakis, J.G.: Digital Communications. McGraw-Hill Book Company, New York (1989)
11. Toumpis, S., Goldsmith, A.J.: New Media Access Protocols for Wireless Ad Hoc Networks Based on Cross-Layer Principles. IEEE Trans. on Wireless Communications 5(8), 2228–2241 (2006)
12. Verdu, S.: Computational Complexity of Optimum Multiuser Detection. Algorithmica 4(3), 303–312 (1989)
13. Verdu, S.: Multiuser Detection. Cambridge University Press, New York (1998)

Author Index